AF690930

ELECTRICAL THEORY

SIMPLY EXPLAINED
THOROUGHLY UNDERSTOOD

ELECTRICAL THEORY

SIMPLY EXPLAINED
THOROUGHLY UNDERSTOOD

Alvin J. Walker

ELECTRICAL THEORY

SIMPLY EXPLAINED—THOROUGHLY UNDERSTOOD

© 2011 Alvin J. Walker
Portions of this material are reproduced from NFPA 70® - 2008 National Electrical Code®, copyright © 2007 National Fire Protection Association. All rights reserved.

Cover and interior design by Ted Ruybal.

All rights reserved. No part of this publication may be reproduced or transmitted in any form or by any means, electronic or mechanical, including photocopy, recording, or any information storage and retrieval system, without permission in writing from the publisher. Printed in the United States of America.

ISBN 13: 978-0-9831358-0-7
LCCN 2010941169
SCI021000 SCIENCE / Electricity
STU026000 STUDY AIDS / Study Guides
STU029000 STUDY AIDS / Vocational
First Edition
1 2 3 4 5 6 7 8 9 10

www.wisdomhousebooks.com

For more information, please contact
Alvin Walker
318-393-6841
www.alvinwalker.com

TABLE OF CONTENTS

PREFACE

If "simply explained" means making it plain and clear, and if that involves the use of examples and illustrations to ensure that what is discussed is "thoroughly understood," then my objective for writing this book was met. In preparing this book, it was very important to communicate (even before opening it) that the contents contained concepts and information that the user could easily comprehend and successfully retain.

The experience of learning the fundamentals of electrical theory can be quite cumbersome. Even after acquiring such valuable knowledge, knowing how to apply it is, without a doubt, a total subject in its own. In this book, you will find that the concepts of learning and applying the fundamentals of electrical theory are simply explained and thoroughly understood.

From the study of the electron to the variations of electrical current, voltage, power, and resistance—along with its counterpart, impedance—to the in-depth analysis of basic and advanced circuits to the conclusion of three-phase systems, transformers, and power factor corrections, the user is assured the discovery of a new or improved approach.

What the user will also discover is the repetitious value of the learning process. When asked what factor contributed to their offensive teams' repeated success with one certain play, legendary football coaches Eddie Robinson and Vince Lombardi both agreed that execution was the secret, and that only came by running that same play over and over in practice until it became second nature to the players during game time.

Based on my many years of hands-on experience in the electrical profession and while instructing students in various subjects of electricity, in particular electrical theory, I can honestly relate to the needs and challenges most students encounter when taking on this ongoing subject. Therefore, I encourage the user to study and develop a sense of discipline when using this book. I know it can be of great help, but you have to use it. In doing so, it will prove to be just like with all of my current and future books to follow—an instructor without actually being in the presence of a classroom. And remember that famous phrase I once heard on a radio station (although it might sound corny and out of touch, it is so true): "Constant review is a student's glue."

Portions of this material are reproduced from NFPA 70® - 2008, *National Electric Code®*, Copyright ©2007, National Fire Protection Association, Quincy, MA 02169. All rights reserved.

All NEC references made in this book are based on the 2008 National Electrical Code.

DISCLAIMER:
Every attempt has been made to ensure that the information and calculations in this book are correct; however, errors can occur. The author of this material does not assume responsibility for errors or omissions and is not liable for the use or misuse of the material contained in this document.

ACKNOWLEDGMENTS

> *Trust in the Lord with all thine heart; and lean not unto thine own understanding. In all thy ways acknowledge him, and he shall direct thy paths.* (Proverbs 3:5–6)

As a young boy, I was taught this verse by a Sunday School teacher who was adamant and persistent about my memorizing it. Little did I realize back then that the verse would one day be a stronghold in my walk with Christ, my life, and how I would learn to treat and respect my fellow man. I am thankful for those individuals who have played an important part in my life when I was a child and to those who have supported me now that I am an adult.

Although the way hasn't been easy and the obstacles have been many, by the grace of God this journey has come to the second of many milestones. And again, for this I say, "Lord, I thank thee—great is your faithfulness."

A special thanks goes out to my late mother and late father, my sisters and late brother-in-law, and to my brother and sister-in-law. I also thank my children, whom I am constantly praying for and love, and my late first-grade teacher for her prayers, love, and encouragement.

Thank you to the publisher, Annette Maxberry-Carrara, and her husband, Jean Louis Carrara; the editors, Deanne Lachner and Tricia Allen; the designer, Ted Ruybal; and to all others who contributed to the success of this book.

And last, to my Christian family and to all of my students and well-wishers who have crossed my path in current and past years, thank you.

INTRODUCTION
ELECTRICAL THEORY—SOMETHING YOU NEED TO KNOW

Can you imagine riding around in a car without a fuel gauge to tell you how much gasoline was left in the tank, or an eye surgeon examining someone's brain without knowing what a brain looks like? Without a doubt, dilemmas such as these spell trouble; nevertheless, this is what working conditions would be like in the electrical profession if you always had to deal with things that were unknown.

Not knowing how to size electrical wire or how to recognize a potential hazard, for instance, or not being aware of electrical terms—these are just a few examples of how a lack of knowledge can either endanger a person's life or compromise one's safety. This is why it is so important to have a reliable source assist you in understanding those things you are not familiar with or just misinformed about. I guarantee it's a lot safer and less costly.

What electrical theory does is provide a means for understanding those things that are either unknown or cannot be perceived, while providing a path for interpreting the end results.

Having a background in electrical theory allows the pursuit of more advanced topics, which leads to comprehending and solving problems beyond basic knowledge.

Considering such advantages, wouldn't you agree that electrical theory is *something you need to know*?

The Basics of Electricity (Understanding the Electron Theory)

Just think what it would be like to live in this world without electricity. Could you imagine living without the convenience of artificial lighting, air-conditioning and heating, the use of modern and advanced technology, or the ability to travel and communicate worldwide? Well, this has the potential to go on and on, so don't let it impede your imagination. To think about it, life would be extremely difficult and perhaps quite primitive because the need for electricity exists everywhere, and that says a whole lot.

Because this vital resource of energy is often taken for granted, very few people ever wonder where it came from. At best, the most conclusive explanation of scientists' findings seems to indicate that electricity is produced by particles called electrons and protons. Although these particles cannot be seen, it is said that they exist in all materials and, therefore, all forms of matter.

What Is Matter? What Are Elements?

Anything that occupies space, or better yet, can be felt or seen, is called *matter*. Matter can be in the form of a solid, liquid, or gas. The basic materials that make up all matter are called *elements*. Compared to matter, one would think there are millions and millions of elements; however, there are only 103 known elements: 92 that are natural elements and 11 that are man-made. Elements cannot be produced or broken up to form other materials. Nitrogen (N), iron (Fe), hydrogen (H), copper (Cu), silicon (Si), and aluminum (Al) are just a few of the most commonly used elements.

The Atom and Its Structure

The smallest particle that an element can be reduced to without losing the properties of that element is called an *atom*. If the atom of an element is broken down or reduced any further, none of the properties of that element would exist.

Regardless of the form of an element, the structure of all atoms contains three subatomic particles: the electron(s), the proton(s), and the neutron(s). These three particles play an important part in the theoretical concepts of electricity.

The Nucleus (The Neutron and the Proton)

Structurally, the center of an atom is called the nucleus. The nucleus is where the protons and neutrons are located, while the electrons circles about the orbit(s) of the nucleus. Because the neutron is an electrically neutral particle, it is not as important as the proton and electron.

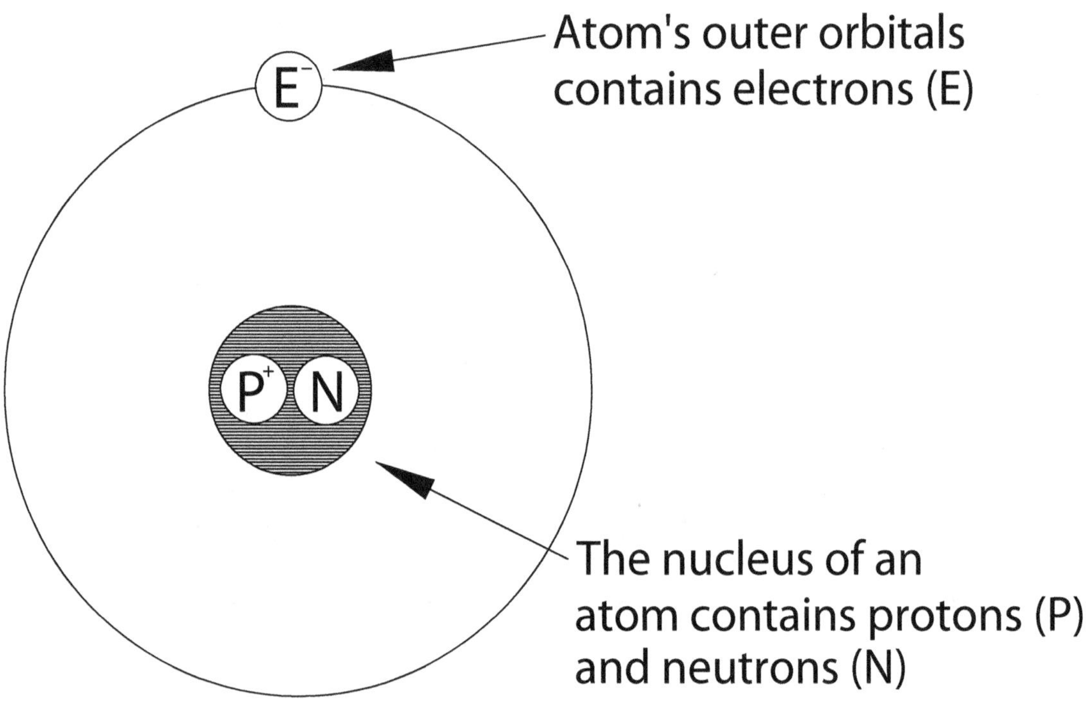

Fig. 1. The structure of an atom.

The atomic number of an element is based on the number of protons located in the nucleus of an atom. Because each element contains a different number of protons, each element can be uniquely identified by its atomic number. Although the diameter of a proton is exceedingly smaller than the electron, the mass (size/weight) of a proton is said be approximately 1840 times that of an electron. As a result, protons are quite stable and play an inactive role in producing electricity. Because protons are positively (+) charged, the same holds true for the nucleus of an atom.

Inversely, the mass of an electron is approximately 1840 times lighter than a proton, yet its diameter is three times that of a proton. Unlike the proton, electrons are negatively (-) charged and can be moved about rather easily because of their near-weightless mass, thus allowing them to play an essential role in producing electricity.

As a review, we know that like charges (positive/positive and negative/negative) repel each other, and unlike charges (positive/negative) attract each other; therefore, it should be clearly understood, as displayed in figure 2, that protons and electrons are attracted to each other.

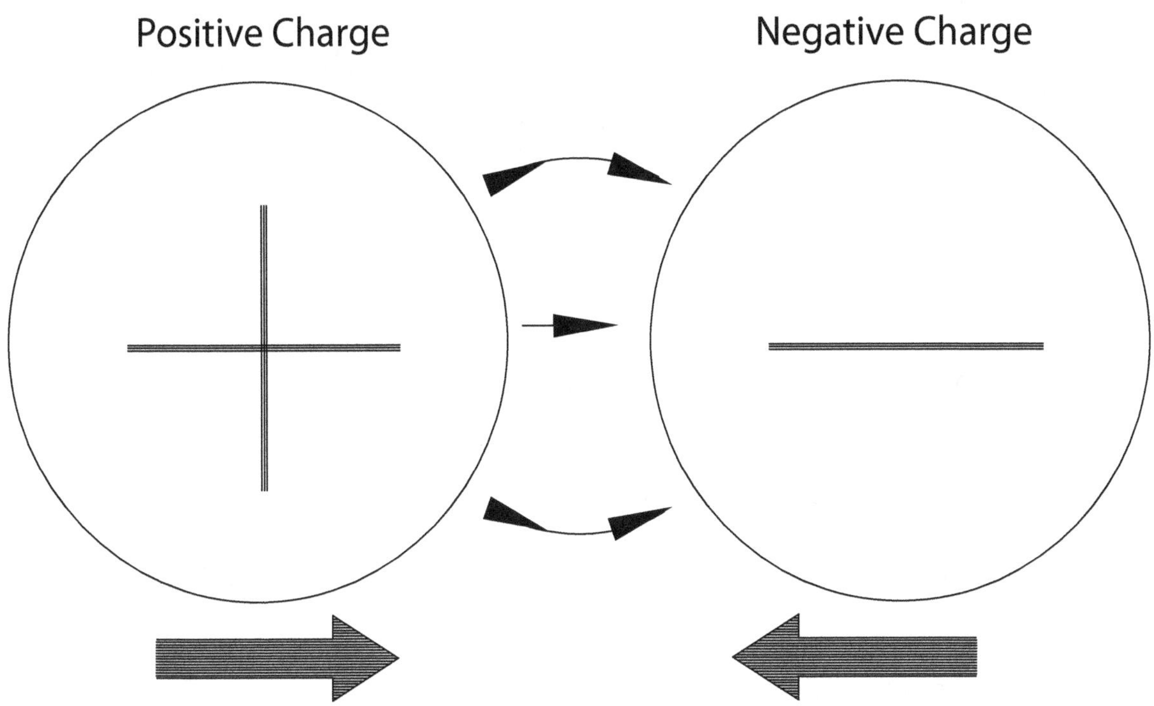

Fig. 2. Unlike charges attract.

In most cases, an atom will contain the same number of protons and electrons, which greatly contributes to the atom's being neutrally charged based on protons and electrons having equal and opposite charges.

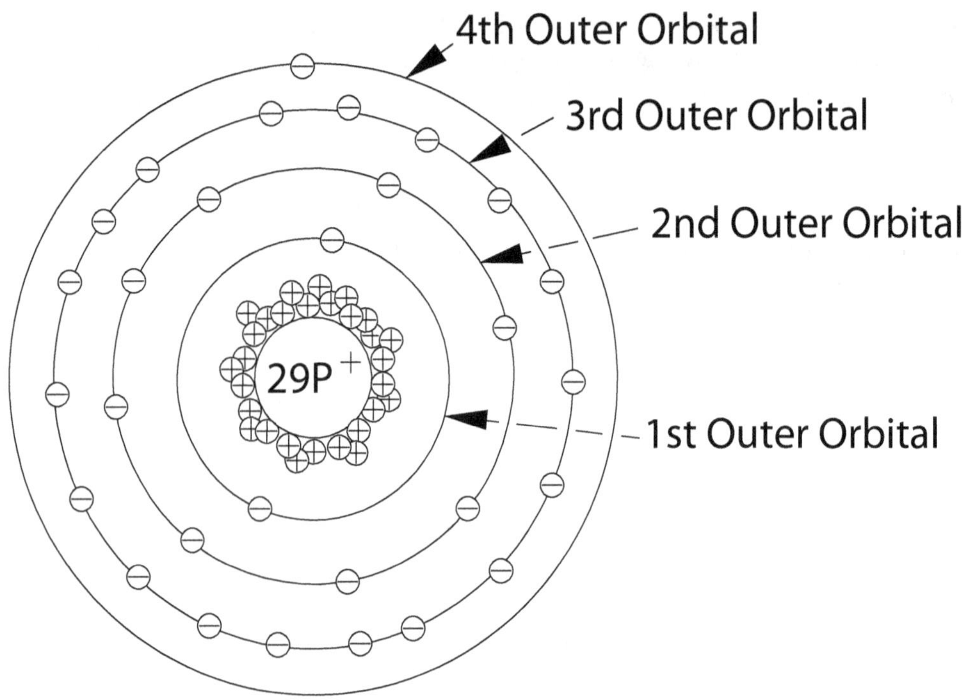

Fig. 3. The copper atom.

Consider the copper atom in figure 3. The copper atom has a total of twenty-nine protons and is surrounded by twenty-nine electrons in its orbital circle. There are four orbital circles surrounding a copper atom. The number of electrons in each orbital circle is as follows: two, eight, eighteen, and one. To say the least, copper material contains millions and millions of copper atoms.*

However, as with all atoms, copper atoms can either gain or lose electrons based on the instability of the atoms' electrons and the electrons' ability to move about freely.

*Except for silver, electrical wire made of copper conducts electricity better than any other natural material.

Valence Electrons/Free Electrons

As previously mentioned, electrons circle about the orbit(s) of the nucleus, which occurs at a high rate of speed. If not for the positively charged nucleus, centrifugal forces would pull the electrons out of the orbit(s) surrounding the nucleus. The electron(s) located the farthest from the nucleus are called valence electrons. If an outside force of significant energy is engaged to assist the centrifugal force, these valence electron(s) can be completely freed from the farthest outer orbit of the nucleus. Once these electrons are freed, they are called free electrons. The movement of free electrons is what produces electricity.

Electron Flow

The movement or flow of electrons is by far the most important concept in producing electricity. However, to understand this concept even better, let's compare the flow of water with the flow of electrons. Consider a dripping water faucet and imagine trying to rinse a dirty dish with single drips of water. Just think, if this was the only means of rinsing the dish, one would be in for a long, long, long wait. Get the point? Now apply a higher pressure to the drips of water by fully turning on the water faucet. These single drips of water are now combined and transformed into a forceful stream, flowing all in the same direction, making the chore of rinsing the dish much easier. In comparison to the single drips of water, if free electrons are to do useful work—that is, producing electricity—they must be flowing under pressure in the same direction.

Electrons under Pressure

Now that we know that there must be an applied pressure to move electrons, let's focus on the type of pressure required. First, pressure is no more than the application of force. Although there are different types of forces other than pressure (that will cause electron flow or movement), such as light, heat, friction, magnetism, and certain types of chemicals, the most recognized or most frequently used force is electrical pressure.

When there is a difference of pressure in a water pipe, water will flow through the pipe. Similarly, when two charges experience a difference of potential, a pressurized reaction or force occurs, which is called an electromotive force, known as *emf* or the most commonly used term, voltage. The unit of measurement used to specify the strength of voltage is the volt. The volt was named after Count Alessandro Volta, an Italian physicist (1745–1827).

Electrical Charge

Because the electrical charge of a single electron is extremely small and insignificant, defining the quantity of measurement of such would be highly impractical. Realistically speaking, deriving the quantity of an electrical charge of a single electron would be as tedious as counting the number of single drips of water required to completely remove a hardened stain from a dirty dish. However, knowing what we now know about electrons and how they move about in massive numbers, it was determined years ago that a given number of electrons would produce an electrical charge significant enough to be given a quantity of measurement. This quantity of measurement, which identifies the electrical charge of a given number of electrons, is called a coulomb. The coulomb was named after Charles-Augustin de Coulomb, a French physicist (1736–1806), who estimated that 6.28×10^{18} electrons or 6.28 billion-billion electrons (6,280,000,000,000,000,000) were needed to produce an electrical charge or one coulomb.

Electrical Current—The Flow of Electrons

By now, you should realize that the flow of electricity is no more than the flow or movement of electrons. The flow of electrons is called electrical current. In comparison, the speed of electron flow is not quite equal in velocity to the speed of light, which is about 186,000 miles per second; nevertheless, the speed of flowing electrons is still far beyond the grasp of the human eye.

If electrons are always flowing in the same direction, the process is called *direct current*, whereas flowing electrons that are continuously changing directions are called *alternating current.*

To initiate the flow of electrons, it is necessary to have an electrical source or power supply to provide electrical pressure or voltage to push electrons through an electrical conductor. Once electrical pressure or voltage is applied to an electrical conductor, the valence electrons that encircle the nucleus of the atoms within the conductor are freed and they move about within the path of the conductor, flowing from atom to atom at a high rate of speed. This type of interaction takes place repeatedly and continuously throughout the entire path of the conductor as long as electrical pressure or voltage is applied.

Once again, using water as a model, the flow of current is measured by the number of electrons that pass a point in an electrical conductor in one second, just as the flow of water is measured in quantities of gallons per minute. The quantity or unit of measurement for current is called amperes. It is said that one ampere of current represents a flow of one coulomb of electrons passing a point in a conductor in one second. The ampere was named after Andre Marie Ampere, a French chemist and physicist (1775–1836).

<u>The Opposition to the Flow of Electrons</u>

Imagine a drinking straw completely stuffed with sand, where each grain of sand represents the number of atoms found in an electrical conductor.

The physical contents of all electrical conductors contain an endless number of atoms, which are very closely compressed, just like the grains of sand in a drinking straw. Because of this, once a valence electron is freed from the outer orbit of an atom, it encounters an enormous amount of opposition in its attempt to move from one atom to another. This type of opposition, which opposes the flow of electrons in an electrical conductor, is called resistance. As the movement of electrons is reduced, resistance affects the amount of current flowing in a conductor.

Resistance is measured in units of ohms. This quantity of measurement was named after Georg Simon Ohm, a German physicist (1789–1854).

OHM'S LAW

The V-C-R Relationship

In every electrical circuit, there are three basic quantities: voltage, current, and resistance.

Before connecting a circuit, it is important to know something about each electrical quantity. By knowing a circuit's operating voltage, the correct size transformer or power supply can be obtained. By knowing the amount of current flowing through a wire, the proper size wire can be selected. It would be awfully frustrating, time-consuming, and hazardous if every electrical circuit required trial and error before being able to determine the circuit's allowable operating voltage, current flow, and total resistance. If electrical work is to be performed safely and without hazards, one must be able to calculate the electrical loads of any given circuit before connecting it.

There is a definite relationship between voltage, current, and resistance in every direct current (*dc*) and purely resistive alternating current (*ac*) circuit. The relationship between voltage, current, and resistance is known as Ohm's Law. Ohm's Law was derived based on experimental evidence gathered by Georg Simon Ohm.

The best approach toward understanding Ohm's Law and its concept is to become familiar with the discussed quantities and other relative information. Therefore, it is recommended that the information to follow is thoroughly reviewed to grasp such vital knowledge.

Definition	**Symbol**	**Units of Measurement**
Voltage—an electromotive force (pressure)	E	volts (V)
Current—the flow of electrons	I	amperes (A)
Resistance—opposition to the flow of current	R	ohms (Ω*)

*Symbolized by the Greek letter Omega.

Using the formula E = I x R, Ohm's Law proves that current (I) is directly proportional to voltage (E) when the resistance (R) remains constant, and inversely proportional to the resistance (R) when the voltage (E) remains constant while the resistance (R) increases or decreases. Check it out:

Directly proportional—Where R remains constant (the same); if E increases or decreases, I does likewise.

Proof: If two different voltage sources (E) (120V, 240V) were applied to two similar electrical conductors (wires) having the same resistance (R), the voltage source with the highest rating (240V) would have the potential to push more electrons through the corresponding conductor, causing a larger flow of current (I) and vice versa.

Inversely proportional—Where E remains constant (the same); if R increases or decreases, I does the opposite.

> Proof: If the same voltage source (E) were applied to two electrical conductors (wires) having different resistances (R), the conductor with the lowest resistance would permit a larger flow of current (I), while the conductor with the highest resistance would permit a smaller flow of current (I).

Based on the given information and the relationship that exists between these three quantities, each quantity can be mathematically calculated using Ohm's Law. Take a look at the formula:

E = I x R, states that the value of E can be found by multiplying I and R

With this formula two other formulas can be derived:

I = E/R which states that the value of I can be found by dividing R into E

and

R = E/I which states that the value of R can be found by dividing I into E

However, to use any of these formulas, two of the quantities' values must be known.

As a useful tool to assist one in memorizing these formulas, the following "three piece (Ohm's Law) pie" chart is provided in figure 4.

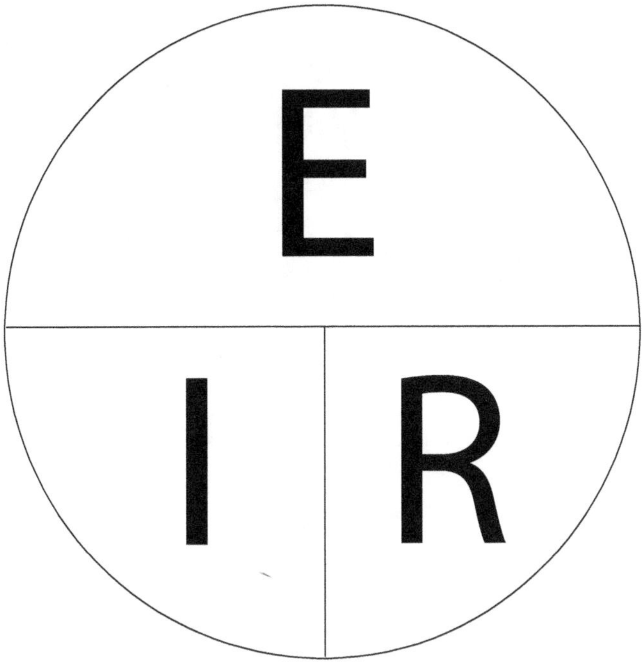

Fig. 4. Ohm's Law pie chart.

To provide further assistance with understanding Ohm's Law, observe the following applications.

If 5 amperes of current flows through a wire having a resistance of 24Ω, what is the applied voltage?

Using the formula E = I x R,

$$5A \times 24\Omega = 120V \text{ (120 volts)}$$

How much current flows when a voltage of 78 volts and a resistance of 39 ohms exist?

Using the formula I = E/R,

$$78V/39\Omega = 2A \text{ (2 amperes)}$$

What is the resistance when 63 volts and 9 amperes of current are utilized?

Using the formula R = E/I,

$$63V/9A = 7\Omega \text{ (7 ohms)}$$

Electrical Power

Electrical power is defined as the amount of work that can be done by an electrical load in some standard period of time. Electrical power is required to produce heat and light or to operate electrical motors and transformers. The amount of work consumed in an electrical load can be either useful or wasted. Useful work comes in the form of heating elements in electric dryers, ranges, or water heaters and in the operation of electric motors. Wasted work results from the heating of electrical wires or poor connections between electrical wires or equipment. The heat(ing) performs no useful work; therefore, the resulting heat that could be used to produce work is lost or wasted.

Electrical power is measured in the unit watts. The numbers of watts consumed indicates the electrical power used by an electrical load. Take for example the light bulb; it is marked according to the number of watts used. The greater the rating of the light bulb in watts, the greater the amount of light (as well as heat) produced. Consider a 5000-watt electric furnace compared to a 10,000-watt electric furnace. A 5000-watt electric furnace will not produce the heat that a 10,000-watt electric furnace will, mainly because less current is required to produce 5000 watts of heat compared to 10,000 watts of heat. In another example, we all are familiar with the unit that is used to identify electrical motors: horsepower. A 200-horsepower motor will produce more torque (rotating force) than a 100-horsepower motor; therefore, it is easy to conclude that the 200-horsepower motor does more work, and more electrical power is needed to operate it.

Similar to the relationship between voltage, current, and resistance in Ohm's Law, there is also a relationship between voltage, current, and electrical power. This relationship is expressed in the definition given for a watt—that is, the electrical power used when 1 ampere of current flows about a potential difference of 1 volt. Take a look at the information provided below before proceeding.

Definition	**Symbol**	**Units of Measurement**
Power—the ability to do work	P	watts (W)
Voltage—an electromotive force (pressure)	E	volts (V)
Current—the flow of electrons	I	amperes (A)

The relationships that exist between these three quantities can also be mathematically calculated similarly to Ohm's Law. Again, consider the definition of a watt and then take notice of the derived formula P = E x I:

P = E x I, states that the value of P can be found by multiplying E and I

With this formula two other formulas can be derived:

E = P/I, which states that E can be found by dividing I into P

and

I = P/E, which states that I can be found by dividing E into P

Once again, to use any of these formulas, two of the quantities' values must be known.

As a useful tool to assist one in memorizing these formulas, the "three-piece (P-E-I) pie" chart, is provided in figure 5.

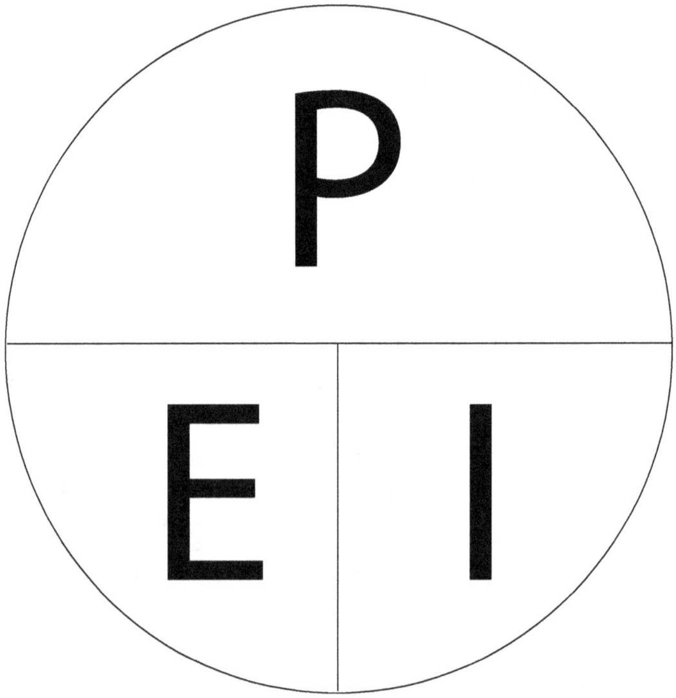

Fig. 5. The P-E-I pie chart.

Formula to Formula

The Ohm's Law and the P-E-I pie charts gave us six formulas to work with, and just like a good pie, the more slices there are, the better it is. First we started with one pie, and then there were two pies. Now if both pies are combined, the results are even better:

Ohm's Law	Power
E = I x R (1)	P = E x I (4)
I = E/R (2)	E = P/I (5)
R = E/I (3)	I = P/E (6)

Because the E and I quantities are no different in the formulas

$$E = I \times R \text{ and } P = E \times I$$

one formula can be substituted into another formula. The formula E = I x R can be substituted for the quantity E in the formula P = E x I, resulting in

$$P = \overset{E}{\boxed{\mathbf{I} \text{ x } \mathbf{R}}} \text{ x I} \quad \text{or} \quad P = \text{I x} \boxed{\mathbf{I} \text{ x } \mathbf{R}} (\text{where I x I} = I^2)$$

to derive another formula,

$$P = I^2 \text{ x R (7)}$$

By substitution, other formulas can be derived for voltage (E), current (I), resistance (R), and power (P). From the formula $P = I^2$ x R, two other formulas can be derived. First,

$$I^2 = P/R, \text{ where } I = {*}\sqrt{P/R} \text{ (8)}$$

*To derive the formula $I = \sqrt{P/R}$, let's start with the formula $P = I^2$ x R. The first thing that has to occur is to get I to one side by itself. This is done by first taking the reciprocal of R, which is 1/R, and multiplying R by its reciprocal 1/R. To eliminate R, observe,

$$P = I^2 \text{ x } \cancel{R} \text{ x } 1/\cancel{R} \text{ (where R x 1/R =1)}$$

Since the reciprocal 1/R was placed on the right side of the formula, it has to be placed on the opposite (left) side as well:

$$\text{P x 1/R (where P x 1/R = P/R)} = I^2$$

Now that I is all to itself, where,

$$P/R = I^2,$$

the only thing left to do is to have I stand completely on its own by eliminating the exponent 2. This is done also by multiplying the exponent 2 by its reciprocal 1/2. Again, just remember whatever is done on one side has to be done on the opposite side; observe,

$$(P/R)^{1/2} = I^{2 \text{ x } 1/2} \text{ (where 2 x 1/2 = 1)}$$

Just as the exponent 2 is also expressed as 2 to the second power, the exponent ½ can also be expressed as ½ to the half power. Any time an exponent is raised to the half power, it becomes exactly the same as the square root of a number; therefore,

$$(P/R)^{1/2} \text{ is the same as } \sqrt{P/R}$$

and

$$I^{2 \text{ x } 1/2} = I$$

As a result,

$$\sqrt{P/R} = I \quad \text{or} \quad I = \sqrt{P/R}$$

and the second formula that's derived from the formula $P = I^2 \times R$ is

$$R = P/I^2 \quad (9)$$

Next, the formula $I = E/R$ can be substituted for the quantity I in the formula $P = E \times I$, resulting in

$$P = E \times \overset{I}{\boxed{\mathbf{E/R}}} \quad \text{or} \quad P = E \times \mathbf{E/R} \quad (\text{where } E \times E = E^2)$$

To derive yet another formula,

$$P = E^2/R \quad (10)$$

And finally, from the formula $P = E^2/R$, we can derive two final formulas:

$$E^2 = P \times R, \text{ where } {**}E = \sqrt{P \times R} \quad (11)$$

**The same approach is used to derive the formula $E = \sqrt{P \times R}$ as was used with the formula $I = \sqrt{P/R}$.

and

$$R = E^2/P \quad (12)$$

As you can see, combining both pies was not a bad idea at all. Instead of six formulas, we now have twelve, and of the twelve, we have three formulas for each quantity.

Voltage	Current	Resistance	Power
$E = I \times R$	$I = E/R$	$R = E/I$	$P = E \times I$
$E = P/I$	$I = P/E$	$R = P/I^2$	$P = I^2 \times R$
$E = \sqrt{P \times R}$	$I = \sqrt{P / R}$	$R = E^2/P$	$P = E^2/R$

Finally, as a useful tool to assist one in developing an understanding of the relationships among the four given quantities, the "twelve (P-E-R-I) piece pie" chart is provided in figure 6. With this chart, any problem involving two of the mentioned quantities can be solved if their values are known.

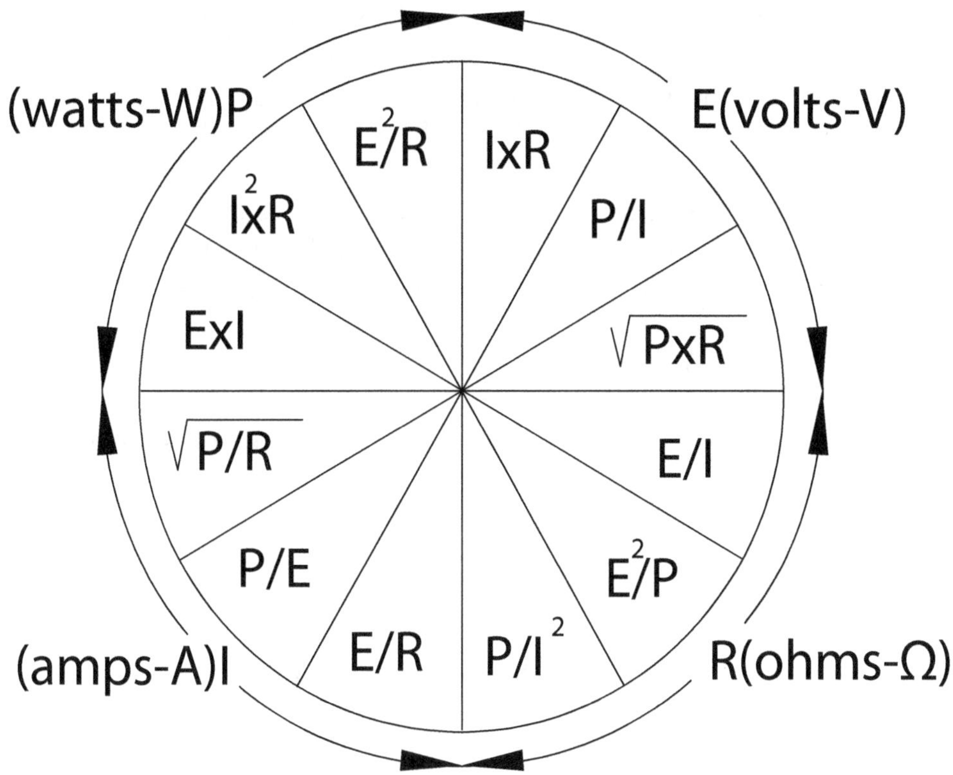

Fig. 6. The P-E-R-I pie chart.

And there you have it: "The more slices, the better the pie."

To build confidence and further enhance one's understanding and ability in deriving unknown electrical quantities, observe the following applications:

How much current flows through a 1600W load that has a resistance of 12Ω?

The question asks for the amount of current flow in this particular application. Therefore, the first step is to locate the three formula that involves calculating current (I).

Afterward, consider the known quantities of the question. They are the load, given in quantities of watts, and the resistance given in quantities of ohms (Ω). Of the three formulas, only one involves the use of both given quantities, $I = \sqrt{P / R}$, where P, the power consumed by the load is measured in quantities of watts (W) and R, the resistance is measured in quantities of ohms (Ω).

Using the formula $I = \sqrt{P / R}$,

$$\sqrt{1600W / 12\Omega} = 11.55A \text{ (11.55 amperes)}$$

Determine the power loss where 8A of current flows through a 15Ω resistor.

The question now asks to calculate the power loss. The next step is to locate the formulas that involve calculating power (P).

Again, consider the known quantities of the question. The two given quantities are 8A of current and 15Ω of resistance. Of the three power formulas, only one involves the inclusion of both given quantities, $P = I^2 \times R$. Therefore, using this formula, the answer is determined to be

$$(8A)^2 \times 15\Omega = 960W \text{ (960 watts)}$$

How much power is consumed when 8A of current flows through a load resistance of 9Ω?

Use the formula $P = I^2 \times R$,

$$P = (8A)^2 \times 9\Omega = 576W \text{ (576 watts)}$$

A 240 volts unit pulls 45A of current. Calculate the power consumption.

Use the formula $P = E \times I$,

$$P = 240V \times 45A = 10{,}800W \text{ (10,800 watts)}$$

A 13Ω heating iron is operated by a 117-volts source. How much power does the heating iron consume?

Use the formula $P = E^2/R$,

$$P = (117V)^2/13\Omega = 1053W \text{ (1053 watts)}$$

Three incandescent light bulbs consume 450W of power. If the bulbs utilize 3.6A of current, how much voltage is supplied to the bulbs?

Use the formula $E = P/I$,

$$E = 450W/3.6A = 125V \text{ (125 volts)}$$

A piece of equipment has a resistance of 3.4 ohms and a 4070W power rating. Determine the equipment's operating voltage.

Use the formula $E = \sqrt{P \times R}$,

$$E = \sqrt{4070W \times 3.4\Omega} = 117.64V \text{ (117.64 volts)}$$

How much current flows in a circuit that has a 2238W load and a resistance of 31.3Ω?

Use the formula $I = \sqrt{P/R}$,

$$I = \sqrt{2238W/31.3\Omega} = 8.46A \text{ (8.46 amperes)}$$

A 480V source supplying a 7650W load produces how much current?

Use the formula $I = P/E$,

$$I = 7650W/480V = 15.94A \text{ (15.94 amperes)}$$

Determine the resistance of a 1868W load that draws 17.39A.

Use the formula $R = P/I^2$,

$$R = 1868W/(17.39A)^2 = 6.18\Omega \text{ (6.18 ohms)}$$

A 36.47kW load is energized by a 265V supply. Determine the resistance of the load.

Use the formula $R = E^2/P$,

$$R = (265V)^2/36{,}470W = 1.93\Omega \text{ (1.93 ohms)}$$

What Is an Electrical Circuit?

An electrical circuit (Fig. 7) is an electrical formation that consists of a voltage source, a load, and connecting wires (conductors), where the voltage source provides energy in the form of an electromotive force that causes current to flow, and where current is opposed by the resistance or impedance (to be discussed) of the load and the connecting wires.

Up until now, the term *circuit* has only been vaguely mentioned. Although a definition was provided, it didn't say anything about the various *types* of circuits. Unknowingly, one could only imagine a circuit as being an arrangement of electrical components. How the electrical components of a circuit are arranged determines the type of electrical circuit the circuit is. Generally speaking, there are three types of electrical circuits: the series circuit, the parallel circuit, and the combination circuit. Just as the names of these circuits differ, so do the operating characteristics and the problem-solving methods used to analyze them. To fully understand and analyze each of these circuits, one must first become familiar with the applicable rules for each type circuit. To learn more about these circuits, let's take a more in-depth look at each individual circuit.

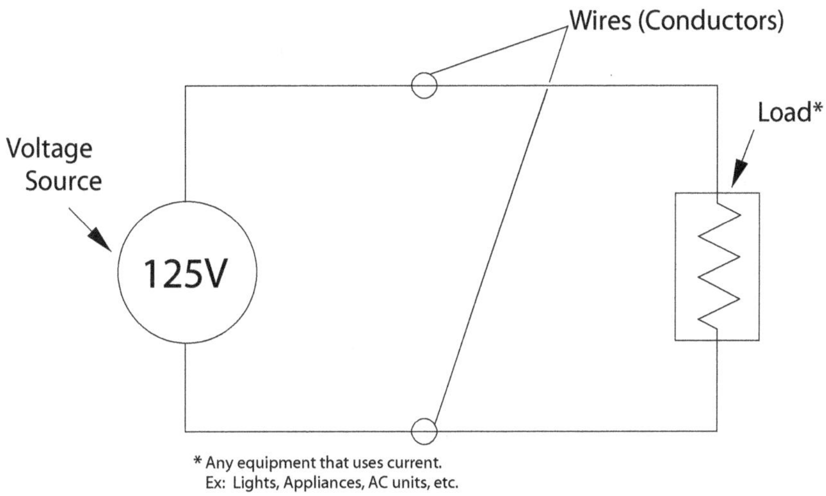

Fig. 7. The electrical circuit.

An active electrical circuit will always include the four electrical quantities previously discussed: voltage, current, resistance, and power.

The Series Circuit

What Is a Series Circuit?

A series circuit is the type of circuit in which only one source of current exists, and all related resistances are connected one after the other. The same current that leaves the voltage source flows through every resistance of a series circuit before returning to the voltage source.

Rules That Apply to Series Circuits

Rule 1. The total resistance (R_T) of a series circuit is equal to the sum of all individual resistances,

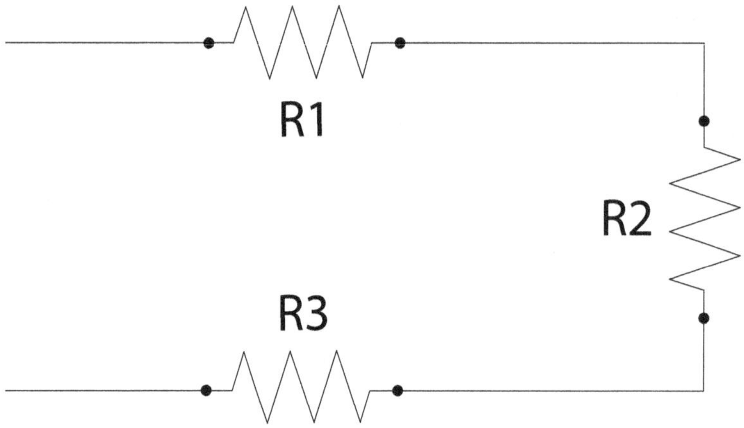

where

$$R_T = R_1 + R_2 + R_3 + \ldots\ldots\ldots\ldots$$

Rule 2. Current (I) is the same (constant), meaning the same current flows through each individual resistance,

where

$$I = E_T/R_T$$

Rule 3. The sum of the voltage drops across each individual resistance is equal to the voltage source (E_T),

where

$$E_T = E_1 + E_2 + E_3 + \ldots\ldots\ldots\ldots$$
$$\text{and } (E_{1,2,3\ldots} = [R_{1,2,3\ldots}] \times I)$$

Rule 4. The sum of the power consumed by each individual resistance is equal to the circuit's total power (P_T),

where

$$P_T = P_1 + P_2 + P_3 + \ldots\ldots\ldots\ldots$$
$$\text{and } (P_{1,2,3\ldots} = [E_{1,2,3\ldots}] \times I)$$

APPLYING THE RULES

Series Resistance (R)

1. What is the total resistance of a 21Ω, 45Ω, 78Ω, 3.5Ω, and 36Ω resistor when connected in series?

   With a series of resistors, the total resistance is determined by adding the resistance of each individual resistor together. Therefore,

$$R_T = 21\Omega + 45\Omega + 78\Omega + 3.5\Omega + 36\Omega = 183.5\Omega$$

Series Resistance Circuits (R)

2. If a 125-volts source were placed across the resistors in question No. 1, how much current would flow through each resistor?

   Using the formula $I = E_T/R_T$

$$I = 125V/183.5\Omega = .681A$$

3. A series circuit energized by a 120V source consists of the following resistors,

   $R_1 = 6\Omega$, $R_2 = ¾\Omega$, $R_3 = 4\Omega$, $R_4 = 9\Omega$, $R_5 = 3.5\Omega$, $R_6 = 13\Omega$

   Determine the following:

   a. The circuit's total resistance, current, and the total power consumed by the circuit.
   b. The voltage drop and power consumed by each resistor.

   a. $R_T = 6\Omega + ¾\Omega + 4\Omega + 9\Omega + 3.5\Omega + 13\Omega = 36.25\Omega$

   With the circuit's total resistance and the given circuit's voltage, the current is

$$I = 120V/36.25\Omega = 3.31A \text{ (rounded off)}$$

The total power consumed by the circuit can be determined by any one of the three given formulas for determining power (P). For verification, we'll use all three formulas.

$$P_T = E \times I = 120V \times 3.31A = 397.2W$$
$$P_T = I^2 \times R = (3.31A)^2 \times 36.25\Omega = 397.16W$$
$$P_T = E^2/R = (120V)^2/36.25\Omega = 397.24W$$

With the exception of the decimal values, the results are all the same, which proves that any of the three formulas could be used to determine the total power consumed by the circuit if related quantities are known.

b. In a series circuit, there is only one source of current and that current flows through each individual resistor. Using the resistance of each resistor along with the source of current, the voltage drop can be determined by using Ohm's Law original formula, E = I x R. Because the same amount of current will flow through each resistor and the value of each resistor varies between ¾Ω and 13Ω, the calculated voltage drop across each resistor will decrease from the largest to the smallest resistance. With this in mind, the results of these calculations and any voltage drop calculation that's involved in a series circuit can be determined before defining a definite answer. As a result, the individual voltage drops in this circuit will conclude in the following order, E_6, E_4, E_1, E_3, E_5, and then E_2.

Voltage drop (E) across each resistor:

(3) $E_1 = 6\Omega \times 3.31A = 19.86V$
(4) $E_3 = 4\Omega \times 3.31A = 13.24V$
(5) $E_5 = 3.5\Omega \times 3.31A = 11.59V$
(6) $E_2 = 3/4\Omega \times 3.31A = 2.48V$
(2) $E_4 = 9\Omega \times 3.31A = 29.79V$
(1) $E_6 = 13\Omega \times 3.31A = 43.03V$

As determined, the result of each individual voltage drop does fall in a like manner, which is validated by the calculations. Because each individual voltage drop, when totaled, should equal the circuit's source voltage (E_T), let's finalize this part of the question through verification.

$$E_T = 19.86V + 2.48V + 13.24V + 29.79V + 11.59V + 43.03V = 119.99V$$

As you can see, the result is approximately 120V.

The power consumed by each resistor should follow in the same order as the results of the voltage drops because the resistance values are the deciding factors.

Earlier, the three formulas for determining power (P) were used to determine the circuit's total power (P_T). Although the same approach could still be used in calculating the power consumed by each individual resistor, for brevity the only formula that will be used is P = E x I, where E is equal to the individual voltage drop (E_R) across each resistor.

Power consumed (P) by each resistor:

(3) $P_1 = 19.86V \times 3.31A = 65.74W$
(4) $P_3 = 13.24V \times 3.31A = 43.82W$
(5) $P_5 = 11.59V \times 3.31A = 38.36W$
(6) $P_2 = 2.48V \times 3.31A = 8.21W$
(2) $P_4 = 29.79V \times 3.31A = 98.6W$
(1) $P_6 = 43.03V \times 3.31A = 142.43W$

The power consumed by each individual resistor when totaled should equal the total power (P_T) consumed by the circuit. Observe:

$$P_T = 65.74W + 8.21W + 43.82W + 98.6W + 38.36W + 142.43W = 397.16W$$

Compared to earlier results, the totaled power consumption is exactly the same as one calculation, 397.16W, and approximately the same as the others, 397.2W and 397.24W.

4. Using the circuit in question No. 3 again, let's consider an alternate method to determine the voltage drop and power consumed by each resistor.

   Based on the circuit's total resistance, let's determine the percentage each resistor contributes to the circuit's total resistance. This will provide an alternate method for analyzing the circuit. The percentage each resistor contributes to the circuit's total resistance is determined by dividing the total resistance into each individual resistance (R/R_T). Observe:

$$\%_1 = \frac{6\Omega}{36.25\Omega} = .166\ (16.6\%) \qquad \%_2 = \frac{3/4\Omega}{36.25\Omega} = .021\ (2.1\%)$$

$$\%_3 = \frac{4\Omega}{36.25\Omega} = .110\ (11.0\%) \qquad \%_4 = \frac{9\Omega}{36.25\Omega} = .248\ (24.8\%)$$

$$\%_5 = \frac{3.5\Omega}{36.25\Omega} = .097\ (9.7\%) \qquad \%_6 = \frac{13\Omega}{36.25\Omega} = .359\ (35.9\%)$$

   Now that the percentage values are determined, they should all total up to 100%.

$$\%_T = 16.6\% + 2.1\% + 11\% + 24.8\% + 9.7\% + 35.9\% = 100.1\%$$

   As you can see, the total extended just a hair over 100% and that's okay. Anytime numbers are rounded off (as was each individual percentage value), the accuracy of the results is always compromised; however, the calculated percentage values are just fine because the total is approximately (more or less) 100%.

   Because these percentage values represent a certain part of a whole value, which in this case is the circuit's total resistance, they can also be used to represent other parts of a whole value, such as the circuit's source (total) voltage (E_T) and the total power consumed (P_T) by the circuit.

   Remember the individual voltage drops and the power consumed by each resistor calculated in question No. 3. As an alternate method, the same or approximate results can now be obtained by multiplying the total voltage (E_T) and the total power (P_T) by each individual percentage value (in decimal form). The alternate method provides an alternate means for checking answers quickly and independently.

   Voltage drops (per question No. 3):

   $E_1 = 19.86V$ $E_2 = 2.48V$ $E_3 = 13.24V$
   $E_4 = 29.79V$ $E_5 = 11.59V$ $E_6 = 43.03V$

   Voltage drops ($E_T = 120V$):

$E_1 = 120V \times .166 = 19.92V$
$E_2 = 120V \times .021 = 2.52V$
$E_3 = 120V \times .11 = 13.2V$
$E_4 = 120V \times .248 = 29.76V$
$E_5 = 120V \times .097 = 11.64V$
$E_6 = 120V \times .359 = 43.08V$

$$E_T = 19.92V + 2.52V + 13.2V + 29.76V + 11.64V + 43.08V = 120.12V$$

Power consumed (per question No. 3):

$P_1 = 65.74W$
$P_2 = 8.21W$
$P_3 = 43.82W$
$P_4 = 98.6W$
$P_5 = 38.36W$
$P_6 = 142.43W$

Power consumed ($P_T = 397.2W$ [median])

$P_1 = 397.2W \times .166 = 65.94W$
$P_2 = 397.2W \times .021 = 8.34W$
$P_3 = 397.2W \times .11 = 43.69W$
$P_4 = 397.2W \times .248 = 98.51W$
$P_5 = 397.2W \times .097 = 38.53W$
$P_6 = 397.2W \times .359 = 142.59W$

$$P_T = 65.94W + 8.34W + 43.69W + 98.51W + 38.53W + 142.59W = 397.6W$$

Compared to the results obtained in question No. 3, the results in both cases are approximately the same. Again, just remember the difference is in the rounded off percentage values.

5. Observe the series circuit below. Each resistor in the circuit is paralleled with a single-pole switch that's displayed in the open position. Use the figure to evaluate the circuit.

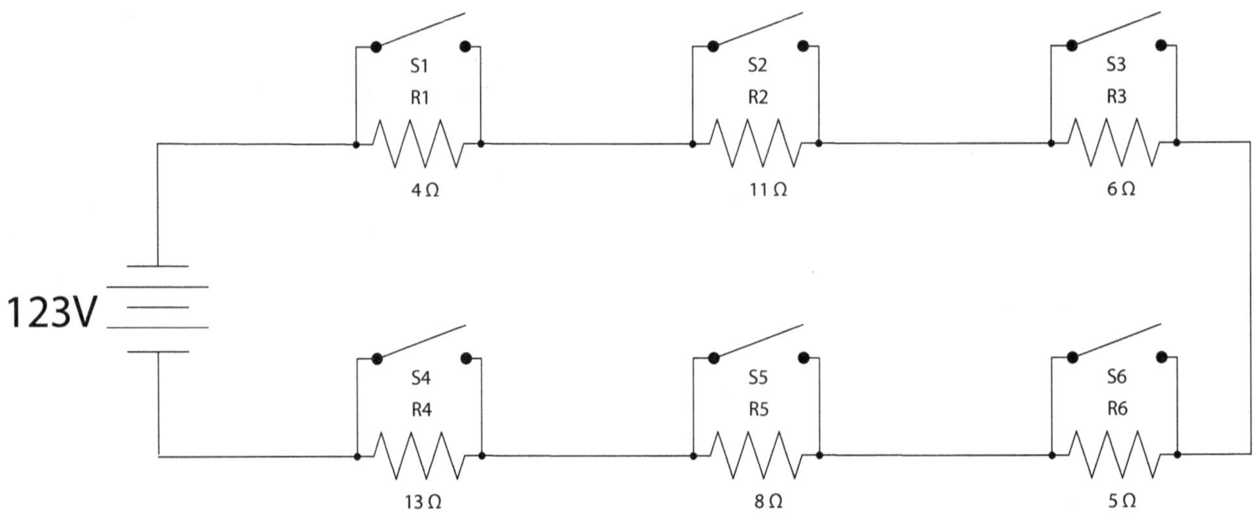

a. Determine the circuit's total resistance with all switches positioned as is.

With all switches open, the circuit's total resistance is 47Ω. That is,

$$R_T = 4\Omega + 11\Omega + 6\Omega + 13\Omega + 8\Omega + 5\Omega = 47\Omega$$

b. How much current flows through the circuit with all switches open?

$$I = 123V/47\Omega = 2.62A \text{ (rounded off)}$$

c. If switches S_1, S_4, and S_5 were closed, how much current would flow through the circuit?

With switches S_1, S_4, and S_5 closed, resistors R_1, R_4, and R_5 would be bypassed or shorted out, meaning those resistors would be removed from the circuit, leaving only R_2, R_3, and R_6 in the circuit, thus leaving the total resistance at

$$R_T = 11\Omega + 6\Omega + 5\Omega = 22\Omega$$

Therefore,

$$I = 123V/22\Omega = 5.59A$$

d. How much current would flow through the circuit if all switches were closed?

With all switches closed, there is zero (0) resistance in the circuit because the switches, when closed, bypass each resistor. With zero resistance, the circuit is short-circuited, thus creating an infinite amount of current flow through the circuit. Anytime there is zero resistance in a live circuit, the amount of current flowing through the circuit goes to infinity (∞) and remains as such until the flow of current is interrupted by means of a circuit meltdown or by the opening of an overcurrent device, if present. Infinity means, in this case, "an unlimited amount of current flow."

$$I = 123V/0\Omega = \infty$$

To get an ideal of how much current would flow through the circuit in a situation like this, let's use a number that is as close to 0Ω as possible. In this case let's use .01Ω. With this resistance, an ideal of the total current and the meaning of infinity can be realized:

$$I = 123V/.01\Omega = 12{,}300A$$

As you can see, a small amount of resistance produces an extreme amount of current at the given voltage, and when "0" resistance exists, you can't imagine how much current is then produced. This is why it is said that current goes to infinity. In this case, the more zeroes that are added after the decimal point (.001Ω, .0001Ω) to reduce the resistance, the more the current increases and dangerously flows to infinity.

The Parallel Circuit

What Is a Parallel Circuit?

A parallel circuit is the type of circuit in which the total current leaves the voltage source and divides between each individual resistance according to the value of the resistance. In a parallel circuit, the voltage drop across each resistance is the same as the voltage source. Just remember, with a parallel circuit, current always takes its least resistive path.

Rules That Apply to Parallel Circuits

Rule 1. The total resistance (R_T) of a parallel circuit is always less than the smallest resistance and can be found by using one of the three methods.

Method 1. Equal resistance.

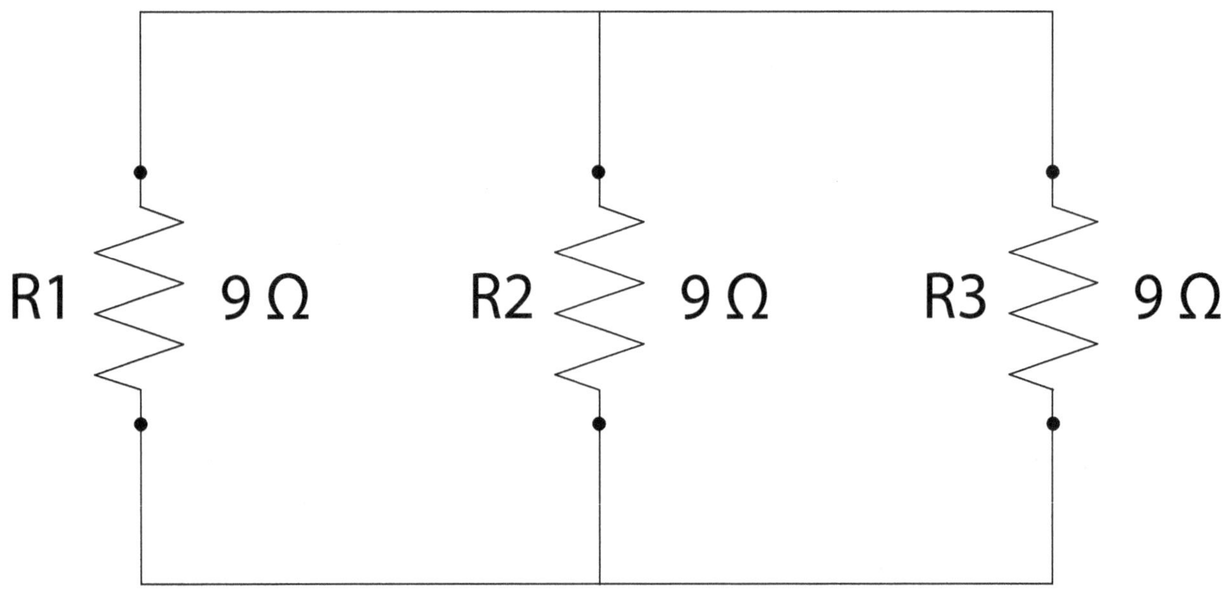

When the individual resistance values of a parallel circuit are the same, divide one resistance value by the number of parallel resistors, where

$$R_T = \frac{\text{R (individual resistance value)}}{\text{N (number of resistors)}}$$

Method 2. Two or more resistances (product of the sum method).*

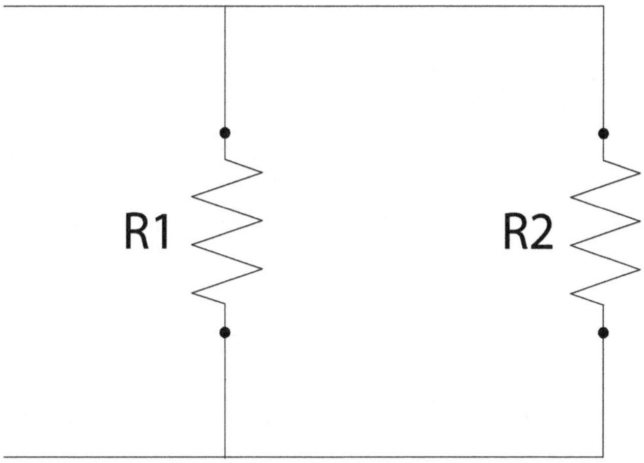

When at least two resistance are in parallel,

$$R_T = \frac{R_1 \text{ x } R_2 \text{ (product)}}{R_1 + R_2 \text{ (sum)}}$$

* The product of the sum method only allows two resistances to be evaluated at one time. When more than two resistances exist, always start with the last two resistances and work toward the first. When a voltage source exists, always start with the two resistances that are farthest away from the source, and work toward the resistance that is nearest the source.

Method 3. Two or more resistances (Reciprocal Method).**

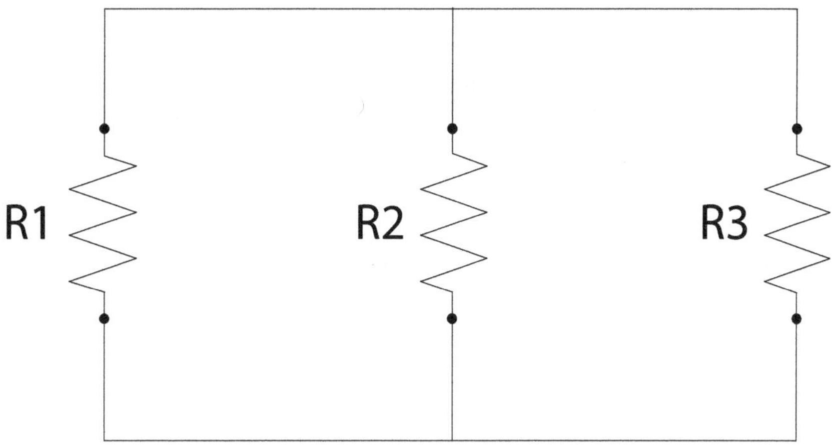

When at least two or more resistances are in parallel,

$$R_T = \frac{1}{1/R_1 + 1/R_2 + 1/R_3 + \ldots}$$

** This method allows an indefinite number of resistances to be evaluated at one time.

Rule 2. The total current (I_T) is equal to the sum of each individual current path through each individual resistance,

where

$$I_T = I_1 + I_2 + I_3 + \dots\dots\dots\dots$$
$$\text{and } (I_{1,2,3\dots} = E / [R_{1,2,3\dots}])$$

Rule 3. The voltage across each individual resistance load is equal to the voltage source (E),

where

$$E = E_1 = E_2 = E_3$$

Rule 4. The total power consumption (P_T) is equal to the sum of each individual resistance load power consumption,

where

$$P_T = P_1 + P_2 + P_3 + \dots\dots\dots\dots$$
$$\text{and } (P_{1,2,3\dots} = E \times [I_{1,2,3\dots}])$$

APPLYING THE RULES

Parallel Resistance (R)

As a reminder, when calculating the total resistance of a group of parallel resistance values or parallel circuits, the total resistance will always be less than the smallest resistance value. With this in mind, the concluding results in determining the total resistance of any parallel arrangement can always be determined before defining a definite answer.

1. Five heating coils are connected in parallel. The resistance of the heating coils is 15.87Ω, 32.3Ω, 26.19Ω, 12.5Ω, and 21.6Ω. The total resistance of the parallel circuit is ____Ω.

   a. 108.46 b. 3.87 c. .2585 d. 13.91

   Because 12.5Ω is the smallest resistance of the group, the total resistance will result in being smaller than 12.5Ω. Although the defining answer still has to be calculated, it is understood that the total resistance will be less than the smallest resistance value.

   Using the reciprocal method, where

$$R_T = \frac{1}{1/R_1 + 1/R_2 + 1/R_3 + 1/R_4 + 1/R_5}$$

$$R_T = \frac{1}{1/15.87\Omega + 1/32.3\Omega + 1/26.19\Omega + 1/12.5\Omega + 1/21.6\Omega}$$

$$R_T = \frac{1}{.2585\Omega} = 3.87\Omega$$

The total resistance of the parallel circuit is 3.87Ω, which is less than 12.5Ω, the smallest resistance value.

2. The total resistance of four 17Ω parallel resistors is ____ Ω.

a. 8.5 b. 5.67 c. 4.25 d. 7.5

$$R_T = \frac{17\Omega}{4} = 4.25\Omega$$

The equal resistance value method was used to determine the total resistance of the four parallel resistors, which is 4.25Ω, a value that is less than 17Ω, the smallest resistance value.

3. Use the product of the sum method to determine the total resistance of the following resistors, where

$R_1 = 7\Omega$, $R_2 = 5\Omega$, $R_3 = 6\Omega$, $R_4 = 12\Omega$, $R_5 = 8\Omega$

Using the formula $R_T = \frac{R_1 \text{ x } R_2}{R_1 + R_2}$

Starting with the last two resistances R_4 and R_5,

$$R_{4/5} = \frac{12\Omega \text{ x } 8\Omega}{12\Omega + 8\Omega} = \frac{96\Omega}{20\Omega} = 4.8\Omega$$

Evaluate R_3 and $R_{4/5}$:

$$R_{3/4/5} = \frac{6\Omega \text{ x } 4.8\Omega}{6\Omega + 4.8\Omega} = \frac{28.8\Omega}{10.8\Omega} = 2.67\Omega$$

Evaluate R_2 and $R_{3/4/5}$:

$$R_{2/3/4/5} = \frac{5\Omega \text{ x } 2.67\Omega}{5\Omega + 2.67\Omega} = \frac{13.35\Omega}{7.67\Omega} = 1.74\Omega$$

Evaluate R_1 and $R_{2/3/4/5}$:

$$R_{1/2/3/4/5} = \frac{7\Omega \text{ x } 1.74\Omega}{7\Omega + 1.74\Omega} = \frac{12.18\Omega}{8.74\Omega} = 1.394\Omega$$

Using the product of the sum method yields a total resistance of 1.394Ω, which again is less than 5Ω, the smallest resistance value of the group of resistances.

Parallel Resistance (R) Circuits

4. If the resistors in question No. 3 above were energized by a 50V source,
   a. How much current flows through each resistor?
   b. What is the voltage drop across each resistor?
   c. How much power is consumed by each resistor?

a. Because current always takes its least resistive path, more current will flow through the resistor having the smallest resistance, as opposed to the least amount of current flowing through the resistor having the greatest resistance. Again, by knowing this, the results can be determined before defining a definite answer. Therefore, as expected, most of the circuit's current will flow through R_2 and the least amount through R_4.

In a parallel circuit, the voltage (drop) across each resistor will always be the same as that of the source voltage, which is 50V. Using the formula I = E/R, the individual current flow through each resistor can be determined.

(3) $1_1 = 50V/7\Omega = 7.14A$
(2) $I_3 = 50V/6\Omega = 8.33A$
(1) $I_2 = 50V/5\Omega = 10A$
(5) $I_4 = 50V/12\Omega = 4.17A$
(4) $I_5 = 50V/8\Omega = 6.25A$

As expected, the results are validated based on the order of each calculated answer.

Just as a series circuit in which the individual voltage drop across each resistor when totaled is equal to the source voltage (E_T), in a parallel circuit when the current flowing through each individual resistor is totaled, the sum is equal to the circuit's total current (I_T):

$$I_T = 7.14A + 10A + 8.33A + 4.17A + 6.25A = 35.89A$$

Using the formula $I_T = E/R_T$ (R_T; see question No. 3) will produce results that are approximately the same.

$$I_T = 50V/1.394\Omega = 35.87A \text{ (rounded off)}$$

b. As stated earlier, the voltage (drop) across each resistor in a parallel circuit will always be the same as that of the source voltage. Therefore, the voltage (drop) across each resistor is 50V.

c. The power consumed by each resistor can also be determined by any one of the three power (P) formulas. However, due to the majority of the individual calculated current values above being rounded off, the formula $P = E^2/R$ would provide the most accurate results because it does not involve the use of I. Since the power consumption of any circuit is dependent upon the amount of current flow, the resistor having the largest current flow will consume the greatest amount of power followed down to the resistor having the smallest current flow consuming the least amount of power.

As an analogy, consider your monthly power bill. The watt/hour meter, which measures the amount of power consumed by the occupancy, is actually a current transformer. As current flows through the service conductors, the amount of current is measured through the watt/hour meter, which translates the amount of power consumed based on the service operating voltage, which is fairly constant compared to the current flow, which varies frequently. The amount of current flow (load) through the watt/hour meter is indicative of the rate at which the measuring disk revolves. In other words, the higher the current flow, the faster the disk revolves, the more power consumed, the more you pay, and vice-versa.

Therefore, the results of the power consumed by each resistor in this circuit can also be determined based on the individual current values per resistance before defining a definite answer.

(3) $P_1 = (50V)^2/7\Omega = 357.14W$ (1) $P_2 = (50V)^2/5\Omega = 500W$
(2) $P_3 = (50V)^2/6\Omega = 416.67W$ (5) $P_4 = (50V)^2/12\Omega = 208.33W$
(4) $P_5 = (50V)^2/8\Omega = 312.5W$

Adding these calculated results will provide the circuit's total power consumption.

$$P_T = 357.14W + 500W + 416.67W + 208.33W + 312.5W = 1794.64W$$

Again, by using the formula $P_T = E/R_T$ (R_T—see question No. 3) approximate results are produced.

$$P_T = (50V)^2/1.394\Omega = 1793.4W$$

5. Using the circuit in question No. 4 again, let's consider the same alternate method as earlier used with the series circuit to determine the current flow and power consumed by each resistor.

Again, based on the circuit's total resistance, let's determine the percentage each resistance contributes to the circuit's total resistance. This will provide an alternate method for analyzing the circuit. Because this is a parallel circuit, which is the opposite of a series circuit, the percentage each resistance contributes to the circuit's total resistance is now determined by dividing each individual resistance into the total resistance (R_T/R). Observe:

$$\%_1 = \frac{1.394\Omega}{7\Omega} = .199\ (19.9\%) \qquad \%_2 = \frac{1.394\Omega}{5\Omega} = .279\ (27.9\%)$$

$$\%_3 = \frac{1.394\Omega}{6\Omega} = .232\ (23.2\%) \qquad \%_4 = \frac{1.394\Omega}{12\Omega} = .116\ (11.6\%)$$

$$\%_5 = \frac{1.394\Omega}{8\Omega} = .174\ (17.4\%)$$

Now that the percentage values are determined, they should all total up to 100%.

$$\%_T = 19.9\% + 27.9\% + 23.2\% + 11.6\% + 17.4\% = 100\%$$

As you can see, in this situation, the total is exactly 100%. However, just remember that the percentage values were rounded off, which again will cause a compromise in accuracy.

These percentage values can be used in like fashion as those used with the series circuit. Remember the current flow and the power consumed by each resistance calculated in question No. 4. As an alternate method, the same or approximate results can now be obtained by multiplying the total current (I_T) and the total power (P_T) by each individual percentage value (in decimal form). Again, the alternate method provides an alternate means for checking answers quick and independently.

Current flow (per question No. 4)

$$I_1 = 7.14A \quad I_2 = 10A \quad I_3 = 8.33A \quad I_4 = 4.17A \quad I_5 = 6.25A$$

Current flow ($I_T = 35.89A$)

$I_1 = 35.89A \times .199 = 7.14A$

$I_2 = 35.89A \times .279 = 10A$

$I_3 = 35.89A \times .232 = 8.33A$

$I_4 = 35.89A \times .116 = 4.16A$

$I_5 = 35.89A \times .174 = 6.24A$

$I_T = 7.14A + 10A + 8.33A + 4.16A + 6.24A = 35.87A$

Power consumed (per question No. 4):

$P_1 = 357.14W$
$P_2 = 500W$
$P_3 = 416.67W$
$P_4 = 208.33W$
$P_5 = 312.5W$

Power consumed ($P_T = 1794.67W$)

$P_1 = 1794.67W \times .199 = 357.14W$
$P_2 = 1794.67W \times .279 = 500.7W$
$P_3 = 1794.67W \times .232 = 416.36W$
$P_4 = 1794.67W \times .116 = 208.18W$
$P_5 = 1794.67W \times .174 = 312.27W$

$P_T = 357.14W + 500.7W + 416.36W + 208.18W + 312.27W = 1794.65W$

Compared to the results obtained in question No. 4, the results in both cases are approximately the same. Again, just remember the difference is in the rounded off percentage values.

6. Observe this parallel circuit. Each resistor in the circuit is paralleled with a single-pole switch that's displayed in the open position. Use the parallel configuration to evaluate the circuit.

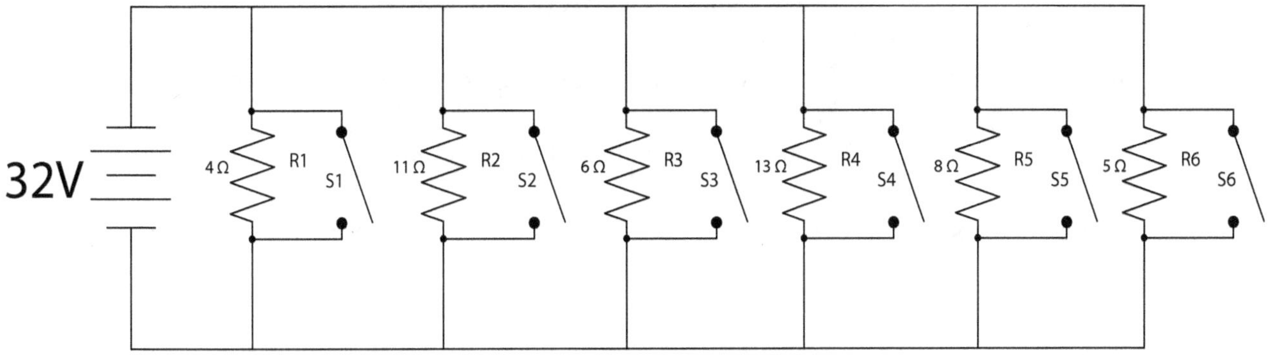

a. Determine the circuit's total resistance with all switches positioned as is.

With all switches open, the circuit's total resistance is 1.1Ω. That is,

$$R_T = \frac{1}{1/4\Omega + 1/11\Omega + 1/6\Omega + 1/13\Omega + 1/8\Omega + 1/5\Omega} = 1.1\Omega \text{ (rounded off)}$$

b. How much current flows through the circuit with all switches open?

$$I_T = 32V/1.1\Omega = 29.1A \text{ (rounded off)}$$

c. If switches S_2, S_3, and S_6 were closed, how much current would flow through the circuit?

With S_2, S_3, and S_6 closed, R_2, R_3, and R_6 would be bypassed or shorted out, leaving only R_1, R_4, and R_5 in the circuit, leaving the total resistance at

$$R_T = \frac{1}{1/4\Omega + 1/13\Omega + 1/8\Omega} = 2.21\Omega \text{ (rounded off)}$$

Therefore,

$$I_T = 32V/2.21\Omega = 14.48A \text{ (rounded off)}$$

d. If switches S_1, S_4, S_5, and S_6 were closed, how much current would flow through the circuit?

With S_1, S_4, S_5, and S_6 closed, R_1, R_4, R_5 and R_6 would be bypassed or shorted out, leaving R_2, and R_3 in the circuit, leaving the total resistance at

$$R_T = \frac{1}{1/11\Omega + 1/6\Omega} = 3.88\Omega \text{ (rounded off)}$$

Therefore,

$$I_T = 32V/3.88\Omega = 8.25A \text{ (rounded off)}$$

e. How much current will flow through the circuit if all switches were closed?

$$I_T = 32V/0\Omega = \infty \text{ (the current goes to infinity)}$$

With all switches closed, the results will be the same as that in question No. 5d., Series Resistance Circuits.

The Combination Circuit

What Is a Combination Circuit?

A combination circuit is a circuit that contains both series and parallel configurations, where the configurations can be either series-parallel or parallel-series connected or combined.

When one or more series resistances are connected in parallel, a configuration is said to be series-parallel connected, and when one or more parallel resistances are connected in series a configuration is said to be parallel-series connected. Again both configurations can be combined. The best way to view either of the two is to always remember,

1. Where resistances are aligned to share the same current the resistances are in series

and

2. Where resistances are aligned to receive different amounts of current, the resistances are in parallel.

Now observe the three type configurations below.

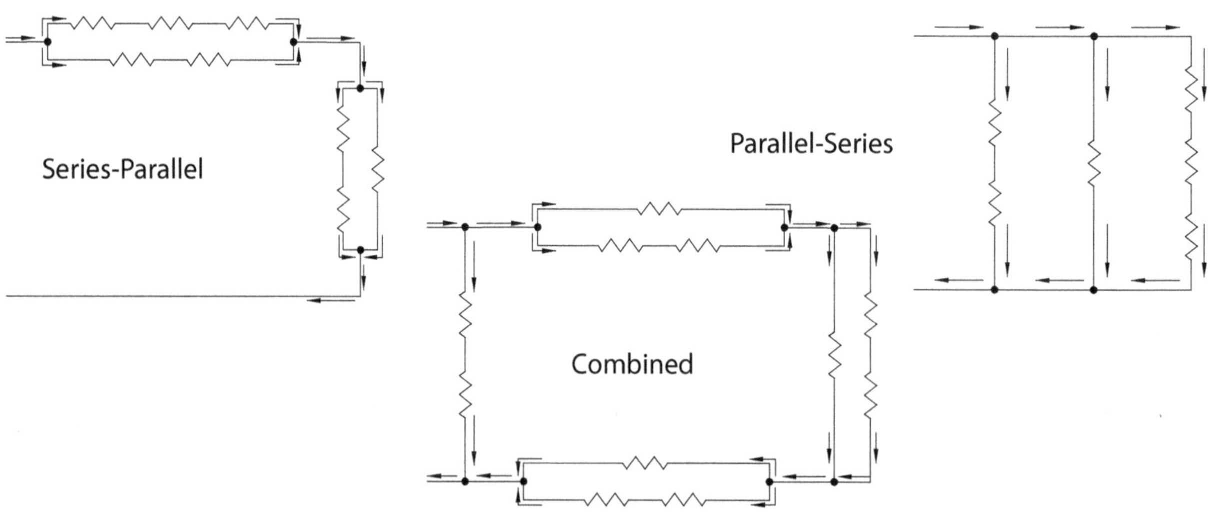

Since the resistance of $R_{6/7}$ (13Ω) is the smallest of the two resistors, the results of the total resistance can be predicted before deriving an answer; therefore, the total resistance between $R_{6/7}$ and $R_{8/9}$ should be less than 13Ω.

Either the product of the sum or the reciprocal method can be used to determine the resistor's total resistance. For this step alone, let's use both methods because they will both provide the same answer.

Product of the Sum Method

$$R_{T(R6/7)\,||\,(R8/9)} = \frac{13\Omega \times 22\Omega}{13\Omega + 22\Omega} = \frac{286\Omega}{35\Omega} = 8.17\Omega \text{ (rounded off)}$$

Reciprocal Method

$$R_{T(R6/7)\,||\,(R8/9)} = \frac{1}{1/13\Omega + 1/22\Omega} = \frac{1}{.1224\Omega} = 8.17\Omega \text{ (rounded off)}$$

As predicted, the total resistance between the two resistors is less than 13Ω. The equivalent resistance between the two parallel resistors will now be referred to as $R_{6/7/8/9}$.

The original circuit has now become the following circuit:

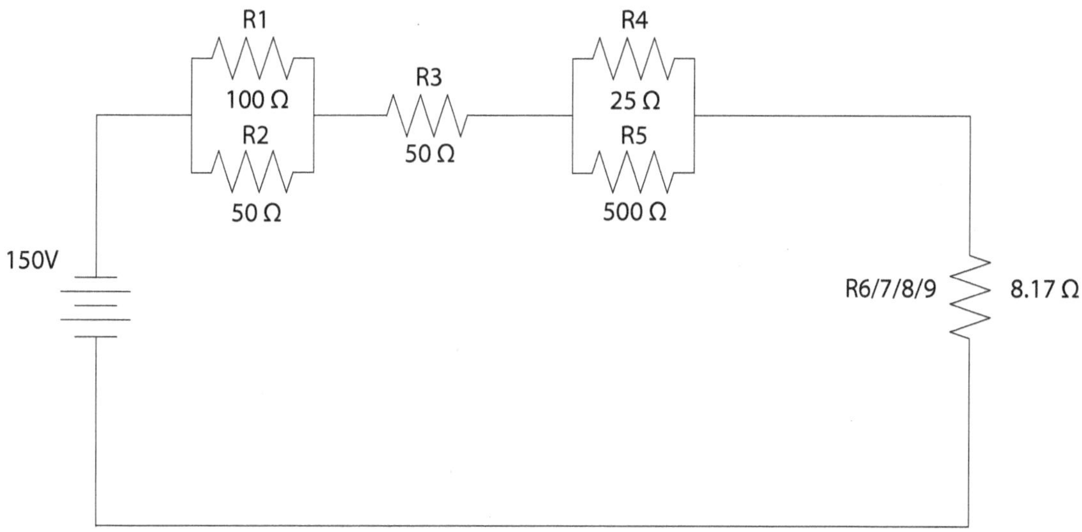

With the exception of the parallel branches R_1 and R_2, and R_4 and R_5, the circuit resembles a series circuit. However, if the total resistance between the parallel branches is determined, the circuit then becomes an equivalent series circuit, which is always the objective when evaluating a series-parallel circuit. Knowing this, let's determine the total resistance between the parallel branches.

Resistors R_1 and R_2 (In parallel)

The results should be less than 50Ω.

$$R_{T(R1\,||\,R2)} = \frac{1}{1/100\Omega + 1/50\Omega} = \frac{1}{.03\Omega} = 33.33\Omega \text{ (rounded off)}$$

Resistors R_1 and R_2 are now referred to as $R_{1/2}$.

Resistors R_4 and R_5 (in parallel)

The results should be less than 25Ω.

$$R_{T(R4\,||\,R5)} = \frac{1}{1/25\Omega + 1/500\Omega} = \frac{1}{.042\Omega} = 23.81\Omega \text{ (rounded off)}$$

Substituting the total resistance of the parallel branches R_1 and R_2 and R_4 and R_5, the circuit now becomes an equivalent series circuit.

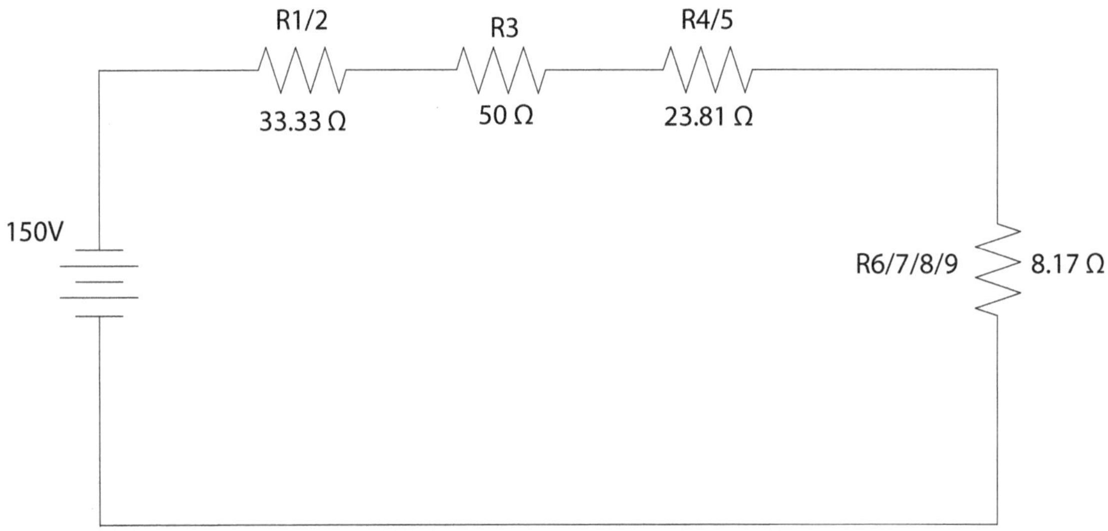

By converting the original combination circuit into an equivalent series circuit, the operating conditions of the original circuit can now be evaluated much easier. Could you imagine trying to determine the total current of the original circuit as is, or the individual voltage drops or current flow through each resistor? With an equivalent series circuit, determining such information is made quite simple. The original circuit's total resistance can now be determined simply by adding the series resistance values together.

$$R_T = 33.33\Omega + 50\Omega + 23.81\Omega + 8.17\Omega = 115.31\Omega$$

With the total resistance, the circuit's total current can be determined.

$$I = 150V/115.31\Omega = 1.3A$$

Because only one source of current flows in a series circuit, the voltage drop across each combination of resistors can be determined. Take a look:

$$E_{1/2} = 33.33\Omega \text{ x } 1.3\text{A} = 43.33\text{V (rounded off)}$$
$$E_3 = 50\Omega \text{ x } 1.3\text{A} = 65\text{V (rounded off)}$$
$$E_{4/5} = 23.81\Omega \text{ x } 1.3\text{A} = 30.95\text{V (rounded off)}$$
$$E_{6/7/8/9} = 8.17\Omega \text{ x } 1.3\text{A} = 10.62\text{V (rounded off)}$$

Notice the consistency of the results: the highest resistance, the largest voltage drop; the lowest resistance, the smallest voltage drop.

Adding each individual voltage drop together, the results should approximate the total source voltage (150V).

$$E_T = 43.33\text{V} + 65\text{V} + 30.95\text{V} + 10.62\text{V} = 149.9\text{V}$$

With these calculated results, the following questions can now be answered based on the original combination circuit.

b. How much current flows through each resistor?

Resistors R_1 and R_2

The circuit's total current will be divided between R_1 and R_2 and will then flow as the total current again once passed through the resistors. Because R_2 has the smallest resistance, most of the total current will flow through R_2.

Current through R_1 and R_2 can be calculated by applying the following formulas: $I_1 = E_{1/2}/R_1$ and $I_2 = E_{1/2}/R_2$. Because both resistors are in parallel, they will share the same voltage drop, which is consistent with the rules applied to parallel circuits; the same will hold for R_4 and R_5. As a result,

$$I_1 = 43.33\text{V}/100\Omega = .433\text{A} \quad I_2 = 43.33\text{V}/50\Omega = .867\text{A}$$

where

$$.433\text{A}(I_1) + .867\text{A}(I_2) = 1.3\text{A}(I_T)$$

Resistor R_3

Because resistor R_3 is the only true series resistor of the circuit, it is the only resistor that will experience the circuit's total current.

$$I_3 = I_T = 1.3\text{A}$$

Resistors R_4 and R_5

$$I_4 = E_{4/5}/R_4 \qquad I_5 = E_{4/5}/R_5$$

$$I_4 = 30.95V/25\Omega = 1.238A \qquad I_5 = 30.95V/500\Omega = .0619A$$

$$1.238A(I_4) + .0619A(I_5) = 1.2999A(I_T) \text{ (1.3A if rounded off)}$$

Resistors R_6 and R_7 and R_8 and R_9

Once the total current flow through R_4 and R_5 is resumed, it is then divided between the parallel-series resistors R_6 and R_7 and R_8 and R_9. Because current always takes its least-resistive path, most of the total current is expected to flow through R_6 and R_7. Since resistors R_6 and R_7 are in series, they will experience the same amount of current as will resistors R_8 and R_9. Because R_6 and R_7 are in parallel with R_8 and R_9, they will share the same voltage drop.

$$I_{6\,\&\,7} = E_{6/7/8/9}/R_{6\,+}R_7 \qquad I_{6\,\&\,7} = 10.62V/13\Omega = .817A$$

$$I_{8\,\&\,9} = E_{6/7/8/9}/R_{8\,+}R_9 \qquad I_{8\,\&\,9} = 10.62V/22\Omega = .483A$$

$$.817A(I_{6\,\&\,7}) + .483A(I_{8\,\&\,9}) = 1.3A$$

Once the calculated current values flow through R_6 and R_7, and R_8 and R_9, they add together again and return to the source and repeat the same process.

c. Determine the individual voltage drops across $R_{6,7,8\&9}$.

Although 10.62V was determined to be the combined voltage drop across R_6 and R_7, and R_8 and R_9, this was based on the overall total resistance of each set of series resistors, not on each individual resistor.

Because the amounts of current flow through R_6 and R_7, and R_8 and R_9 are known, the voltage across each individual resistor can now be determined.

$$E_6 = 10\Omega \text{ x } .817A = 8.17V \qquad E_8 = 7\Omega \text{ x } .483A = 3.38V$$

$$E_7 = 3\Omega \text{ x } .817A = 2.45V \qquad E_9 = 15\Omega \text{ x } .483A = 7.25V$$

Once these individual voltage drops are totaled, they should equal the total voltage drop across each set of series resistors: 10.62V (or approximate).

$$8.17V(E_6) + 2.45V(E_7) = 10.62V$$

$$3.38V(E_8) + 7.25V(E_9) = 10.63V$$

d. How much power is consumed by each resistor?

Because the resistance, current, and voltage drop across each resistor are known, any of the three power formulas can be used; that is, $P = E \times I$, $P = I^2 \times R$, and $P = E^2/R$.

For this question, the formula $P = E \times I$ will be used; however, if you desire to verify the individual results, either of the two remaining formulas can be used.

$$P_1 = 43.33V \times .43A = 18.63W$$
$$P_2 = 43.33V \times .87A = 37.70W$$
$$P_3 = 65V \times 1.3A = 84.50W$$
$$P_4 = 30.95V \times 1.24A = 38.38W$$
$$P_5 = 30.95V \times .062A = 1.92W$$
$$P_6 = 8.17V \times .817A = 6.67W$$
$$P_7 = 2.45V \times .817A = 2W$$
$$P_8 = 3.38V \times .483A = 1.63W$$
$$P_9 = 7.25V \times .483A = 3.50W$$

Make note that with the exception of P_3, all other results were rounded off.

In conclusion, the individual results will be totaled and compared with the circuit's total power consumption using the formula $P = E \times I$, where E is the circuit's source voltage and I is the circuit's total current.

$$P_T = 18.63W + 37.70W + 84.50W + 38.38W + 1.92W + 6.67W + 2W + 1.63W + 3.50W = 194.93W$$

$$P_T = 150V \times 1.3A = 195W$$

As you can see, both calculations yield results that are approximately the same.

2. Observe the circuit below.

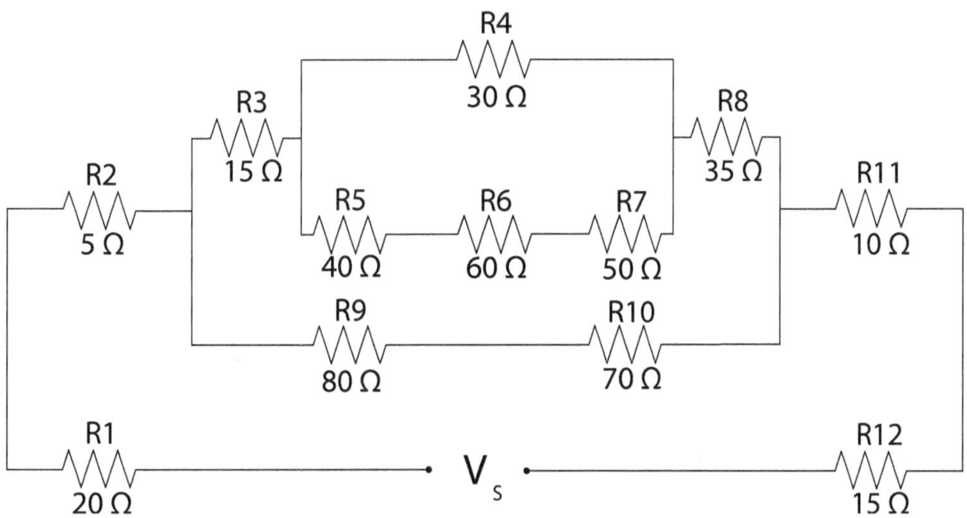

Perhaps this circuit is similar to one that you are familiar with. And just like that circuit, the same question is still probably asked, and that's, "How is the total resistance determined?" Because the circuit appears to be somewhat complex, it can be easily said that the question is valid. Again, this is a combination circuit, and the objective is to convert it into an equivalent series resistance circuit.

Let's assume a voltage source exists where marked V_s. With this in mind, the resistance farthest away from the would-be source is resistor R_4, which is in parallel with the series resistors R_5, R_6, and R_7.

Because the total series resistance of R_5, R_6, and R_7 is 150Ω, the next step is to find the total parallel resistance of R_4, R_5, R_6, and R_7.

$$R_{T\ (R4,\ R5,\ R6||R7)} = \frac{1}{1/30\Omega + 1/150\Omega} = 25\Omega$$

As you can see, the results are consistent with the rule; 25Ω is less than the smallest parallel resistor, 30Ω.

The circuit now resembles the following, where R_4, R_5, R_6 and R_7 are now referred to as $R_{4/5/6/7}$:

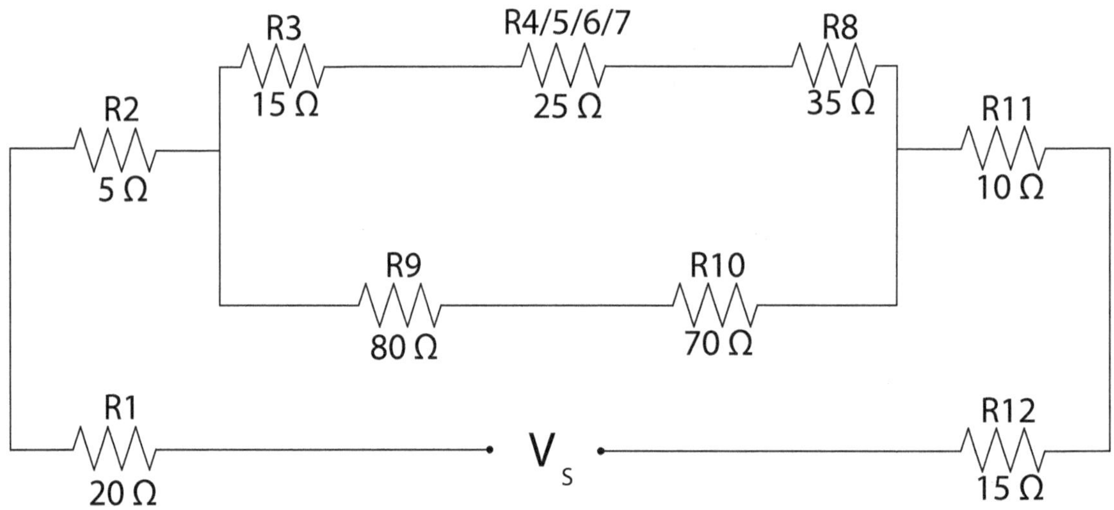

Because $R_{4/5/6/7}$ are in series with resistors R_3 and R_8, the resistance values can be added together, where

$$R_{T\ (R3,\ R4/5/6/7,\ R8)} = 75\Omega$$

The circuit now resembles the following, where R_3, $R_{4/5/6/7}$, and R_8 are now referred to as $R_{3/4/5/6/7/8}$:

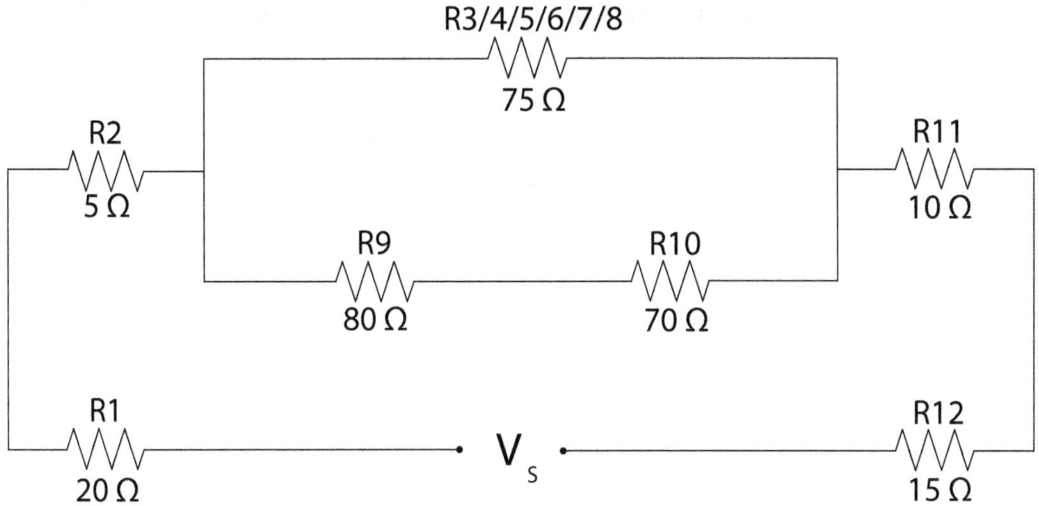

Because $R_{3/4/5/6/7/8}$ are parallel to the series resistors R_9 and R_{10}, which totals 150Ω, the total resistance between the parallel resistors is

$$R_{T\ (R3/4/5/6/7/8\ ||\ R9\ \&\ R10)} = \frac{1}{1/75\Omega + 1/150\Omega} = 50\Omega$$

Again, the results prove consistent with the rule; 50Ω is less than the smallest parallel resistor, 75Ω.

$R_{3/4/5/6/7/8}$ and R_9 and R_{10} can now be referred to as $R_{3/4/5/6/7/8/9/10}$.

By inserting the total resistance of $R_{3/4/5/6/7/8/9/10}$ between resistors R_1, R_2, R_{11}, and R_{12}, the circuit

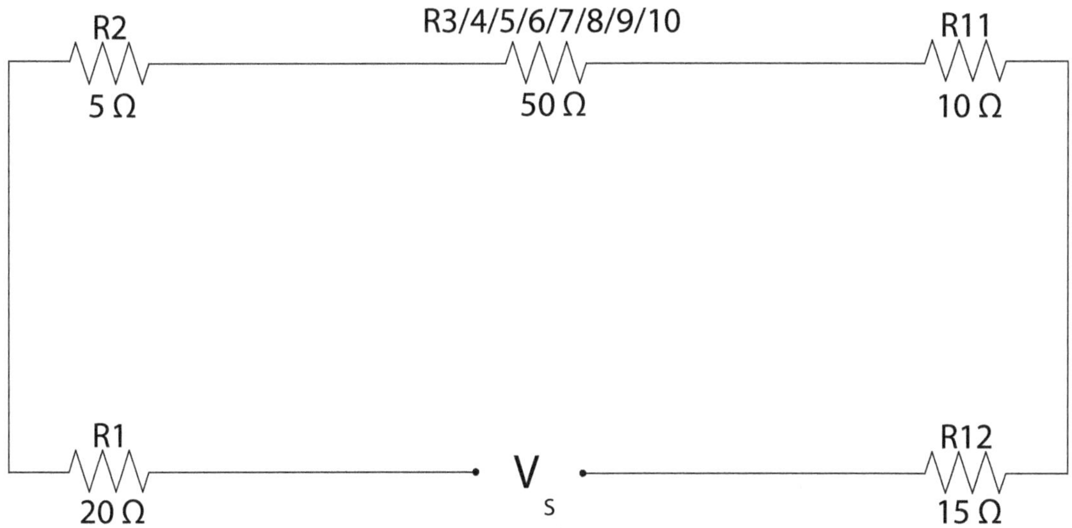

has now been converted into an equivalent series circuit, and the circuit's total resistance can be determined, where

$$R_T = 20\Omega + 5\Omega + 50\Omega + 10\Omega + 15\Omega = 100\Omega$$

In conclusion, the original circuit is now equivalent to the following:

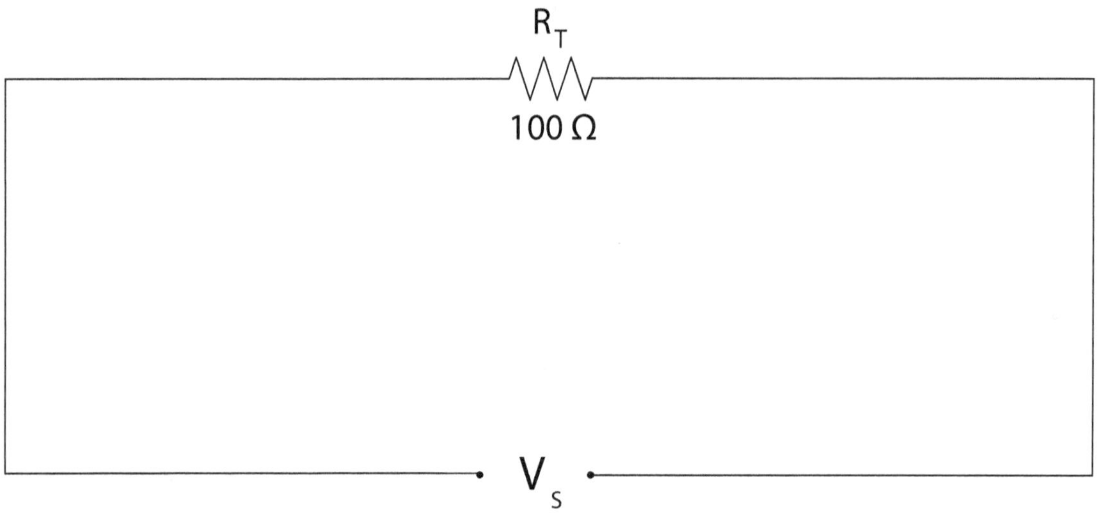

Parallel-Series Resistance (R) Circuit

1. Observe the parallel-series circuit below:

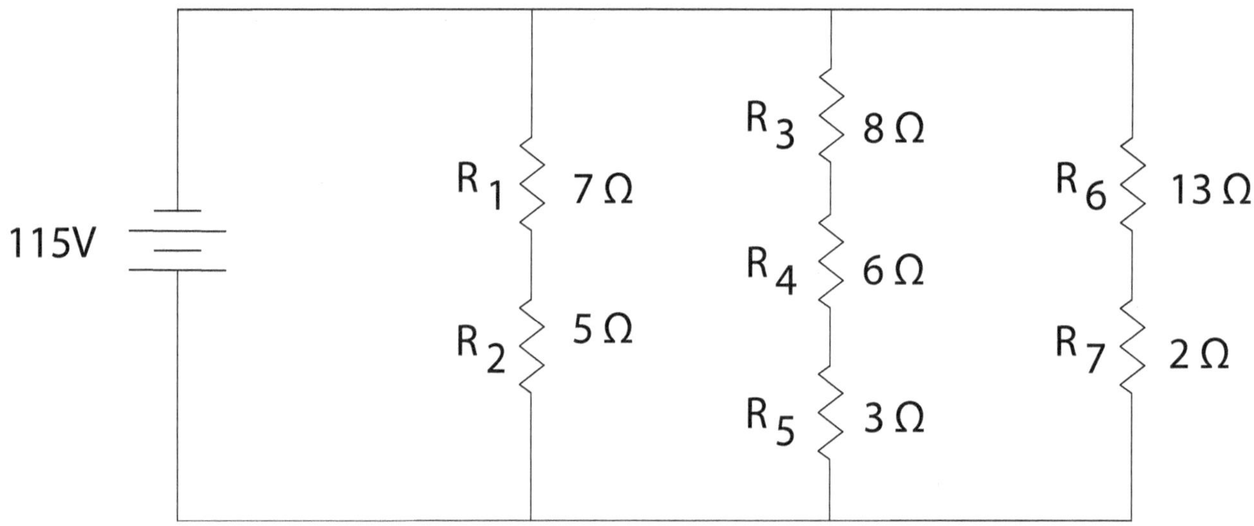

Determine the circuit's total resistance, the current flow through each series resistor, the voltage drop across each resistor, the power consumed by each resistor, and the circuit's total power consumption.

Circuit's Total Resistance

Because the series resistors are in parallel, the circuit's total resistance can be determined just as any other group of parallel resistors. By adding the resistance of each series resistor together, the totaled resistance between each set of series resistors will allow the circuit to appear and operate similar to that of a parallel circuit. Observe this circuit:

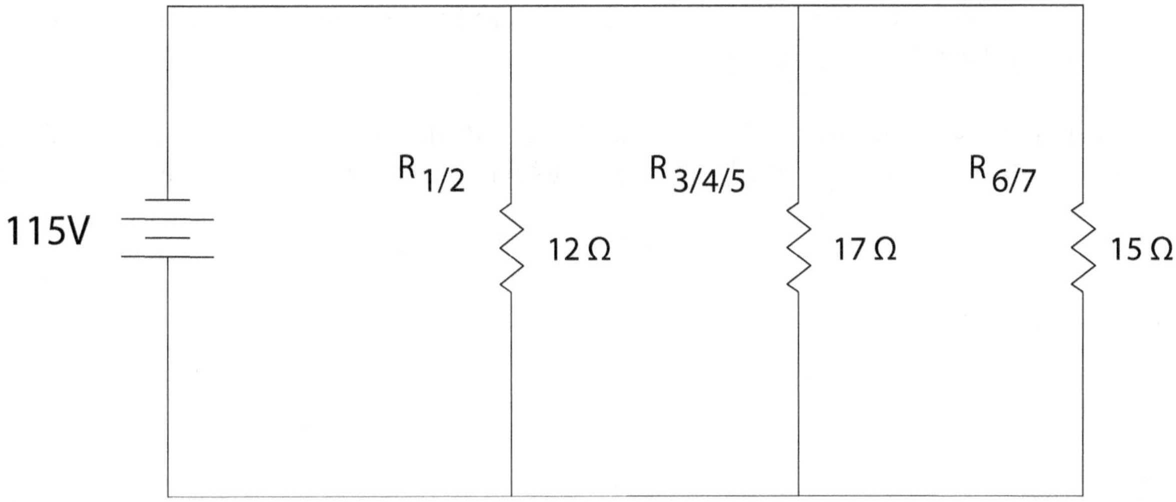

Now the circuit's total resistance can be determined simply by using the product of the sum method or the reciprocal method. In this case, let's apply each method to verify the calculated results.

Using the product of the sum method, only two parallel resistors can be evaluated at a time. Therefore, starting with $R_{6/7}$ and $R_{3/4/5}$, the total resistance $R_{3/4/5/6/7}$ is,

$$\frac{15\Omega \times 17\Omega}{15\Omega + 17\Omega} = 7.97\Omega$$

With this calculated resistance, the remaining resistance $R_{1/2}$ can be used to calculate the circuit's total resistance. Therefore,

$$R_T = \frac{12\Omega \times 7.97\Omega}{12\Omega + 7.97\Omega} = 4.79\Omega$$

Now use the reciprocal method to verify the results found using the product of the sum method. Where

$$R_T = \frac{1}{1/12\Omega + 1/17\Omega + 1/15\,\Omega} = 4.79\Omega$$

Based on the validation of both methods, the total resistance is 4.79Ω.

Individual Currents

Refer to the latter figure. Because the voltage drop across a parallel circuit is the same as the circuit's applied voltage, the individual current flow can be determined using each set of series resistors' total resistance. Therefore,

$I_{R1/2} = 115V/12\Omega$
$= 9.58A$

$I_{R3/4/5} = 115V/17\Omega$
$= 6.76A$

$I_{R6/7} = 115V/15\Omega$
$= 7.67A$

As a result, the current flowing through each resistor can be determined because resistors arranged in series will share the same current.

Although not asked, the circuit's total current (I_T) can be determined by totaling each individual current together and can be verified by using the total resistance and applied voltage. As a result,

$$I_T = 9.58A + 6.76A + 7.67A = 24.01A$$

and

$$I_T = 115V/4.79\Omega = 24.01A$$

Voltage Drops

Referring back to the original parallel-series circuit, clearly it is understood that the voltage drops across each set of parallel resistors are the same as that of the applied voltage. However, because the parallel resistors are connected in series and a given amount of current will flow through each resistor, an individual voltage drop across each resistor will exist. Therefore,

$$E_1 = 7\Omega \times 9.58A = 67.06V$$
$$E_2 = 5\Omega \times 9.58A = \underline{47.90V}$$
$$114.96V$$

$$E_3 = 8\Omega \times 6.76A = 54.08V$$
$$E_4 = 6\Omega \times 6.76A = 40.56V$$
$$E_5 = 3\Omega \times 6.76A = \underline{20.28V}$$
$$114.92V$$

$$E_6 = 13\Omega \times 7.67A = 99.71V$$
$$E_7 = 2\Omega \times 7.67A = \underline{15.34V}$$
$$115.05V$$

Take notice that the voltage drops across each set of series resistors are equal to or approximately the same as the applied voltage because they are in parallel.

Power Consumed

Even though there are three formulas that could be used to determine the power consumed by each resistor, we'll only use one for simplicity: P = E x I.

Therefore,

$$P_1 = 67.06V \times 9.58A = 642.43W$$
$$P_2 = 47.90V \times 9.58A = 458.88W$$
$$P_3 = 54.08V \times 6.76A = 365.58W$$
$$P_4 = 40.56V \times 6.76A = 274.19W$$
$$P_5 = 20.28V \times 6.76A = 137.09W$$
$$P_6 = 99.71V \times 7.67A = 764.78W$$
$$P_7 = 15.34V \times 7.67A = \underline{117.66W}$$
$$2760.61W$$

Circuit's Total Power Consumed

The circuit's total power based on each resistor's individual power consumption has already been determined by summation; however, to verify the results, we'll use the same formula to calculate the circuit's total power based on the applied voltage and the circuit's total current. As a result,

$$P_T = 115V \times 24.01A = 2761.15W$$

As you can see, regardless of whether the circuit's total power consumption is derived by summation or formula, the results are approximately the same.

ALTERNATING CURRENT

Without alternating current (*ac*), it would be impossible to operate induction motors and transformers, and without induction motors and transformers, we would still be living in the eighteenth century.

In the late 1800s, a naturalized American citizen named Nikola Tesla proved that alternating current was far more versatile and superior to direct current (*dc*). He discovered that with alternating current, it was possible to vary the magnitude of voltage through transformation, and electric motors could be made to rotate by means of induction.

In this country, electricity is generated at various voltages that alternate at a frequency of 60 cycles per second, or 60 hertz (Hz). The number of cycles (rate of change) that an *ac* sine wave completes in one second is referred to as the frequency. Hertz is the unit of measurement for frequency—named after German physicist Heinrich Hertz—and is abbreviated Hz where this abbreviation is used in place of the expression *cycles per second.* A hertz signifies that alternating current (voltage) goes through a given number of complete cycles of alternations in one second. In other words, if an alternating current or voltage goes through five complete cycles of alternations in one second, it is said that such alternations occur at a frequency of 5Hz.

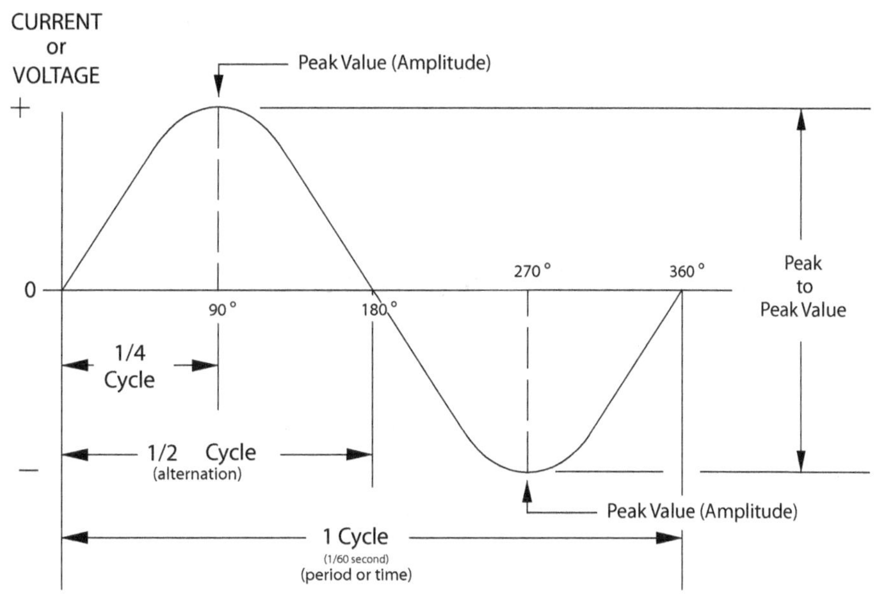

Fig. 8. The sine wave.

As displayed in figure 8, the current (voltage) wave starts at zero, reaches a peak on the positive (+) side of the zero axis, returns to zero, continues to another peak on the negative (-) side of the zero axis, and then returns to zero again. One positive and one negative loop represents one complete or full cycle, or 1Hz. On the other hand, a half-cycle or half of a complete cycle is called an alternation. Therefore, a 60Hz cycle goes through 60 complete sets of positive and negative loops in one second.

Another way of looking at it is this: it takes one-sixtieth of a second (1/60s, or .0167s, or 16.7 ms [milliseconds]) to complete one cycle. The time it takes alternating current (voltage) to complete one full cycle is known as the period, or simply the time. If the operating frequency is known, the period of a cycle can be calculated using the given formula:

$$\text{period} = \frac{1}{\text{frequency}}$$

Because AC voltages are generated by round rotors rotating inside a circular stator, the resulting waveform is called a sine wave. It is called a sine wave because the voltage or current source at any instant is proportional to the sine of the angle at that instant. The maximum height of a sine wave is called the peak value, or amplitude. The peak value can be either above the zero (0) axis, which is referred to as the positive (+) peak value, or below the zero (0) axis, which is referred to as the negative (-) peak value. When combined, the positive and negative peaks are known as the peak-to-peak value.

PRACTICAL APPLICATIONS

An *ac* sine wave completes a full cycle in .005 seconds. Determine the frequency of the cycle.

The frequency of the cycle can be determined by using the converted formula frequency (Hz) = 1/period (time).

$$\text{frequency} = 1/.005\text{s} = 200\text{Hz}$$

The frequency of the cycle is 200 hertz (Hz).

How much time does it take a 50Hz cycle to complete one full cycle?

Use the formula period (time) = 1/frequency (Hz)

$$\text{Period (time)} = 1/50\text{Hz} = .02\text{s}$$

A 50Hz cycle takes .02 seconds to complete one full cycle.

AC Voltage/Current Values

Just as in a *dc* circuit in which *dc* voltage is used to calculate current, the effective value of *ac* voltage is used likewise in a purely *ac* resistive circuit to calculate current. In an *ac* system in which voltage and current values are always changing, there are other parts of these values that must be considered to understand the basic concepts of alternating voltage and current. That being the case, let's take an in-depth look at four values that go hand in hand with *ac* systems, although most of them are never directly applied nor mentioned. Observe the illustrated graph in figure 9.

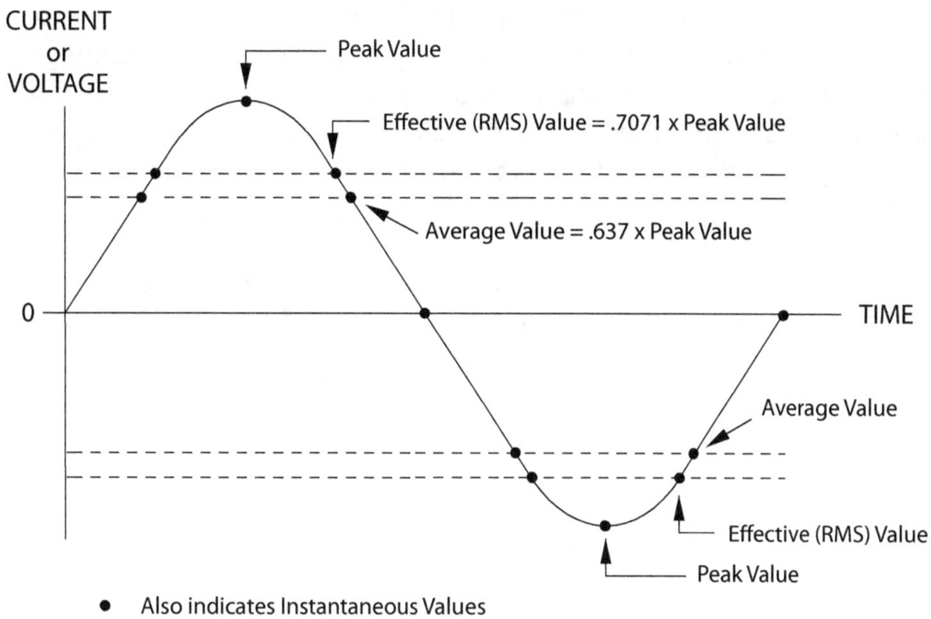

Fig. 9. The *ac* values.

Peak Value: In addition to that which has already been displayed, the peak value is the magnitude of the applied voltage or current that ranges between zero and the highest point of a sine wave for one half-cycle.

PRACTICAL APPLICATION

If the average voltage of an ac circuit is 114 volts, determine the circuit's peak voltage.

The peak voltage of the circuit can be determined by using the formula

$$E_{PEAK} = 1.57 \times E_{AVG}$$

$$E_{PEAK} = 1.57 \times 114V = 178.98V$$

The peak voltage is 178.98V or approximately 179V.

Instantaneous Value: With respect to time or phase angle, the instantaneous value provides a particular voltage or current value at any point about the sine wave. By use of trigonometry (reference the topic Adding Vectors That Forms Right Angles), the given trigonometric functions,

$$\sin\theta = v/V_{max} \text{ and } \sin\theta = i/I_{max}$$

can be used to derive the following instantaneous voltage (v) and current (i) values:

$$v = V_{max} \times \sin\theta \text{ or } i = I_{max} \times \sin\theta$$

about a sine wave, where the small letters v and i represent the calculated instantaneous values, the capital letters V and I represent the maximum (max) applied voltage and operating current of an *ac* system, and θ represents a given phase angle. An instantaneous value can be determined at any time or phase angle about a 360° radius. The maximum value of an *ac* voltage (V_{max}) or an *ac* current (I_{max}) occurs twice during each full cycle. At 90°, the positive maximum values occurs, and at 270° the negative maximum values occurs.

In applying the given formulas, if the instantaneous values of a 100A-240V *ac* system were desired at 35° and 235°, the instantaneous voltages would be

$$v = 240V \text{ x } \sin 35° = 137.66V$$

and

$$v = 240V \text{ x } \sin 235° = -196.60V$$

while the instantaneous currents would be

$$i = 100A \text{ x } \sin 35° = 57.36A$$

and

$$i = 100A \text{ x } \sin 235° = -81.92A$$

Where the results of the calculations yielded a negative value, the negative sign only means a change of direction of each alternating source has occurred. When a phase angle falls between 180° and 360°, instantaneous values will always be negative. This explains why the each instantaneous value yielded a negative value at 235°.

With the given formulas, instantaneous voltage and current values can be determined at any phase angle between 0° and 360°.

Average Value: The average value of alternating voltage or current is the average of all instantaneous voltage and current values occurring during a half-cycle. At any rate, the average value must only be determined between 0 and either of the two peak values. Because a half-cycle ranges from 0 to the positive peak value and from 0 to the negative peak value, if these two half-cycles were combined, the average between the two half-cycles would be zero (same positive peak values plus same negative peak values). With respect to the peak value of either the applied voltage or current, the average value (AV) represents 63.7% (.637) of the respective peak value, where

$$V(I)_{AV} = .637 \text{ x } V(I)_{PK}$$

PRACTICAL APPLICATION

An *ac* current sine wave displays a peak current measuring 234 amps. What is the average value of the measured current?

Using the formula $I_{AVG} = .637 \times I_{PEAK}$

$$I_{AVG} = .637 \times 234A = 149.1A$$

The average current value is 149.1A.

Effective Value: Because alternating current and voltage alternate back and forth from 0 to peak, the amount of energy produced by alternating current (*ac*) is not the same as for direct current (*dc*). Unlike alternating current, direct current provides a constant voltage and a constant flow of current, thus providing an effective (measurable) output. Again, with respect to the peak value of either the applied voltage or current, the effective value (EFF) represents 70.7% (.707) of the respective peak value, where

$$V(I)_{EFF} = .707 \times V(I)_{PK}$$

Alternating current measuring devices are designed to read the effective value of the applied voltage or operating current to assist the user in determining the amount of energy that is being produced in the form of heat. For instance, *ac* ammeters really measure the heating effect of ac current.

To better understand this concept, let's briefly discuss current and resistance. We know that when *dc* current is passed through a given resistance, heat is produced. Well, *ac* current also produces heat when it is passed through the same resistance. In both cases, this heating effect is proportional to I^2R. That is, this heating effect varies with the square of the current (I^2) flow through a specific resistance (R). The larger the current, the greater the heat produced in a given circuit. Therefore, an *ac* ampere can be defined as that current flowing through a given resistance that produces heat at the same rate as a *dc* ampere.

Now getting back to the formula(s), at a peak value of 170 volts, an *ac* voltage would reflect an effective value of 120 volts (170V x .707). This implies that if a resistor were placed across a 170 volt *ac* source, the resistor would produce the same amount of heat as if it were placed across a 120 volt *dc* source.

The effective value is also known as the root-means-square or RMS. The root-means-square provides a method for calculating the effective value of either voltage or current about a sine wave. One half of a cycle is divided into a number of equal parts, and the instantaneous values are calculated at each point. These values are then squared, and an average value is determined. The square root of the derived average value is equal to the effective value of a sine wave. As a result of evaluating several sine waves at different peak values, the conclusion always amounts to the effective value of a sine wave being equal to 70.7% of the different peak values.

PRACTICAL APPLICATION

What is the root-mean-square voltage when an *ac* sine wave reaches a 148 volts peak value?

The root-mean-square (RMS) voltage also referred to as the effective value can be determined by using the formula $E_{EFF(RMS)} = .707 \times E_{PEAK}$,

$$E_{EFF(RMS)} = .707 \times 148V = 104.64V$$

The root-mean-square (RMS) voltage is 104.64V.

AC Phase Angles

In an alternating current circuit where only resistance exists, the circuit's applied voltage and current are in phase with each other. In this case, the expression *in phase* implies a simultaneous occurrence; that is, the applied voltage and current increase and decrease in magnitude, reaching their maximum peaks at the same time although their magnitudes may differ. Observe figure 10.

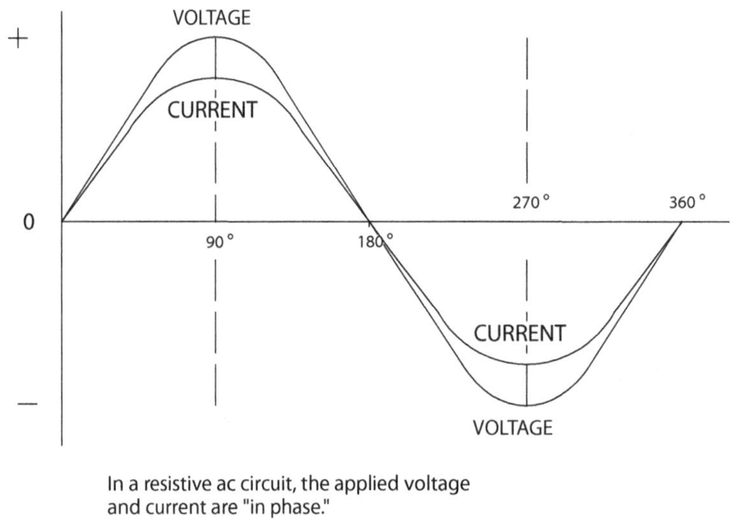

Fig. 10. Purely resistive *ac* circuit.

Because of this in phase relationship between the applied voltage and current, an alternating current circuit that contains only resistance can be approached in the same manner as that for a direct current circuit because there are no phase differences to deal with.

Now when the applied voltage and current are not in phase or out of phase, as with inductive and capacitive alternating current circuits, the resulting effects causes a phase difference to occur, which is recognized as the phase angle. The phase angle, which is identified by the Greek alphabet letter θ, (which is referred to as "theta"), references either the electrical angle at any point about a voltage or current waveform or the displacement of two waveforms with respect to time, which in this case would be both voltage and current. Observe figures 11(a) and (b):

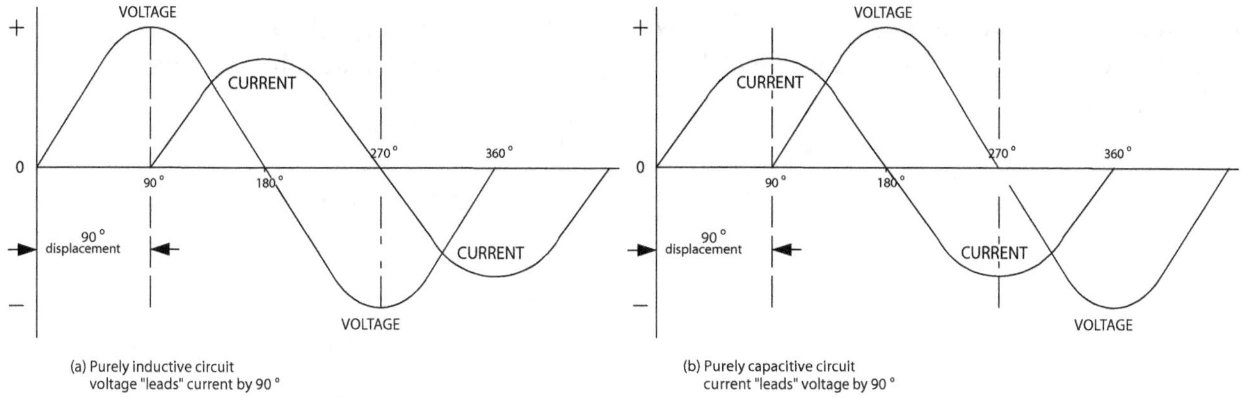

Figs. 11(a) and (b). Purely inductive and capacitive *ac* circuits.

In figure 11(a), the voltage waveform is displaced 90° ahead of the current waveform; in other words, the voltage waveform starts at 0°, while the current waveform starts at 90°. When this type of displacement occurs, it is said that voltage leads current by 90°, or current lags voltage by 90°. Figure 11(a) identifies the phase angle difference that takes place in a purely inductive circuit containing zero resistance.

In figure 11(b), the current waveform is displaced 90° ahead of the voltage waveform; in other words, the current waveform starts at 0°, while the voltage waveform starts at 90°. When this type of displacement occurs, it is said that current leads voltage by 90° or voltage lags current by 90°. Figure 11(b) identifies the phase angle difference that takes place in a purely capacitive circuit containing zero resistance.

In later topics, you will learn that the phase angle is also a function of the power factor in an electrical system.

Vectors

Just as voltage and current go hand in hand in an electrical system, so do phase angles and vectors. Now that an understanding of a phase angle has been developed, to move further in the study of alternating current, the composition and application of a vector must also be understood.

In simple terms, a vector is a quantity that reflects both magnitude (size, amount) and direction (up-down, east-west, right-left, 30°–60°). Quantities such as the number of pounds a box of nails weighs (15 lbs), the wire gauge of an electrical conductor (No. 14, 500kcmil), and the pressure of water (38 psi) only describe the magnitudes of each item. Quantities such as these are called scalars because they are independent of direction. As for vectors, if someone told you to walk three blocks, these instructions would only provide a magnitude of the number of blocks, which would still be a scalar. However, if someone told you to walk three blocks to the north, then the instructions would include magnitude and direction, which now would distinguish the difference between a scalar and a vector.

With alternating current, vectors are used to show the magnitude of voltage (240V) or current (35A) at a particular phase angle (29°) or the magnitudes of both voltage (208V) and current (400A) at difference phase angles or times (43° leading or lagging).

Vectors are also used with alternating current to graphically illustrate the magnitude and phase angles of voltage and current as they relate to their waveforms. Figure 12(a), (b), and (c) display how the voltage and current waveforms in figures 10 and 11(a) and (b) are translated into vectors.

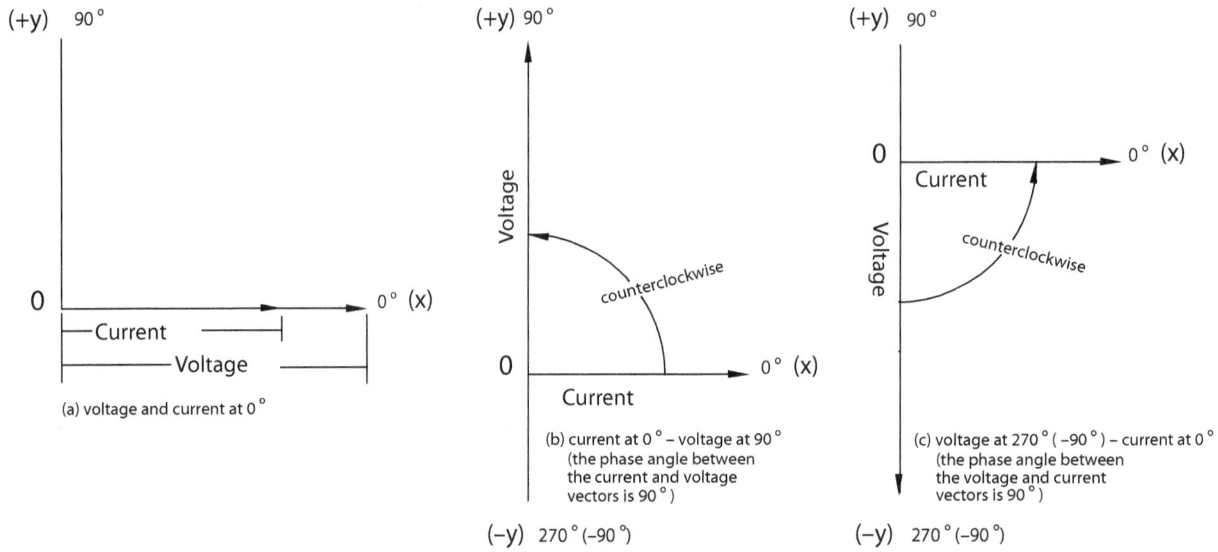

Figs. 12(a), (b), and (c). Translated voltage and current vectors.

Observe figure 12(a). Because the voltage and current waveforms are in phase in a purely resistive circuit, the voltage and current vectors are drawn in the same direction about the horizontal x-axis (0°) starting at zero (0). In all cases, the x-axis at point 0 will serve as the reference point.

Figure 12(b) displays the vector relationship between the voltage and current as translated from their waveforms for a purely inductive circuit. The current vector is at 0° with respect to the x-axis, and the voltage vector is at 90° with respect to the vertical y-axis. The voltage vector is now displaced 90° from the current vector or, as previously stated, voltage leads current by 90°. Because of their 90° displacement, the two vectors form a right triangle. Notice how the angle of rotation is counterclockwise. This angular direction is the common path for most vector displacements.

Figure 12(c) displays the vector relationship between the current and voltage as translated from their waveforms for a purely capacitive circuit. The voltage vector is at 270° (or –90°) with respect to the negative (–) y-axis, while the current vector is at 0° with respect to the x-axis. The current vector is now displaced 90° from the voltage vector or, as previously stated, current leads voltage by 90°. Notice how these two vectors also form a right triangle.

Adding Vectors That Form Right Angles

Perhaps at this point, you have thought more than once how good it would be if all *ac* circuits were resistive. If this were the case, there would be no need to consider phase angles, right triangles, or vectors, not to mention the need for adding vectors. Since this is not the case, the study of vectors must continue, but the goal and objective still remains the same: to keep the discussion of vectors and other associated discussions as simple as possible. Figures 12(b) and (c) clearly reflect the 90° displacement or right angles that takes place between voltage and current. Because there are mathematical concepts for analyzing and solving quantities involving right angles, these concepts will play an important role in evaluating inductive and capacitive circuits, which likewise share 90° phase differences.

When two vectors are displaced by 90° to form a right angle, they can be added together by means of a technique called the triangle method, which is also referred to as the "head to tail" method. This method will be used exclusively based upon its simple approach.

Before getting started, we'll reuse figures 12(b) and (c) to illustrate the head and tail of each vector. Now observe figure 13.

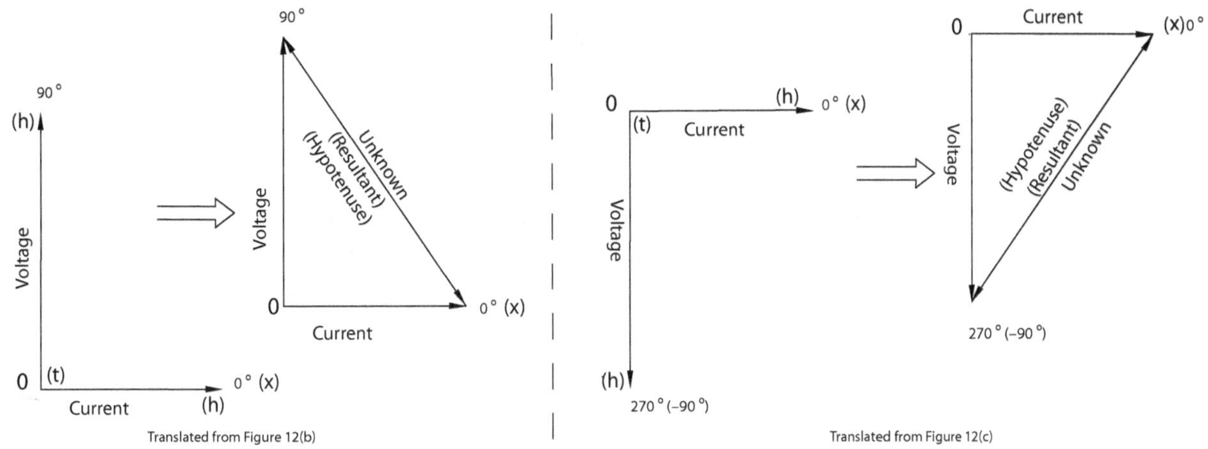

Fig. 13. Head and tail of vectors.

The heads (h) of each vector are the ends that have arrowheads while the tails (t) of each vector are the opposite ends. In opposition to figures 12(b) and (c), if the tails of each voltage vector were positioned at the heads of each current vector, this would prove to be inconsistent with the counterclockwise rotation, although both vectors would still form a 90° angle. When positioned as shown in figure 13, the voltage and current vectors can be added together vectorially to determine the magnitude and direction of the unknown vectors positioned between the heads of the voltage and current vectors. To define the direction of the unknown vectors, arrowheads are placed at the ends of the unknown vectors opposite the 0 reference points at 90° and 0°. The unknown vectors are called the resultant(s) based on the results of the voltage and current vectors being added together vectorially. The resultant is also known as the hypotenuse of a right triangle. Because the voltage and current vectors have magnitudes and are displaced 90° apart, the magnitude of the hypotenuse can be determined by a proven method called the Pythagorean Theorem.

The Pythagorean Theorem is a mathematical formula that was derived as a result of a demonstrated truth. As illustrated in figure 14, the Greek mathematician Pythagoras discovered that the sum of the squares ($a^2[9ft^2] + b^2[16ft^2]$) of the length of the sides (a[3ft] and b[4ft]) of a right triangle is equal to the square ($c^2[25ft^2]$) of the length of the hypotenuse (c[5ft]). Take a look:

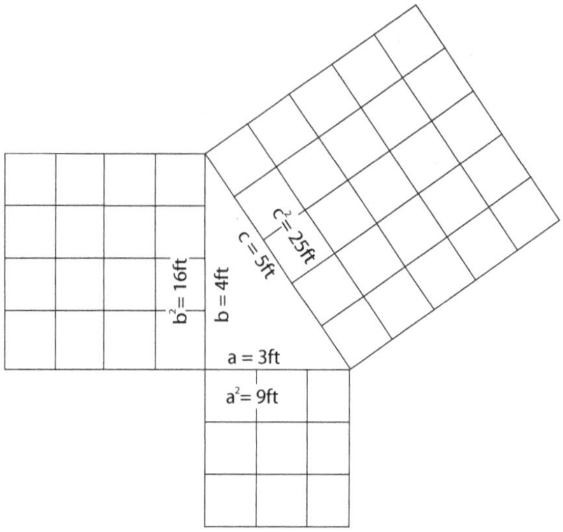

Fig. 14. Illustration of the Pythagorean Theorem.

The two sides of the triangle are 3ft and 4ft in length, while the hypotenuse is 5ft. If the two sides are added together, the sum does not equal the hypotenuse; that is, 3ft + 4ft does not equal 5ft. However, if the two sides and the hypotenuse are squared ($[3ft]^2 = 9ft^2$, $[4ft]^2 = 16ft^2$, and $[5ft]^2 = 25ft^2$), and the square of the two sides are added together,

$$9ft^2 + 16ft^2$$

the result is equal to $25ft^2$, the square of the hypotenuse. Get it?

As a result, the derived formula

$$a^2 + b^2 = c^2 \quad \text{or} \quad c^2 = a^2 + b^2$$

can be converted to derive three additional formulas for determining the lengths of each side of a right triangle, where

$$\text{side a} = \sqrt{c^2 - b^2}, \quad \text{side b} = \sqrt{c^2 - a^2}, \quad \text{side c} = \sqrt{a^2 + b^2}$$

Three very important things to always remember about a right triangle are (1) the hypotenuse will always be the longest side, (2) the lengths of any two sides must be known to determine the length of an unknown side, and (3) these formulas can be used for any right triangle.

In figure 14, the magnitude of the hypotenuse is given (5ft). Now let's determine the direction of the hypotenuse. However, before moving on, just let me remind you again that the hypotenuse is a vector because it has magnitude and direction. The direction of the hypotenuse is simply the phase angle (θ) between the x-axis and the hypotenuse. Observe figures 15(a) and (b). In both illustrations, the phase angle identifies the angular difference between the current and hypotenuse vectors.

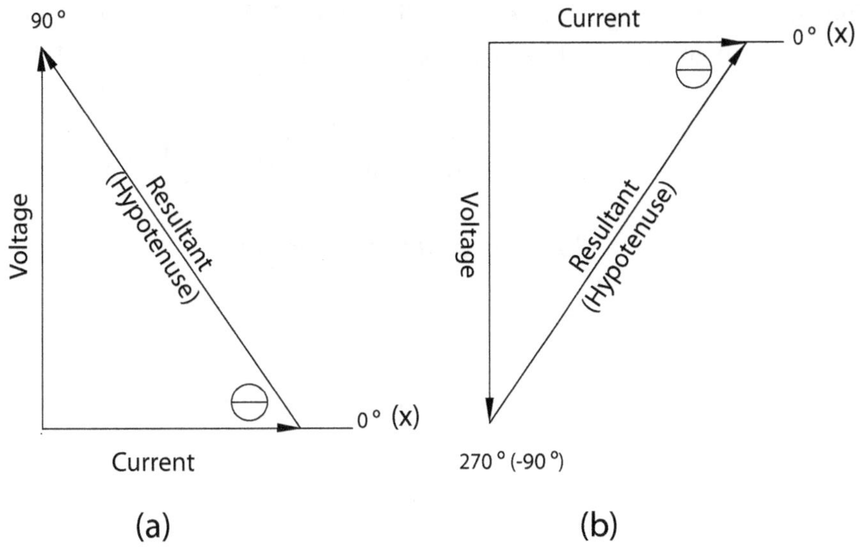

Fig. 15. Directions of hypotenuse.

With the use of trigonometry, the direction of the hypotenuse can be determined by use of three trigonometric functions: the cosine (cos), the sine (sin), and the tangent (tan). Referring to figures 16(a) and (b), you will see that the sides of the right triangles have all now been given specific names that are commonly associated with each trigonometric function.

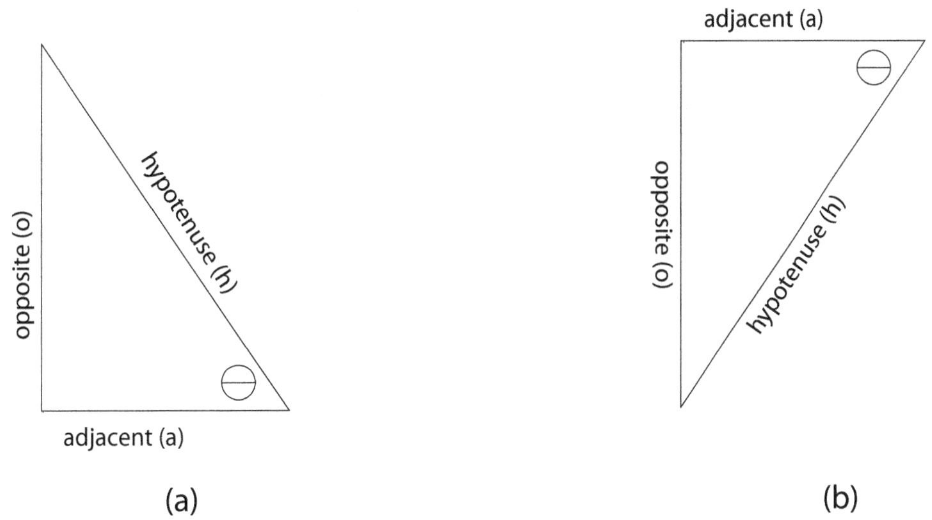

Fig. 16. Named sides of right triangles.

This now eliminates the need to refer to a particular side as either the x-axis or the y-axis, or simply as side a, b, or c, as displayed in figure 14. For clarity, each function will be analyzed individually using the same magnitudes (lengths) also used in figure 14.

cosine (cos) – The *cosine* is the ratio between the adjacent (a) side of the triangle and the hypotenuse (h), where the cosine of the angle (cos θ) is equal to such ratio. That is,

$$\cos \theta = \frac{\text{adjacent side (a)}}{\text{hypotenuse (h)}}$$

The angle θ can be determined by the following *converted* formula:

$$0 = \cos^{-1} \frac{a}{h}$$

$\cos^{-1}$ is also expressed as the arc cosine (arccos) and the inverse cosine (invcos).

AS APPLIED

$$\theta = \cos^{-1} \frac{3'}{5'}$$
$$= \cos^{-1} \ .6$$
$$= 53.13°$$

sine (sin) – The *sine* is the ratio between the opposite (o) side of the triangle and the hypotenuse (h), where the sine of the angle (sin θ) is equal to such ratio. That is,

$$\sin \theta = \frac{\text{opposite side (o)}}{\text{hypotenuse (h)}}$$

The angle θ can be determined by the following *converted* formula:

$$\theta = \sin^{-1} \frac{o}{h}$$

$\sin^{-1}$ is also expressed as the arc sine (arcsin) and the inverse sine (invsin).

AS APPLIED

$$\theta = \sin^{-1} \frac{4'}{5'}$$
$$= \sin^{-1} \ .8$$
$$= 53.13°$$

tangent (tan) – The *tangent* is the ratio between the opposite (o) side of the triangle and the adjacent (a) side of the triangle, where the tangent of the angle (tan θ) is equal to such ratio. That is,

$$\tan \theta = \frac{\text{opposite side (o)}}{\text{adjacent (a)}}$$

The angle θ can be determined by the following *converted* formula:

$$\theta = \tan^{-1}\frac{o}{a}$$

$\tan^{-1}$ is also expressed as the arc tangent (arctan) and the inverse tangent (invtan).

AS APPLIED

$$\theta = \tan^{-1}\ \frac{4'}{3'}$$
$$= \tan^{-1}\ 1.3333$$
$$= 53.13°$$

As you can see, the use of all three functions yielded the same angle.

Finally, instead of using the Pythagorean Theorem to find the magnitudes of the adjacent (a) and opposite (o) sides, the trigonometric functions cos θ and sin θ can be converted to yield the same results. Therefore, if

$$\cos\theta = \frac{a}{h} \text{ and } \sin\theta = \frac{o}{h}$$

then,

$$a = h \times \cos\theta \text{ and } o = h \times \sin\theta$$

AS APPLIED

$$a = 5' \times \cos 53.13° \text{ and } o = 5' \times \sin 53.13°$$
$$= 5' \times .6 \qquad\qquad = 5' \times .8$$
$$= 3' \qquad\qquad = 4'$$

Just as all three functions yielded the same angle, the converted functions also yielded the same magnitudes as in figure 14.

In further topics, you will see how the hypotenuse will be used with similar right triangles to identify the apparent power ([k]VA), the impedance (Z), and the total voltage (E_T) and current (I_T) quantities of a resistive-inductive (RL) circuit and a resistive-capacitive (RC) circuit.

Inductance

In alternating current (*ac*) circuits, coiled conductors or inductors are used primarily to introduce inductance into a circuit.

An inductor is a passive electrical component that stores energy in the form of a magnetic field, which is produced about a conductor when electrical current flows through the conductor, causing the magnetic field to expand and collapse. The term "passive" in the preceding sentence implies such a component consumes energy but never produces energy; that is, it may store energy momentarily but cannot deliver a source of energy continuously like that of a battery. Resistors and capacitors are also passive electrical components, where a resistor dissipates (spreads wastefully) its energy as heat, while a capacitor stores energy in the form of an electric field that charges and discharges.

The term inductance can be defined as that part of an electrical circuit that opposes any change of current flowing through the circuit, thus causing voltage changes to lead current changes or current changes to lag behind voltage changes.

The capital letter L is used to symbolize an inductor. The unit of measurement for an inductor is expressed in henrys and is abbreviated using the capital letter H. Because henrys are rather large units, inductors are usually measured in millihenrys. Millihenrys means one one-thousandth (1/1,000) of a henry and is abbreviated mH.

Compared to resistors, the use of inductors in a circuit is fairly common. Inductors can also be connected either in series or parallel, and the total inductance of each connection can be determined in similar fashion to that of a resistor.

The total inductance (L_T) of a combination of series inductors is calculated in the same manner as that for resistors; that is,

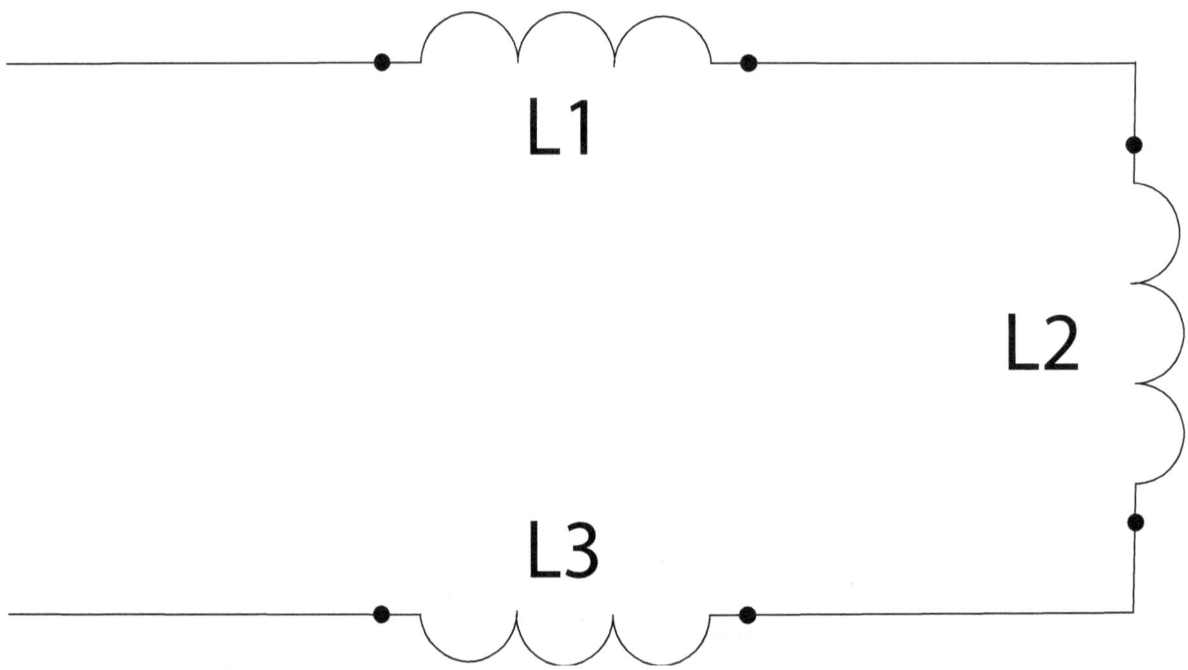

$$L_T = L_1 + L_2 + L_3 + \dots\dots\dots\dots\dots\dots$$

As for parallel inductors, the total inductance (L_T) of a combination of parallel inductors is also calculated in the same manner as that for resistors; that is,

Method 1. Equal inductors.

$$L_T = \frac{\text{L (individual inductor value)}}{\text{N (number of inductors)}}$$

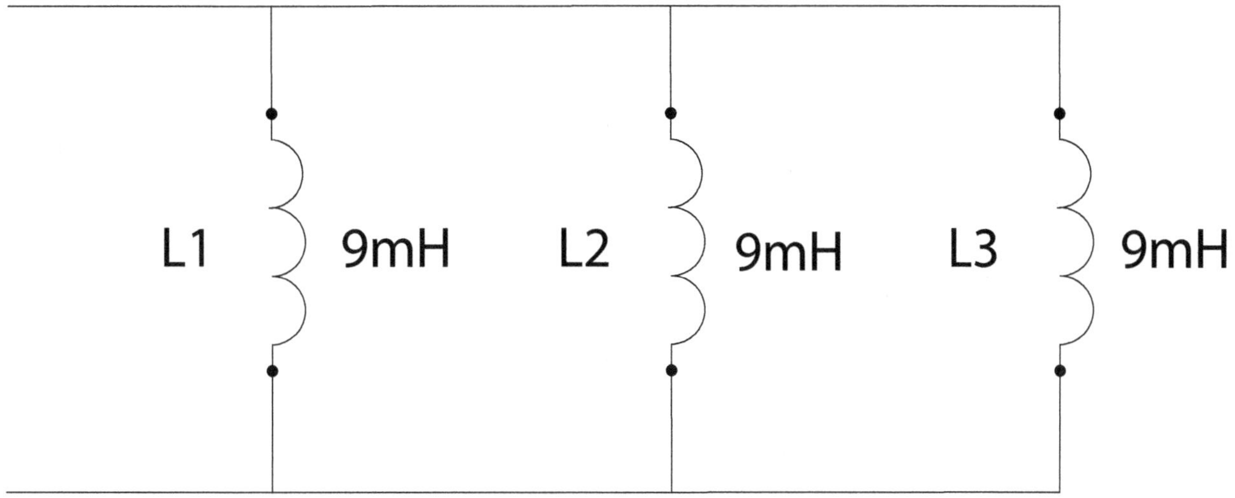

Method 2. Two inductors (product of the sum).*

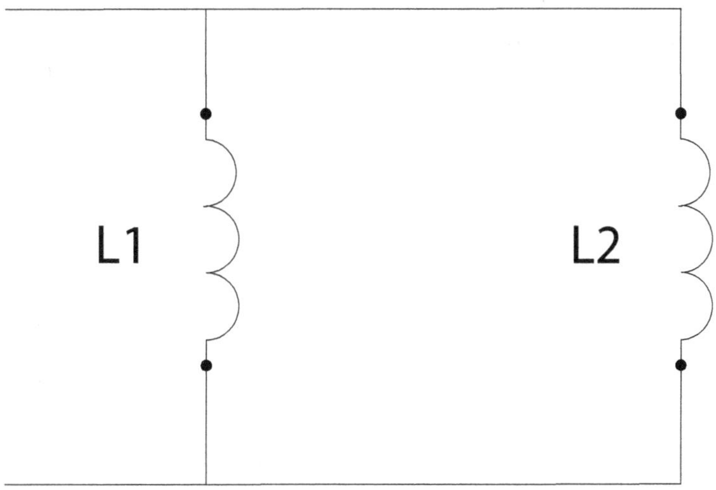

When two inductors are in parallel,

$$L_T = \frac{L_1 \text{ x } L_2 \text{ (product)}}{L_1 + L_2 \text{ (sum)}}$$

* This method only allows two inductors to be evaluated at one time.

Method 3. Two or more inductors (reciprocal method).**

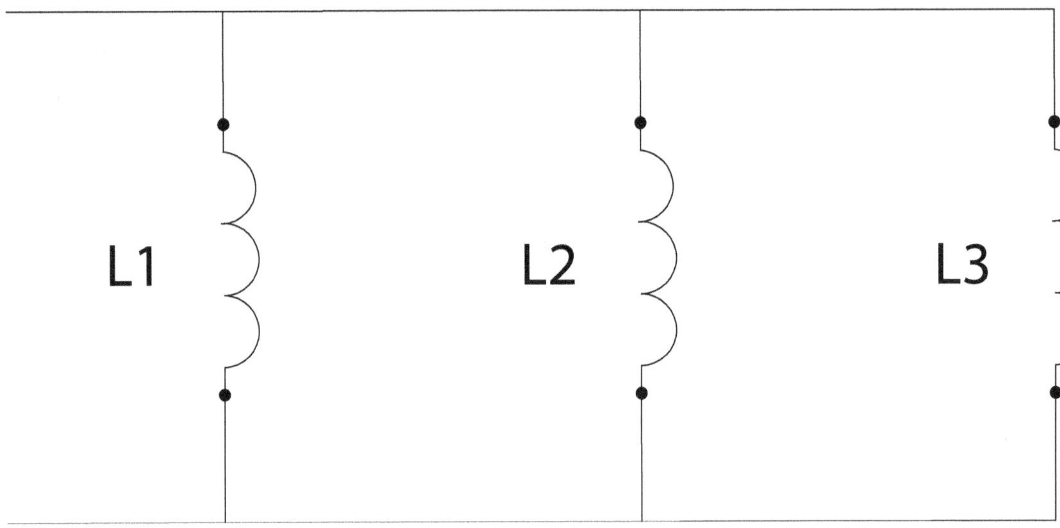

When two or more inductors are in parallel,

$$L_T = \frac{1}{1/L_1 + 1/L_2 + 1/L_3 + \ldots.}$$

** This method allows an indefinite number of inductors to be evaluated at one time.

Inductance Reactance

When *ac* current flows through an inductor, it causes a magnetic field to surround the inductor. Because alternating current changes direction periodically it causes the magnetic field to expand and collapse. As the direction of current changes, so does the magnitude of the current, which causes the strength of the magnetic field, to change likewise. The greater the magnitude of current flow through the inductor, the stronger the magnetic field, and vice versa. The effects of this varying magnetic field cause a varying reaction to take place about the inductor, which causes an opposition to the flow of current through the inductor. This opposition to the flow of current is called inductance reactance. Inductance reactance also prevents current flowing through the inductor from being in phase with the applied voltage. As a result, current lags behind the voltage, which causes it to flow. With an ideal inductor, which has only inductance and zero resistance, current will lag behind the voltage by 90°. Observe figures 11(a) and 12(b).

Similar to resistance, inductance reactance is measured in units of ohms (Ω) and is directly proportional to the operating frequency of the source supplying a circuit and to the size of the involved inductor. In other words, the higher the frequency, or the larger the inductor, the more reactance that will take place in the circuit, and vice versa. The formula for inductance reactance is

$$X_L = 2\pi fL$$

where

X_L = inductance reactance in ohms
π = 3.14
f = frequency in hertz and
L = inductance in henrys (H)

In general, consider the following rules and selected formulas that apply to the different types of purely inductance circuits:

Series

Inductance reactance (X_{LT}): The total inductance reactance equals the sum of the individual inductance reactance occurring about each inductor, where

$$X_{LT} = X_{L1} + X_{L2} + \dots\dots\dots\dots\dots\dots$$

Current (I): The same current flows through each inductor, where

$$I = E_T/X_{LT}$$

Voltage source (E_T): The total voltage source equals the sum of each voltage drop across each inductor, where

$$E_T = E_{L1} + E_{L2} + \dots\dots\dots\dots\dots\dots$$
$$\text{and } (E_{L1,2,\dots} = [X_{L1,2,\dots}] \times I)$$

Reactance power ($P_{VARsL(T)}$)*: The total reactance power equals the sum of the individual reactance power occurring about each inductor, where

$$P_{VARsL(T)} = P_{VARsL1} + P_{VARsL2} + \dots\dots\dots\dots\dots\dots$$
$$\text{and } (P_{VARsL1,2,\dots} = [X_{L1,2,\dots}] \times I^2)$$

* In a purely inductance circuit, only reactance power exists, being that the circuit only consists of inductance, which yields only reactance power. As a result, true power and a power factor do not exist. The terms reactance power and reactive power are the same; for clarity, the term reactance power will be used in all contents of this book. See "Related Series Formulas (Inductance—Overall)" and "The Power Triangle, Power Factor and Phase Angle."

Related Series Formulas (Inductance—Overall)

Inductance

$L = X_L/2\pi f$

Total Inductance

$L_T = X_{LT}/2\pi f$

Inductance Reactance

$X_L = 2\pi fL$ (1) $X_L = E_L/I$ (2) $X_L = E_L^2/P_{VARsL}$ (3) $X_L = P_{VARsL}/I^2$ (4)

Total Inductance Reactance

$X_{LT} = 2\pi fL_T$ (1) $X_{LT} = E_{L(T)}/I$ (2) $X_{LT} = E_T^2/P_{VARsL(T)}$ (3) $X_{LT} = P_{VARsL(T)}/I^2$ (4)

Voltage Drop

$E_L = I \times X_L$ (1) $E_L = P_{VARsL}/I$ (2) $E_L = \sqrt{P_{VARsL} \times X_L}$ (3)

Total Voltage

$E_T = I \times X_{LT}$ (1) $E_T = P_{VARsL(T)}/I$ (2) $E_T = \sqrt{P_{VARsL(T)} \times X_{LT}}$ (3)

Current

$I = E_T/X_{LT}$ (1) $I = P_{VARsL(T)}/E_T$ (2) $I = \sqrt{P_{VARsL(T)}/X_{LT}}$ (3)

Individual Reactance Power

$P_{VARsL} = E_L \times I$ (1) $P_{VARsL} = I^2 \times X_L$ (2) $P_{VARsL} = E_L^2/X_L$ (3)

Total Reactance Power

$P_{VARsL(T)} = E_T \times I$ (1) $P_{VARsL(T)} = I^2 \times X_{LT}$ (2) $P_{VARsL(T)} = E_T^2/X_{LT}$ (3)

Parallel

Inductance Reactance (X_{LT}): The total inductance reactance equals at least one of the given formulas per application. These formulas are derivatives of the total inductance formulas, where

$$X_{LT} = \frac{X_L \text{ (individual inductance reactance value)}}{N \text{ (number of inductance reactances)}} \quad \text{or}$$

$$X_{LT} = \frac{X_{L1} \times X_{L2} \text{ (product)}}{X_{L1} + X_{L2} \text{ (sum)}} \quad \text{or}$$

$$X_{LT} = \frac{1}{1/X_{L1} + 1/X_{L2} + \ldots\ldots\ldots\ldots}$$

Voltage source (E): The voltage drop across each inductor (E_{L1}, E_{L2}, ...) is the same as the voltage source, where

$$E = I_T \times X_{LT}$$

Current (I_T): The total current equals the sum of each individual current path through each inductor, where

$$I_T = I_{L1} + I_{L2} + \ldots\ldots\ldots\ldots\ldots\ldots$$
$$\text{and } (I_{L1,2,\ldots} = E/[X_{L\,1,2,\ldots}])$$

Reactance power ($P_{VARsL(T)}$)*: The total reactance power equals the sum of the individual reactance power occurring about each inductor, where

$$P_{VARsL(T)} = P_{VARsL1} + P_{VARsL2} + \ldots\ldots\ldots\ldots\ldots\ldots$$
$$\text{and } (P_{VARsL1,2,\ldots} = E \times [I_{L1,2,\ldots}])$$

* See "Related Parallel Formulas (Inductance—Overall)."

Related Parallel Formulas (Inductance—Overall)

Inductance

$L = X_L/2\pi f$

Total Inductance

$L_T = X_{LT}/2\pi f$

Inductance Reactance

$X_L = 2\pi fL$ (1) $X_L = E/I_L$ (2) $X_L = E^2/P_{VARsL}$ (3) $X_L = P_{VARsL}/I_L^2$ (4)

Total Inductance Reactance

$X_{LT} = 2\pi fL_T$ (1) $X_{LT} = E/I_T$ (2) $X_{LT} = E^2/P_{VARsL(T)}$ (3) $X_{LT} = P_{VARsL(T)}/I_T^2$ (4)

Voltage

$E = I_T \times X_{LT}$ (1) $E = P_{VARsL(T)}/I_T$ (2) $E = \sqrt{P_{VARsL(T)} \times X_{LT}}$ (3)

Individual Current Path

$I_L = E/X_L$ (1) $I_L = P_{VARsL}/E$ (2) $I_L = \sqrt{P_{VARsL}/X_L}$ (3)

Total Current

$I_T = E/X_{LT}$ (1) $I_T = P_{VARsL(T)}/E$ (2) $I_T = \sqrt{P_{VARsL(T)}/X_{LT}}$ (3)

Individual Reactance Power

$P_{VARsL} = E \times I_L$ (1) $P_{VARsL} = I_L^2 \times X_L$ (2) $P_{VARsL} = E^2/X_L$ (3)

Total Reactance Power

$P_{VARsL(T)} = E \times I_T$ (1) $P_{VARsL(T)} = I_T^2 \times X_{LT}$ (2) $P_{VARsL(T)} = E^2/X_{LT}$ (3)

APPLYING THE RULES

Series Inductance (L)

1. Five inductors are connected in series. If the inductors are rated for 12mH, 6.5mH, 28mH, 71.35mH, and 54mH, determine the total inductance.

   Use the formula $L_T = L_1 + L_2 + L_3 + \ldots$

   $$L_T = 12mH + 6.5mH + 28mH + 71.35mH + 54mH = 171.85mH$$

   The total (series) inductance is 171.85mH.

   For clarity, 171.85mH and any quantity involving units of millihenrys can be converted into units of henrys by dividing the millihenrys' quantity by 1000. Therefore, 171.85mH is the same as .17185H (171.85mH/1,000).

Parallel Inductance (L)

2. Four inductors are connected in parallel. If the inductors are rated for 17mH, 51.8mH, 32mH, and 9.6mH, determine the total inductance.

   Use the formula $L_T = \dfrac{1}{1/L_1 + 1/L_2 + 1/L_3 + \ldots}$

   $$L_T = \frac{1}{1/17mH + 1/51.8mH + 1/32mH + 1/9.6mH} = \frac{1}{.2135mH} = 4.684mH$$

   The total (parallel) inductance is 4.684mH or .004684H.

Inductance Reactance (X_L)

3. What is the inductance reactance of a 4.7mH inductor when supplied by frequencies measuring 40Hz, 180Hz, and 300Hz?

   Use the formula $X_L = 2 \times \pi \times f \times L$

At 40Hz:

$$X_L = 2 \text{ x } 3.14 \text{ x } 40\text{Hz x } .0047\text{H} = 1.18\Omega$$

At 180Hz:

$$X_L = 2 \text{ x } 3.14 \text{ x } 180\text{Hz x } .0047\text{H} = 5.31\Omega$$

At 300Hz:

$$X_L = 2 \text{ x } 3.14 \text{ x } 300\text{Hz x } .0047\text{H} = 8.85\Omega$$

Series Inductance Reactance (X_{LT}) Total

4. Determine the total inductance reactance of the five series inductors (12mH, 6.5mH, 28mH, 71.35mH, 54mH) in question No. 1 at 250Hz.

   Use the formulas $X_{LT} = 2 \text{ x } \pi \text{ x f x } L_T$ or $X_{L1} + X_{L2} + X_{L3} + \ldots..$

   (1) When the total inductance is used (L_T = 171.85mH)

   $$X_{LT} = 2 \text{ x } 3.14 \text{ x } 250\text{Hz x } .17185\text{H} = 269.8\Omega$$

   (2) When the individual inductance reactance are determined and totaled

   (at 12mH) $X_{L1} = 2 \text{ x } 3.14 \text{ x } 250\text{Hz x } .012\text{H} = 18.84\Omega$

   (at 6.5mH) $X_{L2} = 2 \text{ x } 3.14 \text{ x } 250\text{Hz x } .0065\text{H} = 10.21\Omega$

   (at 28mH) $X_{L3} = 2 \text{ x } 3.14 \text{ x } 250\text{Hz x } .028\text{H} = 43.96\Omega$

   (at 71.35mH) $X_{L4} = 2 \text{ x } 3.14 \text{ x } 250\text{Hz x } .07135\text{H} = 112.02\Omega$

   (at 54mH) $X_{L5} = 2 \text{ x } 3.14 \text{ x } 250\text{Hz x } .054\text{H} = 84.78\Omega$

   $$X_{LT} = 18.84\Omega + 10.21\Omega + 43.96\Omega + 112.02\Omega + 84.78\Omega = 269.81\Omega$$

Parallel Inductance Reactance (X_{LT}) Total

5. Determine the total inductance reactance of the four parallel inductors (17mH, 51.8mH, 32mH, 9.6mH) in question No. 2 at 425Hz.

   Use the formula $X_{LT} = 2 \text{ x } \pi \text{ x f x } L_T$ or $\dfrac{1}{1/X_{L1} + 1/X_{L2} + 1/X_{L3} + \ldots..}$

(1) When the total inductance is used (L_T = 4.684mH)

$$X_{LT} = 2 \text{ x } 3.14 \text{ x } 425\text{Hz x } .004684\text{H} = 12.5\Omega$$

(2) When the individual inductance reactance is determined and totaled

(at 17mH) $X_{L1} = 2 \text{ x } 3.14 \text{ x } 425\text{Hz x } .017\text{H} = 45.37\Omega$

(at 51.8mH) $X_{L2} = 2 \text{ x } 3.14 \text{ x } 425\text{Hz x } .0518\text{H} = 138.25\Omega$

(at 32mH) $X_{L3} = 2 \text{ x } 3.14 \text{ x } 425\text{Hz x } .032\text{H} = 85.41\Omega$

(at 9.6mH) $X_{L4} = 2 \text{ x } 3.14 \text{ x } 425\text{Hz x } .0096\text{H} = 25.62\Omega$

$$X_{LT} = \frac{1}{1/45.37\Omega + 1/138.25\Omega + 1/85.41\Omega + 1/25.62\Omega}$$

$$= \frac{1}{.08\Omega} = 12.5\Omega$$

Series Inductance Reactance (X_{LT}) Circuit

6. Now that the inductance reactance has been determined for the inductors in question No. 4, if 575 volts were applied to the inductors, determine the circuit's current, the voltage drop across L_1, L_2, and L_4, and the reactance power produced by L_2 and L_5.

   Use the formula $I = E_T/X_{LT}$

$$I = 575\text{V}/269.8\Omega = 2.13\text{A}$$

   Voltage drop (X_L x I) across L_1, L_2, and L_4

$$E_{L1} = 18.84\Omega \text{ x } 2.13\text{A} = 40.13\text{V}$$
$$E_{L2} = 10.21\Omega \text{ x } 2.13\text{A} = 21.75\text{V}$$
$$E_{L4} = 112.02\Omega \text{ x } 2.13\text{A} = 238.6\text{V}$$

   Reactance power (I^2 x X_L) resulting from L_2 and L_5

$$P_{VARsL2} = (2.13\text{A})^2 \text{ x } 10.21\Omega = 46.32\text{VARs}$$
$$P_{VARsL5} = (2.13\text{A})^2 \text{ x } 84.78\Omega = 384.64\text{VARs}$$

   Now that the voltage drop and reactance power for inductor L_2 has been determined, let's use "like formulas" to cross-check the results. See "Related Series Formulas (Inductance—Overall)."

To determine the voltage drop using the two remaining formulas,

$$E_L = P_{VARsL}/I \qquad E_L = \sqrt{P_{VARsL} \times X_L}$$

$$E_{L2} = P_{VARsL2}/I \qquad E_{L2} = \sqrt{P_{VARsL2} \times X_{L2}}$$

$$= 46.32\text{VARs}/2.13\text{A} \qquad = \sqrt{46.32 \times 10.21}$$

$$= 21.75\text{V} \qquad = \sqrt{472.93}$$

$$= 21.75\text{V}$$

To determine the reactance power using the two remaining formulas,

$$P_{VARsL} = E_L \times I \qquad P_{VARsL} = E_L{}^2/X_L$$

$$P_{VARsL2} = E_{L2} \times I \qquad P_{VARsL2} = E_{L2}{}^2/X_{L2}$$

$$= 21.75\text{V} \times 2.13\text{A} \qquad = (21.75)^2/10.21$$

$$= 46.33\text{VARs} \qquad = 473.1/10.21$$

$$= 46.34\text{VARs}$$

In addition to cross-checking the results in both calculations, this also proves that like formulas will always yield the same or approximate results.

Parallel Inductance Reactance (X_{LT}) Circuit

7. If the inductors in question No. 5 were energized by a 47V source,
   a. How much current flows through L_1 and L_4?
   b. What is the voltage drop across L_3?
   c. How much reactance power is produced by L_2 and L_4?

   a. Current (E/X_L) flow through L_1 and L_4

$$I_{L1} = 47\text{V}/45.37\Omega = 1.04\text{A} \qquad I_{L4} = 47\text{V}/25.62\Omega = 1.83\text{A}$$

   b. Voltage drop across L_3 same as source voltage

$$E_{L3} = 47\text{V}$$

   c. Reactance power produced (E^2/X_L) by L_2 and L_4

$$P_{VARsL2} = (47\text{V})^2/138.25\Omega = 15.98\text{VARs}$$
$$P_{VARsL4} = (47\text{V})^2/25.62\Omega = 86.22\text{VARs}$$

Again, let's use like formulas to cross-check the current flow through inductor L_4. See "Related Parallel Formulas (Inductance—Overall)."

To determine the current flow using the two remaining formulas,

$$I_L = P_{VARsL}/E \qquad I_L = \sqrt{P_{VARsL}/X_L}$$

$$I_{L4} = P_{VARsL4}/E \qquad I_{L4} = \sqrt{P_{VARsL4}/X_{L4}}$$

$$= 86.22\text{VARs}/47\text{V} \qquad = \sqrt{86.22/25.62}$$

$$= 1.83\text{A} \qquad = \sqrt{3.37}$$

$$= 1.84\text{A}$$

As you can see, the like formulas yield the same and approximate results.

Capacitance

The measure of a capacitor's ability to store charge or hold electrons defines capacitance. A capacitor, also a passive electrical component, consists of two or more metal plates (also called conductors) that are placed close together without touching, where energy in the form of an electric field is stored. Either air or some other type of insulating material called the dielectric is used to fill the void between the plates of a capacitor.

The amount of energy a capacitor is capable of storing depends upon three things: the area of the plates, the dielectric of the insulating material, and the distance the plates are placed apart. The larger the area of the plates of a capacitor, the more electrons or energy the capacitor is capable of storing. Therefore, as the area of the plates of a capacitor is increased, so is the capacitance and vice-versa. How effective a dielectric is at allowing a capacitor to store more charge generally depends upon the material the dielectric is made of. Lastly, the closer the plates of a capacitor are to each other, the more effect the charge on one plate will have on the other, meaning the closer the plates, the greater the capacitance of the capacitor.

When capacitors are connected in series, the total capacitance decreases. Capacitors connected in series produce the same effect as increasing the distance between the plates of a capacitor. To increase capacitance, capacitors are connected in parallel. Placing capacitors in parallel produces the same effect as increasing the area of the plates of a capacitor.

The capital letter C is used to symbolize a capacitor. The unit of measurement for a capacitor is expressed in farads and is abbreviated using the capital letter F. Because farads are rather large units, capacitors are usually measured in microfarads or picofarads. Microfarad means one-millionth (1/1,000,000) of a farad and is abbreviated µF, while picofarad means one-millionth of one-millionth of a farad and is abbreviated pF.

With alternating current, the voltage across a capacitor is always increasing and decreasing as it repeatedly changes cycles. Likewise, this effect causes a capacitor to constantly charge and discharge at a rate indifferent to that of the voltage. The charging and discharging effect that occurs back and forth about the plates of a capacitor happens at a faster rate than the buildup of the applied voltage across the plates of a capacitor. As a result, this causes current changes to lead voltage changes or voltage changes to lag behind current changes in a capacitor.

When capacitors are either series or parallel connected, the total capacitance of each connection is determined in the opposite manner to that of a resistor.

The total capacitance (C_T) of a combination of series capacitors is calculated in the same manner as the total resistance of parallel resistors.

Method 1. Equal capacitors.

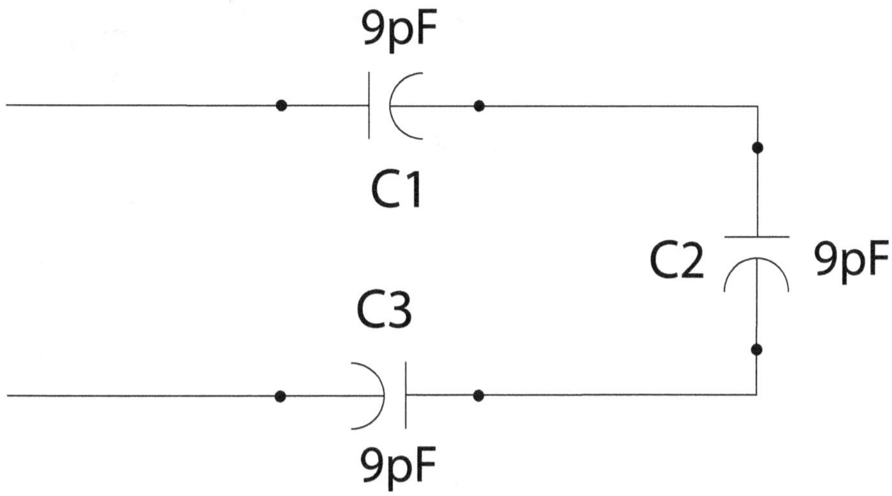

$$C_T = \frac{\text{C (individual capacitor value)}}{\text{N (number of capacitors)}}$$

Method 2. Two capacitors (Product of the Sum).*

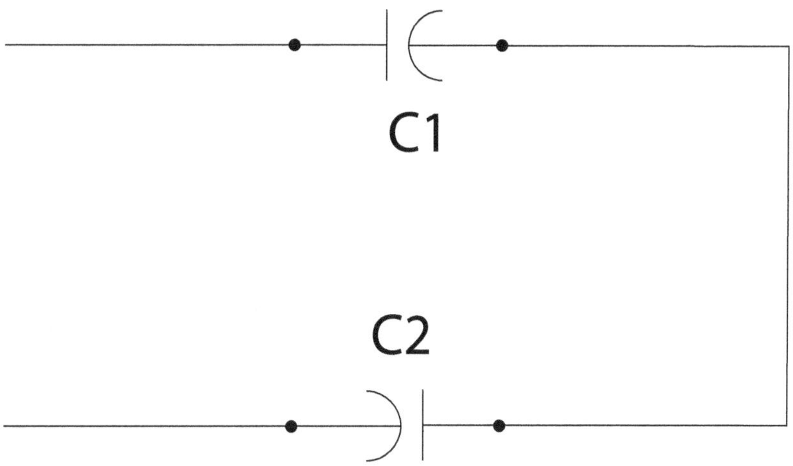

When at least two capacitors are in series,

$$C_T = \frac{C_1 \text{ x } C_2 \text{ (product)}}{C_1 + C_2 \text{ (sum)}}$$

* This method only allows two capacitors to be evaluated at one time.

Method 3. Two or more capacitors (Reciprocal Method).**

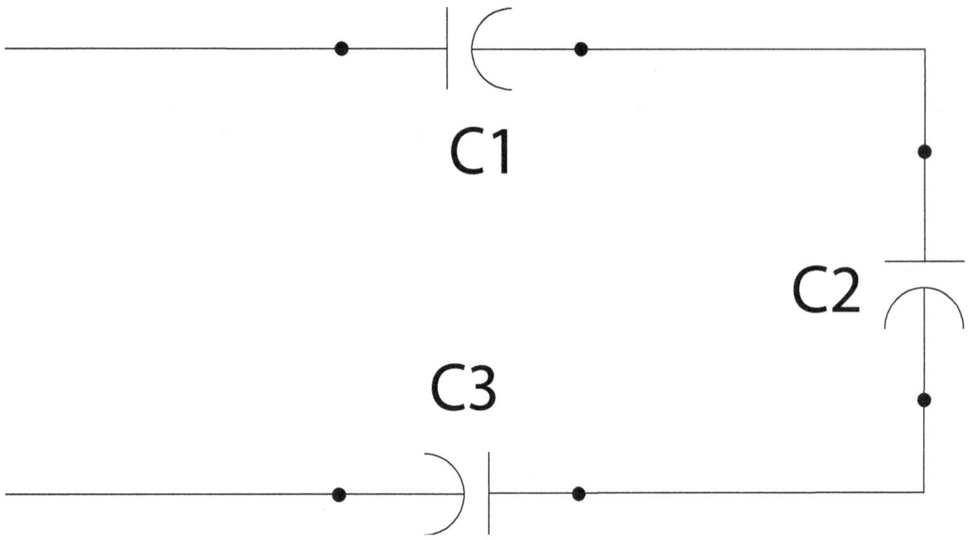

When at least two or more capacitors are in series,

$$C_T = \frac{1}{1/C_1 + 1/C_2 + 1/C_3 + \ldots.}$$

** This method allows an indefinite number of capacitors to be evaluated at one time.

As for parallel capacitors, the total capacitance (C_T) of a combination of capacitors connected in parallel is calculated in the same manner as the total resistance of series resistors:

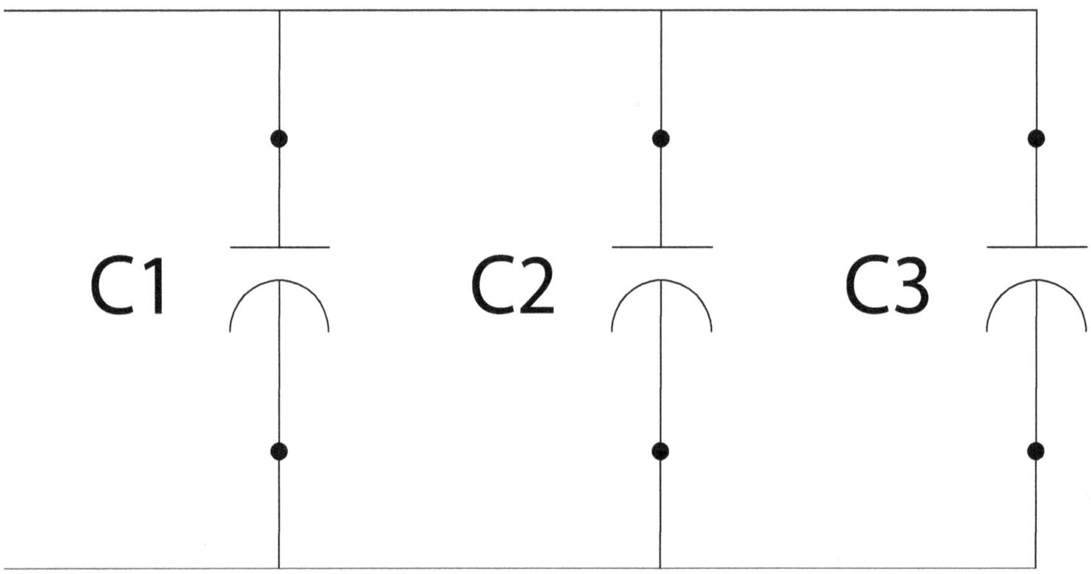

$$C_T = C_1 + C_2 + C_3 + \ldots\ldots\ldots\ldots\ldots\ldots..$$

Capacitance Reactance

When direct current is applied to a capacitor, current ceases to flow when the plates of a capacitor are fully charged, thus leaving the capacitor resembling that of an open circuit.

When a capacitor is placed in an alternating current circuit, it offers opposition to the current flow much like a resistor. Once alternating current is applied to a capacitor, the capacitor will charge during the first half cycle and discharge during the next half cycle. This charging and discharging cycling effect causes a capacitor to produce a counterelectromotive force or reaction that opposes the flow of current. A small capacitance will have more opposition to current flow than a large capacitance. In a similar manner, when a low frequency is applied to a capacitor, it will present a greater opposition than that of a higher frequency, which is the exact opposite of that of an inductor. As a result, the value of a capacitor and the operating frequency determine the amount of current that flows in a circuit.

This opposition to current flow in a capacitor is called capacitive reactance. Capacitive reactance is measured in units of ohms (Ω), just as resistance and inductive reactance. The formula for capacitance reactance is

$$X_C = \frac{1}{2\pi fC}$$

where

X_C = capacitance reactance in ohms
π = 3.14
f = frequency in hertz and
C = capacitance in farads (F)

Contrasted to that of an inductor where current lags the applied voltage by 90°, with a capacitor, the current leads the applied voltage by 90°. In such case, current and voltage are out of phase with each other. Observe figures 11(b) and 12(c).

In general, consider the following rules and selected formulas, which apply to the different types of purely capacitance circuits:

Series

Capacitive reactance (X_{CT}): The total capacitance reactance equals the sum of the individual capacitance reactance occurring about each capacitor, where

$$X_{CT} = X_{C1} + X_{C2} + \dots\dots\dots\dots\dots\dots$$

Current (I): The same current flows about each capacitor, where

$$I = E_T/X_{CT}$$

Voltage source (E_T): The total voltage source equals the sum of each voltage drop across each capacitor, where

$$E_T = E_{C1} + E_{C2} + \dots\dots\dots\dots$$
$$\text{and } (E_{C1,2,\dots} = [X_{C1,2,\dots}] \times I)$$

Reactance power ($P_{VARsC(T)}$)*: The total reactance power equals the sum of the individual reactance power occurring about each capacitor, where

$$P_{VARsC(T)} = P_{VARsC1} + P_{VARsC2} + \dots\dots\dots\dots$$
$$\text{and } (P_{VARsC1,2,\dots} = [X_{C1,2,\dots}] \times I^2)$$

* In a purely capacitance circuit, only reactance power exists, being that the circuit only consists of capacitance, which only yields reactance power. As a result, true power and a power factor do not exist. See "Related Series Formulas (Capacitance—Overall)."

Related Series Formulas (Capacitance—Overall)

Capacitance

$$C = \frac{1}{2\pi f X_C}$$

Total Capacitance

$$C_T = \frac{1}{2\pi f X_{CT}}$$

Capacitance Reactance

$X_C = \frac{1}{2\pi f C}$ (1) $X_C = E_C/I$ (2) $X_C = E_C^2/P_{VARsC}$ (3) $X_C = P_{VARsC}/I^2$ (4)

Total Capacitance Reactance

$X_{CT} = \frac{1}{2\pi f C_T}$ (1) $X_{CT} = E_T/I$ (2) $X_{CT} = E_T^2/P_{VARsC(T)}$ (3) $X_{CT} = P_{VARsC(T)}/I^2$ (4)

Voltage Drop

$E_C = I \times X_C$ (1) $E_C = P_{VARsC}/I$ (2) $E_C = \sqrt{P_{VARsC} \times X_C}$ (3)

Total Voltage

$E_T = I \times X_{CT}$ (1) $E_T = P_{VARsC(T)}/I$ (2) $E_T = \sqrt{P_{VARsC(T)} \times X_{CT}}$ (3)

Current

$I = E_T/X_{CT}$ (1) $I = P_{VARsC(T)}/E_T$ (2) $I = \sqrt{P_{VARsC(T)}/X_{CT}}$ (3)

Individual Reactance Power

$P_{VARsC} = E_C \times I$ (1) $P_{VARsC} = I^2 \times X_C$ (2) $P_{VARsC} = E_C^{\ 2}/X_C$ (3)

Total Reactance Power

$P_{VARsC(T)} = E_T \times I$ (1) $P_{VARsC(T)} = I^2 \times X_{CT}$ (2) $P_{VARsC(T)} = E_T^{\ 2}/X_{CT}$ (3)

Parallel

Capacitive reactance (X_{CT}): The total capacitive reactance equals at least one of the given formulas per application.

$$X_{CT} = \frac{X_C \text{ (individual capacitance reactance value)}}{N \text{ (number of capacitance reactances)}} \quad \text{or}$$

$$X_{CT} = \frac{X_{C1} \times X_{C2} \text{ (product)}}{X_{C1} + X_{C2} \text{ (sum)}} \quad \text{or}$$

$$X_{CT} = \frac{1}{1/X_{C1} + 1/X_{C2} + \dots\dots\dots\dots}$$

Voltage source (E): The voltage drop across each capacitor (E_{C1}, E_{C2}, ...) is the same as the voltage source, where

$$E = I_T \times X_{CT}$$

Current (I_T): The total current equals the sum of each individual current path about each capacitor, where

$$I_T = I_{C1} + I_{C2} + \dots\dots\dots\dots$$
$$\text{and } (I_{C1,2,\dots} = E/[X_{C1,2,\dots}])$$

Reactance power ($P_{VARs(C(T)}$)*: The total reactance power equals the sum of the individual reactance power occurring about each capacitor, where

$$P_{VARsC(T)} = P_{VARsC1} + P_{VARsC2} + \dots\dots\dots\dots$$
$$\text{and } (P_{VARsC1,2,\dots} = E \times [I_{C1,2,\dots}])$$

* See "Related Parallel Formulas (Capacitance—Overall)."

Related Parallel Formulas (Capacitance—Overall)

Capacitance

$$C = \frac{1}{2\pi f X_C}$$

Total Capacitance

$$C_T = \frac{1}{2\pi f X_{CT}}$$

Capacitance Reactance

$X_C = \frac{1}{2\pi f C}$ (1) $X_C = E / I_C$ (2) $X_C = E^2 / P_{VARsC}$ (3) $X_C = P_{VARsC} / I_C^2$ (4)

Total Capacitance Reactance

$X_{CT} = \frac{1}{2\pi f C_T}$ (1) $X_{CT} = E / I_T$ (2) $X_{CT} = E^2 / P_{VARsC(T)}$ (3) $X_{CT} = P_{VARsC(T)} / I_T^2$ (4)

Voltage

$E = I_T \times X_{CT}$ (1) $E = P_{VARsC(T)} / I_T$ (2) $E = \sqrt{P_{VARsC(T)} \times X_{CT}}$ (3)

Individual Current Path

$I_C = E / X_C$ (1) $I_C = P_{VARsC} / E$ (2) $I_C = \sqrt{P_{VARsC} / X_C}$ (3)

Total Current

$I_T = E / X_{CT}$ (1) $I_T = P_{VARsC(T)} / E$ (2) $I_T = \sqrt{P_{VARsC(T)} / X_{CT}}$ (3)

Individual Reactance Power

$P_{VARsC} = E \times I_C$ (1) $P_{VARsC} = I_C^2 \times X_C$ (2) $P_{VARsC} = E^2 / X_C$ (3)

Total Reactance Power

$P_{VARsC(T)} = E \times I_T$ (1) $P_{VARsC(T)} = I_T^2 \times X_{CT}$ (2) $P_{VARsC(T)} = E^2 / X_{CT}$ (3)

APPLYING THE RULES

Series Capacitance (C)

1. Four capacitors are connected in series. If the capacitors are rated for 6µf, 14µf, 23µf, and 35µf, determine the total capacitance.

   Use the formula $C_T = \frac{1}{1/C_1 + 1/C_2 + 1/C_3 + \ldots\ldots}$

$$C_T = \frac{1}{1/6\mu f + 1/14\mu f + 1/23\mu f + 1/35\mu f} = \frac{1}{.310\mu f} = 3.23\mu f$$

The total (series) capacitance is 3.23μf.

For clarity, 3.23μf and any quantity involving units of microfarads can be converted into units of farads by dividing the microfarads' quantity by 1,000,000. Therefore, 3.23μf is the same as .00000323F (3.23μf /1,000,000).

Parallel Capacitance (C)

2. Five capacitors are connected in parallel. If the capacitors are rated for 3μf, 25μf, 43μf, 68μf, and 75μf, determine the total capacitance.

Use the formula $C_T = C_1 + C_2 + C_3 + \ldots..$

$$C_T = 3\mu f + 25\mu f + 43\mu f + 68\mu f + 75\mu f = 214\mu f$$

The total (parallel) capacitance is 214μf or .000214F.

Capacitance Reactance (X_C)

3. What is the capacitive reactance of a 27μf capacitor when supplied by two separate *ac* circuits operating at 50 and 60 hertz frequencies?

Use the formula $X_C = \frac{1}{2 \text{ x } \pi \text{ x } f \text{ x } C}$

At 50Hz:

$$X_C = \frac{1}{2 \text{ x } 3.14 \text{ x } 50\text{Hz x } .000027\text{F}} = \frac{1}{.0085} = 117.65\Omega$$

At 60Hz:

$$X_C = \frac{1}{2 \text{ x } 3.14 \text{ x } 60\text{Hz x } .000027\text{F}} = \frac{1}{.0102} = 98.04\Omega$$

Series Capacitance Reactance (X_{CT}) Total

4. Determine the total capacitance reactance of the four series capacitors (6μf, 14μf, 23μf, 35μf) in question No. 1 at 75Hz.

Use the formulas $X_{CT} = \frac{1}{2 \text{ x } \pi \text{ x } f \text{ x } C_T}$ or $X_{CT} = X_{C1} + X_{C2} + X_{C3} + \ldots..$

(1) When the total capacitance is used ($C_T = 3.23\mu f$)

$$X_{CT} = \frac{1}{2 \text{ x } 3.14 \text{ x } 75Hz \text{ x } .00000323F} = \frac{1}{.00152} = 657.89\Omega$$

(2) When the individual capacitance reactance is determined and totaled,

$$(\text{at } 6\mu f)\ X_{C1} = \frac{1}{2 \text{ x } 3.14 \text{ x } 75Hz \text{ x } .000006F} = \frac{1}{.00283} = 353.36\Omega$$

$$(\text{at } 14\mu f)\ X_{C2} = \frac{1}{2 \text{ x } 3.14 \text{ x } 75Hz \text{ x } .000014F} = \frac{1}{.00659} = 151.75\Omega$$

$$(\text{at } 23\mu f)\ X_{C3} = \frac{1}{2 \text{ x } 3.14 \text{ x } 75Hz \text{ x } .000023F} = \frac{1}{.01083} = 92.34\Omega$$

$$(\text{at } 35\mu f)\ X_{C4} = \frac{1}{2 \text{ x } 3.14 \text{ x } 75Hz \text{ x } .000035F} = \frac{1}{.0165} = 60.61\Omega$$

$$X_{CT} = 353.36\Omega + 151.75\Omega + 92.34\Omega + 60.61\Omega = 658.1\Omega$$

Parallel Capacitance Reactance (X_{CT}) Total

5. Determine the total capacitance reactance of the five parallel capacitors (3µf, 25µf, 43µf, 68µf, 75µf) in question No. 2 at 100Hz.

Use the formula $X_{CT} = \frac{1}{2 \text{ x } \pi \text{ x f x } C_T}$ or $\frac{1}{1/X_{C1} + 1/X_{C2} + 1/X_{C3} + \ldots}$

(1) When the total capacitance is used ($C_T = 214\mu f$)

$$X_{CT} = \frac{1}{2 \text{ x } 3.14 \text{ x } 100Hz \text{ x } .000214F} = \frac{1}{.1344} = 7.44\Omega$$

(2) When the individual capacitance reactance is determined and totaled

$$(\text{at } 3\mu f)\ X_{C1} = \frac{1}{2 \text{ x } 3.14 \text{ x } 100Hz \text{ x } .000003F} = \frac{1}{.00188} = 531.91\Omega$$

$$(\text{at } 25\mu f)\ X_{C2} = \frac{1}{2 \text{ x } 3.14 \text{ x } 100Hz \text{ x } .000025F} = \frac{1}{.0157} = 63.69\Omega$$

$$(\text{at } 43\mu f)\ X_{C3} = \frac{1}{2 \text{ x } 3.14 \text{ x } 100Hz \text{ x } .000043F} = \frac{1}{.0270} = 37.04\Omega$$

$$\text{(at 68}\mu\text{f) } X_{C4} = \frac{1}{2 \text{ x } 3.14 \text{ x } 100\text{Hz x } .000068\text{F}} = \frac{1}{.0427} = 23.42\Omega$$

$$\text{(at 75}\mu\text{f) } X_{C5} = \frac{1}{2 \text{ x } 3.14 \text{ x } 100\text{Hz x } .000075\text{F}} = \frac{1}{.0471} = 21.23\Omega$$

$$X_{CT} = \frac{1}{1/531.91\Omega + 1/63.69\Omega + 1/37.04\Omega + 1/23.42\Omega + 1/21.23\Omega}$$

$$= \frac{1}{.1344}$$

$$= 7.44\Omega$$

Series Capacitance Reactance (X_{CT}) Circuit

6. Now that the capacitance reactance has been determined for the capacitors in question No. 4, if 250 volts were applied to the capacitors, determine the circuit's current, the voltage drop across C_1 and C_3, and the reactance power produced by C_2, C_3, and C_4.

   Use the formula $I = E_T/X_{CT}$ (use most accurate value for X_{CT}):

   $$I = 250\text{V}/658.1\Omega = .38\text{A}$$

   Voltage drop (X_C x I) across C_1 and C_3:

   $$E_{C1} = 353.36\Omega \text{ x } .38\text{A} = 134.28\text{V} \qquad E_{C3} = 92.34\Omega \text{ x } .38\text{A} = 35.09\text{V}$$

   Reactance power produced (I^2 x X_C) by C_2 and C_4:

   $$P_{VARsC2} = (.38\text{A})^2 \text{ x } 151.75\Omega = 21.91\text{VARs}$$
   $$P_{VARsC3} = (.38\text{A})^2 \text{ x } 92.34\Omega = 13.33\text{VARs}$$
   $$P_{VARsC4} = (.38\text{A})^2 \text{ x } 60.61\Omega = 8.75\text{VARs}$$

   Now that the voltage drop and reactance power for capacitor C_3 has been determined, let's use like formulas to cross-check the results. See "Related Series Formulas (Capacitance—Overall)."

   To determine the voltage drop using the two remaining formulas,

   $$E_C = P_{VARsC}/I \qquad E_C = \sqrt{P_{VARsC} \text{ x } X_C}$$
   $$E_{C3} = P_{VARsC3}/I \qquad E_{C3} = \sqrt{P_{VARsC3} \text{ x } X_{C3}}$$
   $$= 13.33\text{VARs}/.38\text{A} \qquad = \sqrt{13.33 \text{ x } 92.34}$$
   $$= 35.08\text{V} \qquad = \sqrt{1230.89}$$
   $$= 35.08\text{V}$$

To determine the reactance power using the two remaining formulas,

$P_{VARsC} = E_C \times I$ $\quad\quad$ $P_{VARsC} = E_C{}^2/X_C$

$P_{VARsC3} = E_{C3} \times I$ $\quad\quad$ $P_{VARsC3} = E_{C3}{}^2/X_{C3}$
$= 35.08V \times .38A$ $\quad\quad$ $= (35.08)^2/92.34$
$= 13.33VARs$ $\quad\quad$ $= 1230.61/92.34$
$= 13.33VARs$

In both calculations, the results are the same.

Parallel Capacitance Reactance (X_{CT}) Circuit

7. If the capacitors in question No. 5 were energized by a 65V source,
   a. How much current flows through C_2 and C_3?
   b. What is the voltage drop across C_4?
   c. How much reactance power is produced by C_1, C_2 and C_5?

   a. Current (E/X_C) flow through C_2 and C_3:

$$I_{C2} = 65V/63.69\Omega = 1.02A \qquad I_{C3} = 65V/37.04\Omega = 1.75A$$

   b. Voltage drop across C_4 same as source voltage:

$$E_{C4} = 65V$$

   c. Reactance power produced (E^2/X_C) by C_1, C_2, and C_5

$$P_{VARsC1} = (65V)^2 / 531.91\Omega = 7.94VARs$$
$$P_{VARsC2} = (65V)^2 / 63.69\Omega = 66.34VARs$$
$$P_{VARsC5} = (65V)^2/21.23\Omega = 199.01VARs$$

Again, let's use like formulas to cross-check the current flow through capacitor C_2. See "Related Parallel Formulas (Capacitance—Overall)."

To determine the current flow using the two remaining formulas,

$I_C = P_{VARsC}/E$ $\qquad$ $I_C = \sqrt{P_{VARsC}/X_C}$

$I_{C2} = P_{VARsC2}/E$ $\qquad$ $I_{C2} = \sqrt{P_{VARsC2}/X_{C2}}$

$= 66.34VARs/65V$ $\qquad$ $= \sqrt{66.34/63.69}$

$= 1.02A$ $\qquad$ $= \sqrt{1.04}$

$= 1.02A$

As you can see, the like formulas yield the same results.

Impedance

The total opposition to the flow of current in an *ac* circuit where resistance (R) and reactance (X_C [capacitors] or X_L [inductors] or both) quantities are combined to produce a total current-limiting opposition defines impedance. Impedance is measured in units of ohms and is symbolized by the capital letter Z.

When an *ac* circuit consists of resistance and either inductance or capacitance reactance or both, whether connected in series, parallel, or combined, the impedance cannot be determined simply by adding the values of these quantities together. This is because the flow of current with respect to resistance and either inductance or capacitance is not in phase. One of the most important steps that must be taken before determining the impedance is to ensure that the resistance and reactance quantities share the same units of measurement in ohms. To obtain the impedance of resistance and reactance quantities, their ohmic values have to be added together vectorially instead of arithmetically. To do this, the following formulas are used to determine the impedance per circuit configuration in figure 17:

	R and X_L	R and X_C	R and (X_L and X_C)*
Series	$Z = \sqrt{R^2 + X_L^2}$	$Z = \sqrt{R^2 + X_C^2}$	$Z = \sqrt{R^2 + (X_{L\text{ or }C} - X_{C\text{ or }L})^2}$
Parallel	$Z = \dfrac{R \times X_L}{\sqrt{R^2 + X_L^2}}$	$Z = \dfrac{R \times X_C}{\sqrt{R^2 + X_C^2}}$	$Z = \dfrac{R \times (X_{L\text{ or }C} - X_{C\text{ or }L})}{\sqrt{R^2 + (X_{L\text{ or }C} - X_{C\text{ or }L})^2}}$
	or	or	or
	$Z = \dfrac{1}{\sqrt{(1/R)^2 + (1/X_L)^2}}$	$Z = \dfrac{1}{\sqrt{(1/R)^2 + (1/X_C)^2}}$	$Z = \dfrac{1}{\sqrt{(1/R)^2 + (1/[X_{L\text{ or }C} - X_{C\text{ or }L}]^2)}}$

* If X_L is larger than X_C, then X_C is subtracted from X_L ($X_L - X_C$).

and

If X_C is larger than X_L, then X_L is subtracted from X_C ($X_C - X_L$).

Fig. 17. Impedance per circuit configuration.

Take notice that the formulas for calculating the impedance for parallel circuits are similar to the product of the sum and the reciprocal formulas for *dc* and purely resistive *ac* circuits.

IMPEDANCE APPLICATIONS

Series Resistance (R) and Capacitance (C)

1. An 8Ω resistor is placed in series with a 100μf capacitor. Determine the impedance of the components if the operating frequency is 85Hz.

   Calculate capacitance reactance:

$$X_C = \frac{1}{2 \times 3.14 \times 85\text{Hz} \times .000100\text{F}} = \frac{1}{.0534} = 18.73\Omega$$

   The capacitance reactance of the capacitor is 18.73Ω at 85Hz.

   To find the impedance,

$$Z = \sqrt{(8\Omega)^2 + (18.73\Omega)^2} = \sqrt{64\Omega + 350.81\Omega} = \sqrt{414.81\Omega} = 20.37\Omega$$

   The impedance of the RC series components is 20.37Ω.

Parallel Resistance (R) and Capacitance (C)

2. A 10Ω resistor is placed in parallel with an 80μf capacitor. Determine the impedance of the components if the operating frequency is 85Hz.

   Calculate capacitance reactance:

$$X_C = \frac{1}{2 \times 3.14 \times 85\text{Hz} \times .000080\text{F}} = \frac{1}{.0427} = 23.42\Omega$$

   The capacitance reactance of the capacitor is 23.42Ω at 85Hz.

   To find the impedance,

$$Z = \frac{1}{\sqrt{(1/10\Omega)^2 + (1/23.42\Omega)^2}} = \frac{1}{\sqrt{.01\Omega + .00182\Omega}} = \frac{1}{\sqrt{.01182\Omega}} = 9.198\Omega$$

   The impedance of the RC parallel components is 9.198Ω.

Series Resistance (R) and Inductance (L)

3. A 14Ω resistor is placed in series with a 22.8mH inductor. Determine the impedance of the components if the operating frequency is 135Hz.

   Calculate inductance reactance:

$$X_L = 2 \text{ x } 3.14 \text{ x } 135\text{Hz x } .0228\text{H} = 19.33\Omega$$

   The inductance reactance of the inductor is 19.33Ω at 135Hz.

   To find the impedance,

$$Z = \sqrt{(14\Omega)^2 + (19.33\Omega)^2} = \sqrt{(196\Omega) + (373.65\Omega)} = \sqrt{569.65\Omega} = 23.87\Omega$$

   The impedance of the RL-series components is 23.87Ω.

Parallel Resistance (R) and Inductance (L)

4. A 9Ω resistor is placed in parallel with a 16mH inductor. Determine the impedance of the components if the operating frequency is 135Hz.

   Calculate inductance reactance:

$$X_L = 2 \text{ x } 3.14 \text{ x } 135\text{Hz x } .016\text{H} = 13.56\Omega$$

   The inductance reactance of the inductor is 13.56Ω at 135Hz.

   To find the impedance,

$$Z = \frac{1}{\sqrt{(1/9\Omega)^2 + (1/13.56\Omega)^2}} = \frac{1}{\sqrt{.0123\Omega + .00544\Omega}} = \frac{1}{\sqrt{.01774\Omega}} = 7.51\Omega$$

   The impedance of the RL parallel components is 7.51Ω.

Series Resistance (R) Inductance (L) and Capacitance (C)

5. A 27Ω resistor is placed in series with a 44.3mH inductor and a 88μf capacitor. Determine the impedance of the components if the operating frequency is 30Hz.

   Calculate inductance and capacitance reactance:

$$X_L = 2 \text{ x } 3.14 \text{ x } 30\text{Hz x } .0443\text{H} = 8.35\Omega$$

$$X_C = \frac{1}{2 \text{ x } 3.14 \text{ x } 30\text{Hz x } .000088\text{F}} = \frac{1}{.0166} = 60.24\Omega$$

The capacitance reactance of the capacitor is 60.24Ω at 30Hz while the inductance reactance of the inductor is 8.35Ω at 30Hz. Therefore, the inductance reactance is subtracted from the capacitance reactance in the formula.

To find the impedance,

$$\begin{aligned} Z &= \sqrt{(27\Omega)^2 + (60.24\Omega - 8.35\Omega)^2} \\ &= \sqrt{(27\Omega)^2 + (51.89\Omega)^2} \\ &= \sqrt{729\Omega + 2692.57\Omega} \\ &= \sqrt{3421.57\Omega} \\ &= 58.49\Omega \end{aligned}$$

The impedance of the RLC series components is 58.49Ω.

Parallel Resistance (R) Inductance (L) and Capacitance (C)

6. A 33Ω resistor is placed in parallel with a 136mH inductor and a 369μf capacitor. Determine the impedance of the components if the operating frequency is 30Hz.

Calculate inductance and capacitance reactance:

$$X_L = 2 \text{ x } 3.14 \text{ x } 30\text{Hz x } .136\text{H} = 25.62\Omega$$

$$X_C = \frac{1}{2 \text{ x } 3.14 \text{ x } 30\text{Hz x } .000369\text{F}} = \frac{1}{.07} = 14.29\Omega$$

The inductance reactance of the inductor is 25.62Ω at 30Hz, while the capacitance reactance of the capacitor is 14.29Ω at 30Hz. As a whole, the inductance reactance is larger than the capacitance reactance; however, when placed in the formula, both become reciprocals. Therefore, the matter of which number (quantity) is larger is reversed when converted to decimals, where .039Ω is smaller than .0699Ω. If left as is, the remainder would be negative, but when squared, the results will always yield a positive value.

To find the impedance:

$$Z = \frac{1}{\sqrt{(1/33\Omega)^2 + (1/[25.62\Omega] - 1/[14.29\Omega])^2}}$$

$$Z = \frac{1}{\sqrt{(1/33\Omega)^2 + (.039\Omega - .0699\Omega)^2}}$$

$$Z = \frac{1}{\sqrt{.00092\Omega + (-.0309\Omega)^2}}$$

$$Z = \frac{1}{\sqrt{.00092\Omega + .000955\Omega}}$$

$$Z = \frac{1}{\sqrt{.001875\Omega}}$$

$$Z = \frac{1}{.0433}$$

$$Z = 23.1\Omega$$

The impedance of the RLC parallel components is 23.1Ω.

The Power Triangle, Power Factor, and Phase Angle

As mentioned earlier, if all electrical loads were purely resistive, there wouldn't be a need to consider the reactions and effects inductors and capacitors have on electrical systems. With the exception of synchronous motors, where used for power factor corrections and perhaps a few other occasional applications, purely capacitive loads are somewhat uncommon. Therefore, because most electrical loads are a combination of resistance (R) and inductance (L), the primary focus for this topic will be as such. As a result, inductance reactance (X_L), impedance (Z), power factor (cos θ), etc. are all products of such loads.

In summary, when resistance is combined with an inductor and current is applied, there is a reaction that takes place about the inductor. By now we should know the results of this reaction, which is easily recognized as inductance reactance. Where resistance and inductance reactance are combined, the presence of impedance is certain. Where impedance exists, so does a resulting power factor. And with a power factor, the three power elements can be derived: apparent power, which is measured in units of volt-amperes (VA) or kilovolt-amperes (kVA); true power, which is measured in units of watts (W) or kilowatts (kW); and reactance power, which is measured in units of vars (VARs) or kilovars (kVARs).

The apparent power is recognized as the power needed to meet the demands of an electrical load or system. The transformer is an electrical power source used to supply an electrical load or system based on electrical demands. Therefore, it is safe to say that an electrical load or system's apparent power can be adequately supplied by a transformer when a transformer is sized to meet or exceed the electrical demands of an electrical load or system. Like apparent power, transformers are sized in units of kVA.

The true power is the portion of power that is drawn from the apparent power that is used to supply and operate an electrical load. In other words, it is the power that's consumed by the load and is therefore considered the power that does work. The true power is associated with the resistive (R) part of a load.

The reactance power is the portion of power that is also drawn from the apparent power but is considered wasted power because it does not produce work like that of its counterpart, the true power. It is the power that magnetizing-related equipment such as motors and transformers needs to produce the magnetizing effect. The reactance power is associated with the reactance ($X_{L\ or}\ X_C$) part of a load. Although the primary focus for this topic is resistance and inductance loads, the same approach is also used for resistance and capacitance reactance loads.

Because the true power is only a portion of the apparent power, that portion of the true power as it relates to the apparent power can be found by dividing the apparent power into the true power:

$$\frac{\text{True power}}{\text{Apparent power}} \quad \begin{matrix}\text{(portion)} \\ \text{(whole)}\end{matrix}$$

Again, because the true power is only a portion of the apparent power, it is only a percentage of the apparent power as a whole. When calculated, the results will usually yield a value less than 1 (a decimal). The results of such calculation is referred to as the power factor:

$$\text{Power factor (PF)} = \frac{\text{true power}}{\text{apparent power}} \quad \begin{matrix}\text{(W) (kW)} \\ \text{(VA) (kVA)}\end{matrix}$$

The power factor is simply the ratio between the true power and the apparent power. When the apparent power and true power are equal, a unity power factor is realized. Unlike the three power elements that have specific units of measurements to uniquely identify them, the power factor is unitless because it relates to two corresponding units of power.

To display the relationship between these three power elements, a right triangle can also be used and will be referred to in this case as the power triangle. Observe figure 18.

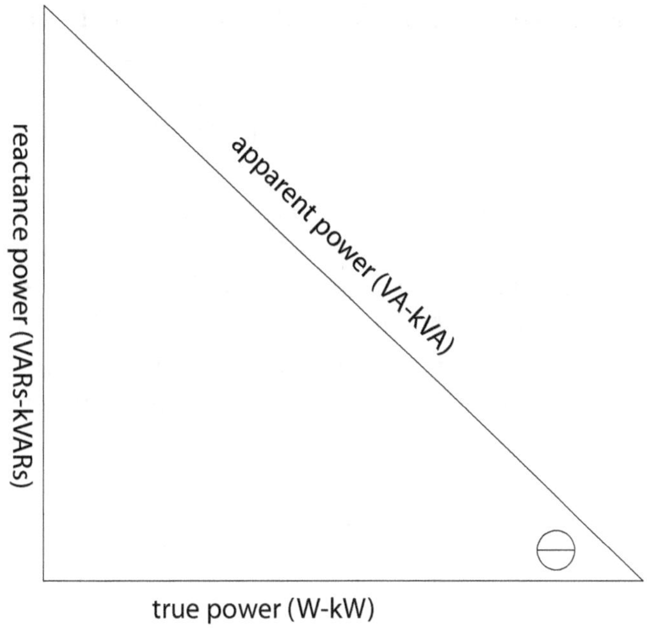

Fig. 18. The power triangle.

Earlier we learned that the trigonometric function cosine theta was equal to the adjacent side of a right triangle over the hypotenuse; that is, $\cos\theta$ = adjacent/hypotenuse. Looking at the power triangle, we see that the adjacent side of the triangle with respect to theta is equivalent to the true power (kW) and the hypotenuse is equivalent to the apparent power (kVA). As a result, an equivalent formula is derived:

$$\cos\theta = kW/kVA$$

If the power factor (PF) = kW/kVA and $\cos\theta$ = kW/kVA, then the power factor is equal to $\cos\theta$; that is,

$$PF = \cos\theta$$

and

$$\theta = \cos^{-1} PF \text{ (derived)}$$

Again, in this situation theta (θ) is no more than the phase angle between the true power and the apparent power and is a function of the power factor.

In a series circuit, when the phase angle falls between 0° and 45°, a load or circuit is said to be more resistive, and when the phase angle falls between 46° and 90°, a load or circuit is said to be more inductive (or capacitive). For a parallel circuit, the previously mentioned is totally opposite. In general, the closer the phase angle is to zero degrees (0°), the better the power factor, and the closer the phase angle is to 90°, the less favorable (desired) the power factor. To maintain an 80% or greater power factor, the phase angle has to be less than 37°.

Now that the power factor has been determined, it can be used to calculate two of the power quantities:

$$\text{kW} = \text{kVA} \times \cos\theta \quad \text{and} \quad \text{kVA} = \text{kW}/\cos\theta$$

And with the phase angle, the reactance power can also be calculated.

Knowing that the trigonometric function sine theta is equal to the opposite side of a right triangle over the hypotenuse ($\sin\theta$ = opposite/hypotenuse), the power triangle shows the opposite side of the triangle with respect to theta being equivalent to the reactance power (kVARs). Because the hypotenuse has already been identified as the apparent power, the concluding formula can be used to derive the reactance power:

$$\sin\theta = \text{kVARs}/\text{kVA}$$

where

$$\text{kVARs} = \text{kVA} \times \sin\theta \quad \text{and} \quad \text{kVA} = \text{kVARs}/\sin\theta$$

Just as the power factor is equal to the ratio kW/KVA, the ratio kVARs/kVA is also recognized as the reactance power factor (RPF).

Therefore, if the reactance power factor (RPF) = kVARs/kVA and $\sin\theta$ = kVARs/kVA, then

$$\text{RPF} = \sin\theta$$

and

$$\theta = \sin^{-1} \text{RPF (derived)}$$

Similar to the power factor, the phase angle θ is a function of the reactance power factor. The use of the reactance power factor is not nearly as common as the power factor and is seldom recognized.

Other Source Triangles

Based on what is understood about a right triangle, other source triangles can also be illustrated as well. Observe figure 19.

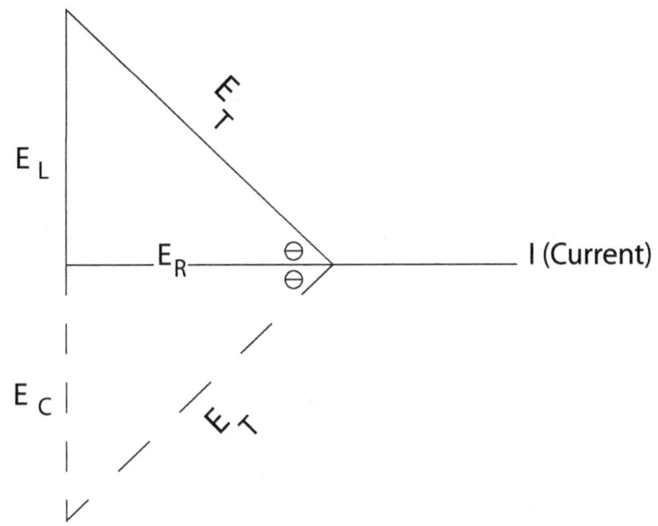

Fig. 19. The voltage triangle (series circuit).

The voltage triangle is relative to the series circuit where there is only one source of current and three elements of voltage (E_T, E_R, and E_L or $_C$).

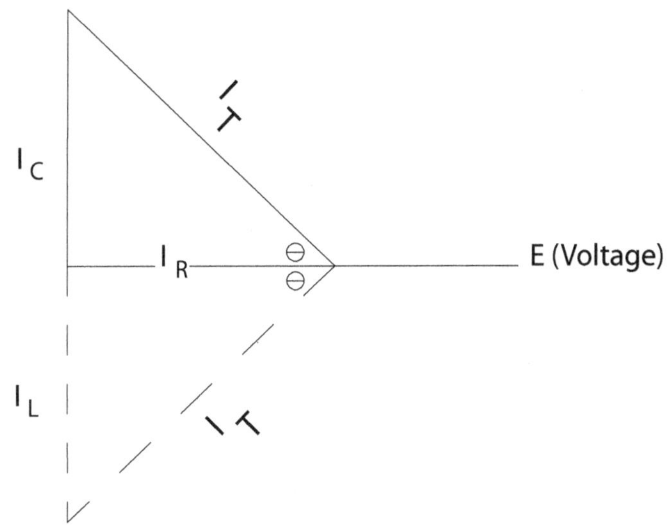

Fig. 20. The current triangle (parallel circuit).

The current triangle is relative to the parallel circuit where there is only one voltage source and three elements of current (I_T, I_R, and I_L or $_C$).

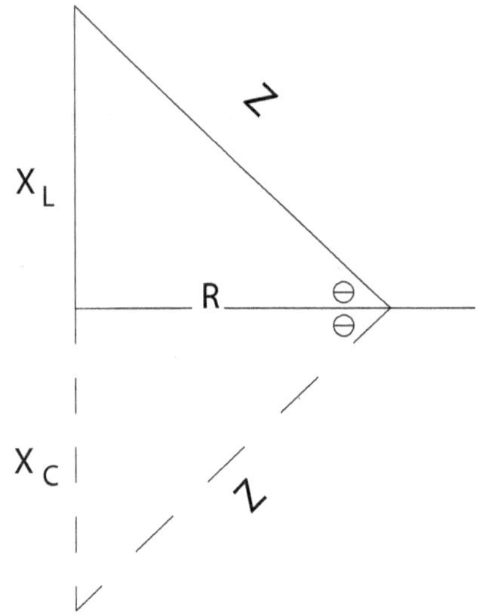

Fig. 21. The impedance triangle.

The impedance triangle consists of four current opposing components (Z, R, X_L, and X_C). Although the components affect time-varying voltage and current quantities of an *ac* circuit, they themselves do not vary with time and therefore are not considered phasors. The application of the impedance triangle is only applicable for series circuits.

Finally, the power triangle as observed in figure 22 can also be re-generated using related formulas opposed to the use of trigonometric functions to calculate the three power elements (figure 22).

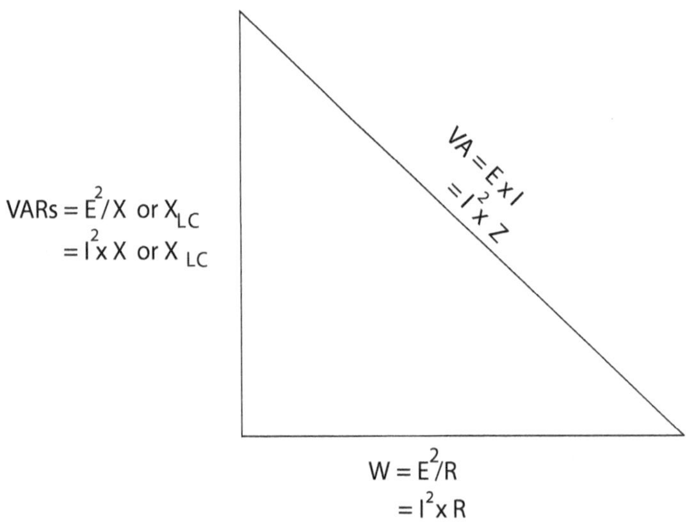

Fig. 22. The power triangle (re-generated).

As follows, each illustrated triangle will be translated into an actual circuit application based on alternating current.

Resistance (R)-Inductance (L) Circuits

Series-RL

In any series circuit, regardless of whether it is being supplied from a *dc* or *ac* source, one thing is certain: the same amount of current will flow through each connected component. Along with this topic and others to follow, detailed information will be used to develop a thorough understanding of all combinations of resistance–reactance circuits. Now let's begin by observing figure 23.

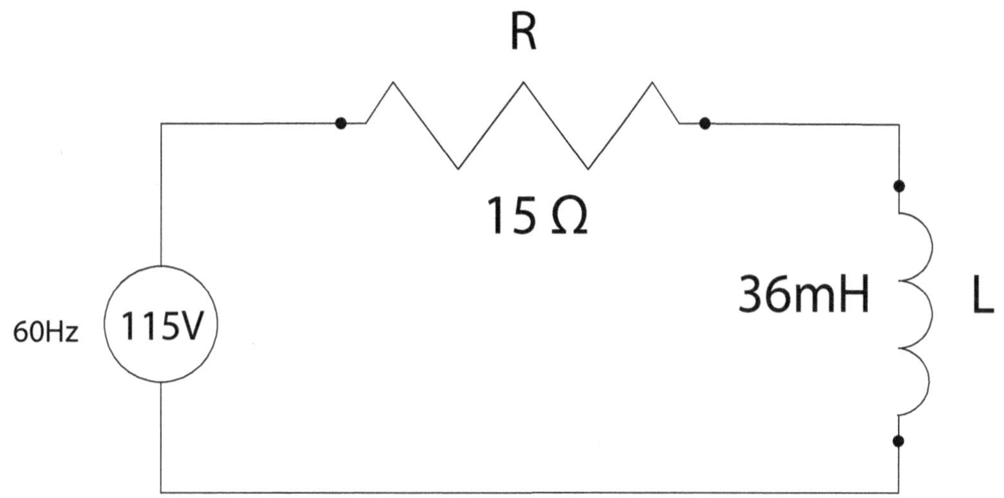

Fig. 23. Series-RL circuit.

Based on the series resistance–inductance circuit's operating conditions and the given components, the following quantities will be calculated: the inductance reactance (X_L), the impedance (Z), the current (I), the voltage drop across each component (E_R and E_L), the total voltage (E_T), the apparent power (VA), the true power (W), the reactance power (VARs), the power factor (PF), and the phase angle (θ).

Because a variation of like quantities will exist while analyzing this circuit, certain calculated quantities will be randomly selected to perform adjoining calculations.

Inductance Reactance (X_L) {Series-RL}

Before the given quantity for the inductor can be used, it must be converted to a henry (H), opposed to a millihenry (mH), to be used in the inductance reactance formula. Therefore,

$$36\text{mH} = \frac{36}{1000} = .036\text{H}$$

Now, based on the converted unit and an operating frequency of 60Hz, the inductance reactance is

$$X_L = 2 \text{ x } 3.14 \text{ x } 60\text{Hz x } .036\text{H} = 13.565\Omega$$

Impedance (Z) {Series-RL}

As you recall, in a purely resistive *ac* circuit, the voltage and current are in phase with each other, and in a purely inductive circuit, the voltage and current are not in phase with each other. To be specific, the voltage (E) leads the current (I) by 90° about the inductor. Because the current flowing through the resistor and the inductor is not in phase with the voltage across each component, the opposition to current for both components cannot be added together similarly to that of a *dc* or purely resistance *ac* circuit. To calculate the total opposition to the flow of current or impedance for this circuit, the vector related formula for a Series RL circuit must be used: $Z = \sqrt{R^2 + X_L^2}$.

Therefore,

$$
\begin{aligned}
Z &= \sqrt{(15)^2 + (13.565)^2} \\
&= \sqrt{225 + 184} \\
&= \sqrt{409} \\
&= 20.224\Omega
\end{aligned}
$$

Now that the circuit's impedance has been determined, other quantities can be determined also.

Current (I) {Series-RL}

In a *dc* or purely resistive *ac* circuit, the Ohm's Law formula I = E/R could be used to calculate the circuit's current. Because this circuit has a combined opposition to the flow of current that is an impedance, the R value can be substituted with Z, causing the current to be calculated based on the converted formula, $I = E_T/Z$. Since E_T is shown to be 115V,

$$
\begin{aligned}
I &= 115V/20.224\Omega \\
&= 5.686A
\end{aligned}
$$

Voltage Drops (E_R and E_L) {Series-RL}

Because current will flow through both components in the same amount, a voltage drop will exist across each component. Again, referring to Ohm's Law, where E = I x R, this formula can be used to calculate the voltage drop across R. However, this formula cannot be used with the inductor, although the inductance reactance is also measured in units of ohms. On the other hand, because the unit of measurement for inductance reactance is ohms and inductance reactance is in opposition to the flow of current, the R value in Ohm's Law can be substituted with X_L. With this substitution, the formula is converted to $E = I \times X_L$, and the voltage drop across the inductor can be calculated. With both formulas in place,

$$
\begin{aligned}
E_R &= I \times R \\
&= 5.686A \times 15\Omega \\
&= 85.29V
\end{aligned}
\qquad \text{and} \qquad
\begin{aligned}
E_L &= I \times X_L \\
&= 5.686A \times 13.565\Omega \\
&= 77.13V
\end{aligned}
$$

Total Voltage (E_T) {Series-RL}

Referring back to that of a *dc* or purely resistive *ac* circuit, where the total voltage is determined by adding each individual voltage drop together, it should be clear that this concept will not work for this type circuit. Once again, it is because the voltage across the inductor is not in phase with the circuit's current. As a result, the total voltage has to be determined vectorially as opposed to the use of basic addition. Because of this the total voltage has to be calculated based on the vector related formula, $E_T = \sqrt{E_R^2 + E_L^2}$.

Therefore,

$$E_T = \sqrt{(85.29)^2 + (77.13)^2}$$
$$= \sqrt{7274.38 + 5949.04}$$
$$= \sqrt{13,223.42}$$
$$= 114.99V$$

As expected, when calculated vectorially, the calculated total voltage should be the same (or approximately the same) as the given total voltage: 115V.

The Apparent Power (VA) {Series-RL}

When an *ac* circuit is purely resistive, the product of E x I produces the same results as a corresponding *dc* circuit and that is power in watts. Unlike a purely resistive *ac* circuit, when such circuit consists of resistance along with an inductor or a capacitor, the product of E x I is not equal to the power in watts but instead is equal to the apparent power. Using that which was given as the circuit's total voltage (E) and the calculated current (I) the circuit's apparent power can be determined by use of the formula $P_{VA} = E_T \times I$ in addition to the formulas, $P_{VA} = I^2 \times Z$ and $P_{VA} = E_T^2/Z$.

$$P_{VA} = 115V \times 5.686A = 653.89VA$$

$$P_{VA} = (5.686A)^2 \times 20.224\Omega = 32.33A \times 20.224\Omega = 653.84VA$$

$$P_{VA} = (115V)^2/20.224\Omega = 13,225V/ 20.224\Omega = 653.93VA$$

The True Power (W) {Series-RL}

The true power is always associated with the resistive part of an *ac* circuit, and because of this, it can be calculated by using the formula $P_W = I^2 \times R$ or $P_W = E_R^2/R$, where E_R is the voltage drop across the 15Ω resistance. The true power can also be calculated by means of the power factor, where the same results are obtained as that of the formulas. The application of the power factor will be used later, so let's now apply the formulas.

$$P_W = I^2 \times R = (5.686A)^2 \times 15\Omega = 484.96W$$

and

$$P_W = E_R^2/R = (85.29V)^2/15\Omega = 484.96W$$

The Reactance Power (VARs) {Series-RL}

The reactance power is always associated with the reactance part of an *ac* circuit, the inductance reactance. Just as the R value was previously substituted in Ohm's Law with X_L, the same substitution can also be made in the formulas used for calculating reactance power.

With this substitution, the formulas are converted to $P_{VARsL} = I^2 \times X_L$ and $P_{VARsL} = E_L^2/X_L$, where E_L is the voltage drop across the 36mH inductor.

The reactance power can also be calculated by means of the phase angle where the same results are obtained as that of the formulas. Such application will also be used later. Now to apply the formulas:

$$P_{VARsL} = I^2 \text{ x } X_L = (5.686A)^2 \text{ x } 13.565\Omega = 438.56\text{VARs}$$

and

$$P_{VARsL} = E_L^{\,2}/X_L = (77.13V)^2/13.565\Omega = 438.56\text{VARs}$$

The Power Factor (PF) {Series-RL}

With the circuit's impedance and resistance alone, the circuit's power factor could be determined solely with those quantities. However, the same could be said about the circuit's apparent power and true power. With each set of quantities the power factor can be calculated, and, best of all, they both will yield the same results or an approximate. Because the power factor is defined as the ratio between the true power and the apparent power and because the true power is relevant to the circuit's resistance and the apparent power is relevant to the impedance, this makes both quantities available for use in determining the power factor. Take a look:

$$PF = \frac{P_W}{P_{VA}} \quad \text{and} \quad PF = \frac{R}{Z}$$

Using the given and calculated quantities,

$$PF = \frac{484.96W}{653.89VA} = .7417\ (74.17\%) \quad \text{and} \quad PF = \frac{15\Omega}{20.224\Omega} = .7417\ (74.17\%)$$

As you can see, when calculated, both sets of quantities yield the same results. Just remember, of the four quantities used, three were derived (P_W, P_{VA}, Z) from calculating and one was given (R), and, where numbers are rounded off (resulting from calculation), the degree of accuracy is always compromised. Because R was given, it can be considered a fixed value, which leaves the degree of accuracy uncompromised. Therefore, if only one formula were selected to determine the power factor, R/Z would prove to yield more accurate results (although both calculations came out to be the same) compared to its counterpart P_W/P_{VA}, where both quantities used were rounded off values.

The Phase Angle (θ) {Series-RL}

In a series RL circuit, formulas for the phase angle can be derived using any of the three trigonometric formulas:

$$\tan\theta = \frac{E_L}{E_R} \text{ or } \frac{X_L}{R} \quad \text{and} \quad \cos\theta = \frac{R}{Z}$$

From the three formulas, the following derived formulas—along with a previously derived formula—will be used to determine the phase angle:

$$\theta = \tan^{-1}\frac{E_L}{E_R} \text{ or } \frac{X_L}{R} \quad \text{and} \quad \theta = \cos^{-1}\frac{R}{Z}$$

$$\theta = \cos^{-1} PF \text{ (previous)}$$

Based on the derived formulas the circuit's phase angle is

$$\theta = \tan^{-1} \frac{77.13V}{85.29V} \text{ or } \frac{13.565\Omega}{15\Omega} \quad \text{and} \quad \theta = \cos^{-1} \frac{15\Omega}{20.224\Omega}$$

$$= \tan^{-1} .9043 \qquad\qquad = \cos^{-1} .7417$$

$$= 42.12° \qquad\qquad = 42.12°$$

$$\theta = \cos^{-1} .7417 \text{ (previous)}$$
$$= 42.12°$$

Using all four formulas, the results are all the same.

Power {Series-RL}
Previously, the three power quantities were calculated using given or calculated values. The quantities will now be determined using an alternate approach. By using the formula cos $\theta = P_W/P_{VA}$, the true power can now be determined by use of the derived formula, $P_W = P_{VA}$ x cos θ, where

$$P_W = P_{VA} \text{ x } \cos\theta$$
$$= 653.89VA \text{ x } \cos 42.12°$$
$$= 653.89VA \text{ x } .7417$$
$$= 484.99W$$

By using the formula sin $\theta = P_{VARsL}/P_{VA}$, the reactance power can now be determined by use of the derived formula, $P_{VARsL} = P_{VA}$ x sin θ, where

$$P_{VARsL} = P_{VA} \text{ x } \sin\theta$$
$$= 653.89VA \text{ x } \sin 42.12°$$
$$= 653.89VA \text{ x } .6707$$
$$= 438.56VARs$$

As you can see, the results are approximately the same as earlier calculated. By using these results, they can be substituted into the Pythagorean theorem, and the apparent power can be verified.

$$P_{VA} = \sqrt{(484.99)^2 + (438.56)^2}$$
$$= \sqrt{235{,}215.3 + 192{,}334.9}$$
$$= \sqrt{427{,}550.2}$$
$$= 653.87VA$$

With alternate methods, calculations can be either cross-checked or verified to ensure correct answers.

The given series resistance–reactance circuit has now been completely analyzed and the objective has been satisfied.

The following display is an overall list of formulas that can be used for solving Series-RL Circuits.

Related Series Formulas (Resistance–Inductance Circuits—Overall)

[Only one source of current (I); therefore, $I = I_R = I_L$.]

Resistance (R) {Series-RL}

$R = E_R/I$ (1) $R = E_R^2/P_W$ (2) $R = P_W/I^2$ (3)

$R = Z \times PF (\cos\theta)$ (4) $R = \sqrt{Z^2 - X_L^2}$ (5) $R = X_L/\tan\theta$ (6)

Inductance (L) {Series-RL}

$L = X_L/2\pi f$

Inductance Reactance (X_L) {Series-RL}

$X_L = 2\pi fL$ (1) $X_L = E_L/I$ (2) $X_L = E_L^2/P_{VARsL}$ (3)

$X_L = P_{VARsL}/I^2$ (4) $X_L = \sqrt{Z^2 - R^2}$ (5) $X_L = Z \times \sin\theta$ (6)

$X_L = R \times \tan\theta$ (7)

Impedance (Z) {Series-RL}

$Z = \sqrt{R^2 + X_L^2}$ (1) $Z = E_T/I$ (2) $Z = E_T^2/P_{VA}$ (3)

$Z = P_{VA}/I^2$ (4) $Z = R/PF (\cos\theta)$ (5) $Z = X_L/\sin\theta$ (6)

Current (I) {Series-RL}

$I = E_R/R$ (1) $I = P_W/E_R$ (2) $I = E_L/X_L$ (3)

$I = P_{VARsL}/E_L$ (4) $I = \sqrt{P_W/R}$ (5) $I = \sqrt{P_{VARsL}/X_L}$ (6)

$I = E_T/Z$ (7) $I = \sqrt{P_{VA}/Z}$ (8) $I = P_{VA}/E_T$ (9)

Voltage Drops (E_R and E_L) {Series-RL}

$E_R = I \times R$ (1) $E_R = P_W/I$ (2) $E_R = \sqrt{P_W \times R}$ (3)

$E_R = \sqrt{E_T^2 - E_L^2}$ (4) $E_R = E_T \times PF (\cos\theta)$ (5) $E_R = E_L/\tan\theta$ (6)

$E_L = I \times X_L$ (1) $E_L = P_{VARsL}/I$ (2) $E_L = \sqrt{P_{VARsL} \times X_L}$ (3)

$E_L = \sqrt{E_T^2 - E_R^2}$ (4) $E_L = E_T \times \sin\theta$ (5) $E_L = E_R \times \tan\theta$ (6)

Total Voltage (E_T) {Series-RL}

$E_T = P_{VA}/I$ (1) $E_T = \sqrt{P_{VA} \times Z}$ (2) $E_T = I \times Z$ (3)

$E_T = \sqrt{E_R^2 + E_L^2}$ (4) $E_T = E_R/\cos\theta (PF)$ (5) $E_T = E_L/\sin\theta$ (6)

The Apparent Power (VA) {Series-RL}

$P_{VA} = E_T \times I$ (1) $P_{VA} = I^2 \times Z$ (2) $P_{VA} = E_T^2/Z$ (3)

$P_{VA} = \sqrt{P_W^2 + P_{VARsL}^2}$ (4) $P_{VA} = P_W/PF (\cos\theta)$ (5) $P_{VA} = P_{VARsL}/\sin\theta$ (6)

The True Power (W) {Series-RL}

$P_W = I^2 \times R$ (1) $P_W = E_R^2/R$ (2) $P_W = E_R \times I$ (3)

$P_W = \sqrt{P_{VA}^2 - P_{VARsL}^2}$ (4) $P_W = P_{VA} \times PF (\cos\theta)$ (5) $P_W = P_{VARsL} / \tan\theta$ (6)

The Reactance Power (VARs) {Series-RL}

$P_{VARsL} = E_L \times I$ (1) $P_{VARsL} = I^2 \times X_L$ (2) $P_{VARsL} = E_L^2 / X_L$ (3)

$P_{VARsL} = \sqrt{P_{VA}^2 - P_W^2}$ (4) $P_{VARsL} = P_{VA} \times \sin\theta$ (5) $P_{VARsL} = P_W \times \tan\theta$ (6)

The Power Factor (PF) {Series-RL}

$PF = R/Z (\cos\theta)$ (1) $PF = E_R/E_T (\cos\theta)$ (2) $PF = P_W/P_{VA} (\cos\theta)$ (3)

The Phase Angle (θ) {Series-RL}

$\theta = \cos^{-1}\frac{E_R}{E_T}$ (1) $\theta = \sin^{-1}\frac{E_L}{E_T}$ (2) $\theta = \tan^{-1}\frac{E_L}{E_R}$ (PF) (3)

$\theta = \cos^{-1}\frac{R}{Z}$ (PF) (4) $\theta = \sin^{-1}\frac{X_L}{Z}$ (5) $\theta = \tan^{-1}\frac{X_L}{R}$ (6)

$\theta = \cos^{-1}\frac{P_W}{P_{VA}}$ (PF) (7) $\theta = \sin^{-1}\frac{P_{VARsL}}{P_{VA}}$ (8) $\theta = \tan^{-1}\frac{P_{VARsL}}{P_W}$ (9)

<u>Parallel-RL</u>

In any parallel circuit, regardless of whether it is being supplied from a *dc* or *ac* source, one thing is for certain: the voltage drop across each parallel component is the same as the applied voltage. Observe figure 24.

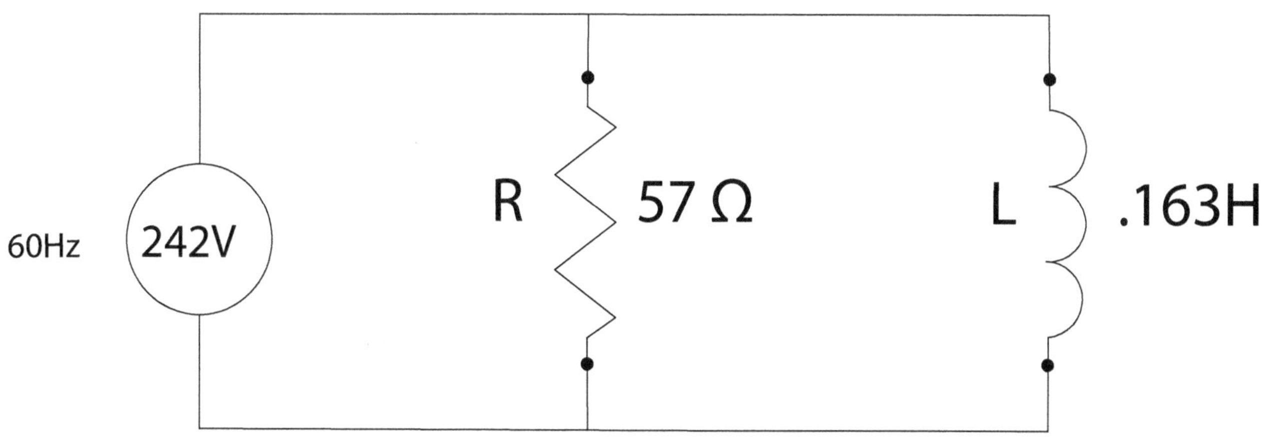

Fig. 24. Parallel-RL circuit.

Based on the parallel resistance–inductance circuit's operating conditions and the given components, the following quantities will be calculated: the inductance reactance (X_L), the impedance (Z), the total current (I_T), the individual current paths through each component (I_R and I_L), the apparent power (VA), the true power (W), the reactance power (VARs), the power factor (PF), and the phase angle (θ).

Because a variation of like quantities will exist while analyzing this circuit, certain calculated quantities will be randomly selected to perform adjoining calculations.

Inductance Reactance (X_L) {Parallel-RL}

Based on the .163H inductor and an operating frequency of 60Hz, the inductance reactance is

$$X_L = 2 \text{ x } 3.14 \text{ x } 60\text{Hz x } .163\text{H} = 61.42\Omega$$

Impedance (Z) {Parallel-RL}

Being in parallel will require the impedance to be calculated differently from that of a series RL circuit. But the fact still remains that the voltage and current are not in phase, and the impedance has to be calculated vectorially. Therefore, to calculate the impedance for this circuit, the vector-related formula for a parallel RL circuit must be used:

$$Z = \frac{R \text{ x } X_L}{\sqrt{R^2 + X_L^2}} \quad \text{or} \quad Z = \frac{1}{\sqrt{1/R^2 + 1/X_L^2}}$$

Applying both formulas,

$$Z = \frac{57 \text{ x } 61.42}{\sqrt{(57)^2 + (61.42)^2}} \quad \text{or} \quad Z = \frac{1}{\sqrt{1/(57)^2 + 1/(61.42)^2}}$$

$$Z = \frac{3500.94}{\sqrt{3249 + 3772.42}} \quad \text{or} \quad Z = \frac{1}{\sqrt{1/\ 3249 + 1/3772.42}}$$

$$Z = \frac{3500.94}{83.79} \quad \text{or} \quad Z = \frac{1}{\sqrt{.00030779 + .0002651}}$$

$$Z = 41.78\Omega \quad \text{or} \quad Z = \frac{1}{.0239351}$$

$$Z = 41.78\Omega$$

Total Current (I_T) {Parallel-RL}

Since E is shown to be 242V, the total current can also be calculated using the formula

$$I_T = E/Z$$

$$I_T = 242\text{V}/41.78\Omega$$

$$= 5.79\text{A}$$

Individual Current (I_R and I_L) {Parallel-RL}

Because this is a parallel circuit, the voltage drop across each component will be equal to that of the applied voltage; however, there will be more than one path for current flow, unlike that of a series circuit. Because the ohmic values of R and X_L are not the same, both components will experience different amounts of current flow: I_R and I_L.

Again, the component with the smallest ohmic value will experience the largest amount of current. The Ohm's Law formula I = E/R can be used to calculate the current flow through R and can also be used, where R is substituted with X_L. That being the case,

$$I_R = E/R = 242V/57\Omega = 4.25A \quad \text{and} \quad I_L = E/X_L = 242V/61.42\Omega = 3.94A$$

Referring to that of a *dc* or purely resistive *ac* circuit where the total current is determined by adding each individual current together, it should be clear that this concept will not work for this type circuit either. Again, it is because the voltage across the inductor is not in phase with the current flowing through it. As a result, the total current must also be determined vectorially by means of the vector-related formula $I_T = \sqrt{I_R^2 + I_L^2}$. Therefore,

$$I_T = \sqrt{(4.25)^2 + (3.94)^2} = \sqrt{33.59} = 5.796A$$

As expected, when calculated vectorially, the total current should be the same (or approximately the same) as that previously calculated ($I_T = 242V/41.78\Omega = 5.79A$).

The Apparent Power (VA) {Parallel-RL}

Using that which was given as the circuit's voltage (E), the calculated current (I), and impedance (Z), the circuit's apparent power can be determined by use of the formulas $P_{VA} = E \times I_T$, $P_{VA} = I_T^2 \times Z$, and $P_{VA} = E^2/Z$.

$$P_{VA} = 242V \times 5.796A = 1402.63VA$$

$$P_{VA} = (5.796A)^2 \times 41.78\Omega = 33.59 \times 41.78\Omega = 1403.39VA$$

$$P_{VA} = (242V)^2/41.78\Omega = 58{,}564V/41.78\Omega = 1401.72VA$$

The True Power (W) {Parallel-RL}

The true power is calculated using the formulas $P_W = I_R^2 \times R$ and $P_W = E^2/R$, where I_R is the current flowing through R.

$$P_W = I_R^2 \times R = (4.25A)^2 \times 57\Omega = 1029.56W \quad \text{and} \quad P_W = E^2/R = (242V)^2/57\Omega = 1027.44W$$

The Reactance Power (VARs) {Parallel-RL}

The reactance power is calculated using the formulas $P_{VARsL} = I_L^2 \times X_L$ and $P_{VARsL} = E^2/X_L$, where I_L is the current flowing through L:

$$P_{VARsL} = I_L^2 \times X_L = (3.94A)^2 \times 61.42\Omega = 953.46VARs \quad \text{and} \quad P_{VARsL} = E^2/X_L = (242V)^2/61.42\Omega = 953.50VARs$$

The Power Factor (PF) {Parallel-RL}

When calculating the power factor in the series-RL circuit, the formula PF = R/Z was used. With a parallel-RL circuit, the formula is the complete opposite:

$$PF = \frac{Z}{R}$$

where the power factor will always result to a value of either 100% (1) or less because the impedance in a parallel circuit will always result to being less than the resistance.

Using the given and calculated quantities along with the above and previously given power factor formula (P_W/P_{VA}),

$$PF = \frac{1027.44W}{1401.72VA} = .733\ (73.3\ \%) \quad \text{and} \quad PF = \frac{41.78\Omega}{57\Omega} = .733\ (73.3\%)$$

As you can see, when calculated, both sets of quantities yield the same results.

The Phase Angle (θ) {Parallel-RL}

In a parallel-RL circuit, formulas for the phase angle can be derived using any of the three trigonometric formulas:

$$\tan\theta = \frac{I_L}{I_R} \text{ or } \frac{R}{X_L} \quad \text{and} \quad \cos\theta = \frac{Z}{R}$$

From the three formulas, the following derived formulas, along with a previously derived formula, will be used to determine the phase angle:

$$\theta = \tan^{-1} = \frac{I_L}{I_R} \text{ or } \frac{R}{X_L} \quad \text{and} \quad \theta = \cos^{-1}\frac{Z}{R}$$

$$\theta = \cos^{-1} PF \text{ (previous)}$$

Based on the derived formulas, the circuit's phase angle is

$$\theta = \tan^{-1}\frac{3.94A}{4.25A} = \tan^{-1} .9271 = 42.83°$$

or

$$\frac{57\Omega}{61.42\Omega} = .9280 = 42.86°$$

and

$$\theta = \cos^{-1}\frac{41.78\Omega}{57\Omega} = \cos^{-1} .733 = 42.86°$$

$$\theta = \cos^{-1} .733 \text{ (previous)} = 42.86°$$

Using all four formulas, the results are all approximately the same.

Power {Parallel-RL}

Previously, the three power quantities were calculated using given or calculated values. The quantities will now be determined using an alternate approach. By using the formula $\cos\theta = P_W/P_{VA}$, the true power can now be determined by use of the derived formula $P_W = P_{VA} \text{ x } \cos\theta$, where

$$\begin{aligned} P_W &= P_{VA} \text{ x } \cos\theta \\ &= 1401.72VA \text{ x } \cos 42.86° \\ &= 1401.72VA \text{ x } .733 \\ &= 1027.46W \end{aligned}$$

By using the formula $\sin\theta = P_{VARsL}/P_{VA}$, the reactance power can now be determined by use of the derived formula $P_{VARsL} = P_{VA} \text{ x } \sin\theta$, where

$$\begin{aligned} P_{VARsL} &= P_{VA} \text{ x } \sin\theta \\ &= 1401.72VA \text{ x } \sin 42.86° \\ &= 1401.72VA \text{ x } .6802 \\ &= 953.45VARs \end{aligned}$$

As you can see, the results are approximately the same as earlier calculated. By using these results, they can be substituted into the Pythagorean theorem, and the apparent power can be verified.

$$\begin{aligned} P_{VA} &= \sqrt{(1027.46)^2 + (953.45)^2} \\ &= \sqrt{1{,}055{,}674.1 + 909{,}066.9} \\ &= \sqrt{1{,}964{,}741} \\ &= 1401.69VA \end{aligned}$$

The given parallel resistance–reactance circuit has now been completely analyzed, and the objective has been satisfied.

The following display is an overall list of formulas that can be used for solving Parallel-RL circuits.

Related Parallel Formulas (Resistance–Inductance Circuits—Overall)

[Only one source of voltage (E); therefore, $E = E_R = E_L$.]

Resistance (R) {Parallel-RL}

$R = E/I_R$ (1) $R = E^2/P_W$ (2) $R = P_W/I_R^{\,2}$ (3)
$R = Z/\cos\theta$ (PF) (4) $R = X_L \times \tan\theta$ (5) $R = \dfrac{1}{\sqrt{(1/Z)^2 - (1/X_L)^2}}$ (6)

Inductance (L) {Parallel-RL}

$L = X_L/2\pi f$

Inductance Reactance (X_L) {Parallel-RL}

$X_L = 2\pi fL$ (1) $X_L = E/I_L$ (2) $X_L = E^2/P_{VARsL}$ (3)
$X_L = P_{VARsL}/I_L^{\,2}$ (4) $X_L = R/\tan\theta$ (5) $X_L = Z/\sin\theta$ (6)
$X_L = \dfrac{1}{\sqrt{(1/Z)^2 - (1/R)^2}}$ (7)

Impedance (Z) {Parallel-RL}

$Z = \dfrac{1}{\sqrt{(1/R)^2 + (1/X_L)^2}} = \dfrac{R \times X_L}{\sqrt{(R)^2 + (X_L)^2}}$ (1) $Z = E/I_T$ (2)
$Z = E^2/P_{VA}$ (3) $Z = P_{VA}/I_T^{\,2}$ (4) $Z = R \times \cos\theta$ (PF) (5)
$Z = X_L \times \sin\theta$ (6)

Individual Current (I_R and I_L) {Parallel-RL}

$I_R = E/R$ (1) $I_R = P_W/E$ (2) $I_R = \sqrt{P_W/R}$ (3)
$I_R = \sqrt{I_T^{\,2} - I_L^{\,2}}$ (4) $I_R = I_T \times \cos\theta$ (PF) (5) $I_R = I_L/\tan\theta$ (6)
$I_L = E/X_L$ (1) $I_L = P_{VARsL}/E$ (2) $I_L = \sqrt{P_{VARsL}/X_L}$ (3)
$I_L = \sqrt{I_T^{\,2} - I_R^{\,2}}$ (4) $I_L = I_T \times \sin\theta$ (5) $I_L = I_R \times \tan\theta$ (6)

Total Current (I_T) {Parallel-RL}

$I_T = E/Z$ (1) $I_T = P_{VA}/E$ (2) $I_T = \sqrt{P_{VA}/Z}$ (3)
$I_T = \sqrt{I_R^{\,2} + I_L^{\,2}}$ (4) $I_T = I_R/\cos\theta$ (PF) (5) $I_T = I_L/\sin\theta$ (6)

Voltage (E) {Parallel-RL}

$E = I_R \times R$ (1) $E = P_W / I_R$ (2) $E = I_L \times X_L$ (3)

$E = P_{VARsL} / I_L$ (4) $E = \sqrt{P_W \times R}$ (5) $E = \sqrt{P_{VARsL} \times X_L}$ (6)

$E = I_T \times Z$ (7) $E = P_{VA} / I_T$ (8) $E = \sqrt{P_{VA} \times Z}$ (9)

The Apparent Power (VA) {Parallel-RL}

$P_{VA} = E \times I_T$ (1) $P_{VA} = I_T^2 \times Z$ (2) $P_{VA} = E^2/Z$ (3)

$P_{VA} = P_W/\cos\theta$ (PF) (4) $P_{VA} = P_{VARsL} / \sin\theta$ (5) $P_{VA} = \sqrt{P_W^2 + P_{VARsL}^2}$ (6)

The True Power (W) {Parallel-RL}

$P_W = E \times I_R$ (1) $P_W = E^2/R$ (2) $P_W = I_R^2 \times R$ (3)

$P_W = P_{VA} \times \cos\theta$ (PF) (4) $P_W = P_{VARsL} / \tan\theta$ (5) $P_W = \sqrt{P_{VA}^2 - P_{VARsL}^2}$ (6)

The Reactance Power (VARs) {Parallel-RL}

$P_{VARsL} = E \times I_L$ (1) $P_{VARsL} = I_L^2 \times X_L$ (2) $P_{VARsL} = E^2/X_L$ (3)

$P_{VARsL} = \sqrt{P_{VA}^2 - P_W^2}$ (4) $P_{VARsL} = P_{VA} \times \sin\theta$ (5) $P_{VARsL} = P_W \times \tan\theta$ (6)

The Power Factor (PF) {Parallel-RL}

PF ($\cos\theta$) = Z/R (1) PF ($\cos\theta$) = I_R/I_T (2) PF ($\cos\theta$) = P_W / P_{VA} (3)

The Phase Angle (θ) {Parallel-RL}

$\theta = \cos^{-1}\dfrac{I_R}{I_T}$ (PF) (1) $\theta = \sin^{-1}\dfrac{I_L}{I_T}$ (2) $\theta = \tan^{-1}\dfrac{I_L}{I_R}$ (3)

$\theta = \cos^{-1}\dfrac{Z}{R}$ (PF) (4) $\theta = \sin^{-1}\dfrac{Z}{X_L}$ (5) $\theta = \tan^{-1}\dfrac{R}{X_L}$ (6)

$\theta = \cos^{-1}\dfrac{P_W}{P_{VA}}$ (PF) (7) $\theta = \sin^{-1}\dfrac{P_{VARsL}}{P_{VA}}$ (8) $\theta = \tan^{-1}\dfrac{P_{VARsL}}{P_W}$ (9)

Resistance (R)–Capacitance (C) Circuits

Series-RC

Just because the topic has changed from that of a series resistance–inductance circuit to a series resistance–capacitance circuit, the same rule still applies in terms of current flow—the same amount of current will flow through and about each connected component. Observe figure 25.

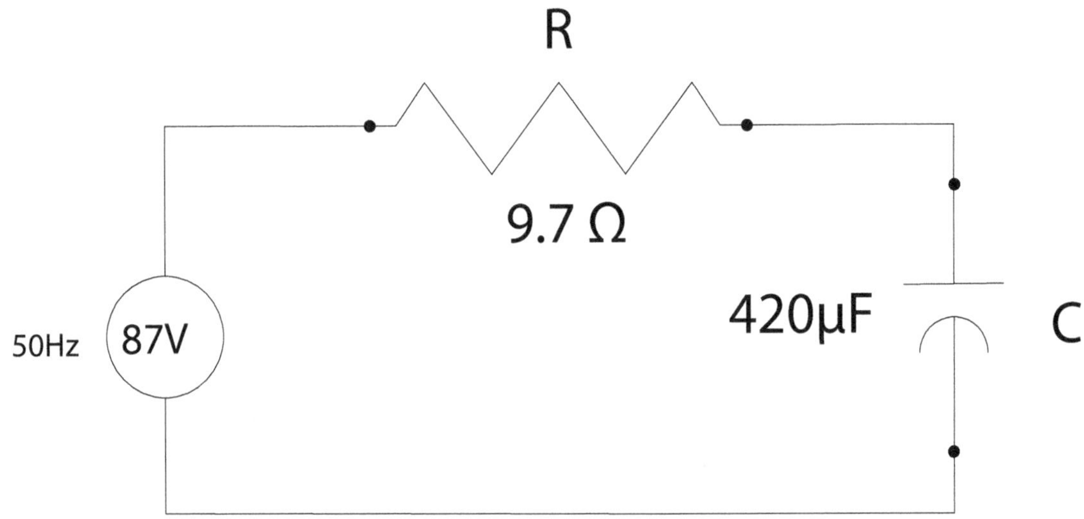

Fig. 25. Series-RC circuit.

Based on the series resistance–capacitance circuit's operating conditions and the given components, the following quantities will be calculated: the capacitance reactance (X_C), the impedance (Z), the current (I), the voltage drop across each component (E_R and E_C), the total voltage (E_T), the apparent power (VA), the true power (W), the reactance power (VARs), the power factor (PF), and the phase angle (θ).

Because a variation of like quantities will exist while analyzing this circuit, certain calculated quantities will be randomly selected to perform adjoining calculations.

Capacitance Reactance (X_C) {Series-RC}

Before the given quantity for the capacitor can be used, it must be converted to a farad (F) opposed to a microfarad (µF) to be used in the capacitance reactance formula. Therefore,

$$420\mu F = \frac{420}{1{,}000{,}000} = .000420F$$

Now based on the converted unit and an operating frequency of 50Hz the capacitance reactance is

$$X_C = \frac{1}{2 \times 3.14 \times 50Hz \times .000420F} = 7.58\Omega$$

Impedance (Z) {Series-RC}

In a purely capacitive circuit the voltage and current are not in phase with each other or to be specific the current (I) leads the voltage (E) by 90° about the capacitor. Because the current flowing about the capacitor is not in phase with the voltage across it, the opposition to current for both components cannot be added together. To calculate the impedance for this circuit, the vector-related formula for a series-RC circuit must be used; that is, $Z = \sqrt{R^2 + X_C^2}$. Therefore,

$$
\begin{aligned}
Z &= \sqrt{(9.7)^2 + (7.58)^2} \\
&= \sqrt{94.09 + 57.46} \\
&= \sqrt{151.55} \\
&= 12.31\Omega
\end{aligned}
$$

Now that the circuit's impedance has been determined, other quantities can be determined also.

Current (I) {Series-RC}

Since E is shown to be 87V,

$$
\begin{aligned}
I &= 87V/12.31\Omega \\
&= 7.07A
\end{aligned}
$$

Voltage Drops (E_R and E_L) {Series-RC}

Because current will flow through the resistance and about the capacitor in the same amount, a voltage drop will exist across each component. Therefore,

$$
\begin{aligned}
E_R &= I \times R \\
&= 7.07A \times 9.7\Omega \\
&= 68.58V
\end{aligned}
\qquad \text{and} \qquad
\begin{aligned}
E_C &= I \times X_C \\
&= 7.07A \times 7.58\Omega \\
&= 53.59V
\end{aligned}
$$

Total Voltage (ET) {Series-RC}

Because the voltage across the capacitor is not in phase with the circuit's current, the total voltage has to be determined vectorially using the formula $E_T = \sqrt{E_R^2 + E_C^2}$. Therefore,

$$
\begin{aligned}
E_T &= \sqrt{(68.58)^2 + (53.59)^2} \\
&= \sqrt{4703.22 + 2871.89} \\
&= \sqrt{7575.11} \\
&= 87.04V
\end{aligned}
$$

As expected when calculated vectorially, the calculated total voltage should be the same (or approximately the same) as the given total voltage; that is, 87V.

The Apparent Power (VA) {Series-RC}

Using that which was given as the circuit's total voltage (E) and the calculated current (I), the circuit's apparent power can be determined by use of the formula $P_{VA} = E_T \times I$ in addition to the formulas $P_{VA} = I^2 \times Z$ and $P_{VA} = E_T^2/Z$.

$$P_{VA} = 87V \times 7.07A = 615.1VA$$

$$P_{VA} = (7.07A)^2 \times 12.31\Omega = 49.98A \times 12.31\Omega = 615.25VA$$

$$P_{VA} = (87V)^2/12.31\Omega = 7569V/\ 12.31\Omega = 614.87VA$$

The True Power (W) {Series-RC}

The true power is calculated using the formula $P_W = I^2 \times R$ or $P_W = E_R^2/R$, where E_R is the voltage drop across the 9.7Ω resistance. The true power can also be calculated by means of the power factor, where the same results are obtained as that of the formulas. The application of the power factor will be used later, so let's now apply the formulas.

$$P_W = I^2 \times R = (7.07A)^2 \times 9.7\Omega = 484.85W$$

and

$$P_W = E_R^2/R = (68.58V)^2/9.7\Omega = 484.87W$$

The Reactance Power (VARs) {Series-RC}

The reactance power is calculated using the formulas $P_{VARsC} = I^2 \times X_C$ and $P_{VARsC} = E_C^2/X_C$, where E_C is the voltage drop across the 420μF capacitor. The reactance power can also be calculated by means of the phase angle where the same results are obtained as that of the formulas. Such application will also be used later. Now to apply the formulas:

$$P_{VARsC} = I^2 \times X_C = (7.07A)^2 \times 7.58\Omega = 378.89VARs$$

and

$$P_{VARsC} = E_C^2/\ X_C = (53.59V)^2/7.58\Omega = 378.88VARs$$

The Power Factor (PF) {Series-RC}

Using the given and calculated quantities along with the formulas, $PF = P_W/P_{VA}$ and $PF = R/Z$ the circuit's power factor is

$$PF = \frac{484.85W}{615.1VA} = .7882\ (78.82\%)$$

and

$$PF = \frac{9.7\Omega}{12.31\Omega} = .788\ (78.8\%)$$

Take note that the quantities used to calculate both results were randomly selected.

The Phase Angle (θ) {Series-RC}

In a series-RC circuit, formulas for the phase angle can be derived using any of the three trigonometric formulas:

$$\tan\theta = \frac{E_C}{E_R} \text{ or } \frac{X_C}{R} \quad \text{and} \quad \cos\theta = \frac{R}{Z}$$

From the three formulas, the following derived formulas—along with the previously used formula—will be used to determine the phase angle:

$$\theta = \tan^{-1}\frac{E_C}{E_R} \text{ or } \frac{X_C}{R} \quad \text{and} \quad \theta = \cos^{-1}\frac{R}{Z}$$

$$\theta = \cos^{-1} PF \text{ (previous)}$$

Based on the derived formulas, the circuit's phase angle is

$$\theta = \tan^{-1}\frac{53.59V}{68.58V} \text{ or } \frac{7.58\Omega}{9.7\Omega} \quad \text{and} \quad \theta = \cos^{-1}\frac{9.7\Omega}{12.31\Omega}$$

$$= \tan^{-1} .7814 \qquad\qquad = \cos^{-1} .788$$

$$= 38° \qquad\qquad = 38°$$

$$\theta = \cos^{-1} .788 \text{ (previous)}$$
$$= 38°$$

Using all four formulas, the results are all the same.

Power {Series-RC}

Previously, the three power quantities were calculated using given or calculated values. Again, the quantities will be determined using an alternate approach. By using the formula $\cos\theta = P_W/P_{VA}$, the true power can now be determined by use of the derived formula, $P_W = P_{VA} \text{ x } \cos\theta$, where

$$\begin{aligned} P_W &= P_{VA} \text{ x } \cos\theta \\ &= 615.1VA \text{ x } \cos 38° \\ &= 615.1VA \text{ x } .7880 \\ &= 484.70W \end{aligned}$$

By using the formula $\sin\theta = P_{VARsC}/P_{VA}$, the reactance power can now be determined by use of the derived formula, $P_{VARsC} = P_{VA} \text{ x } \sin\theta$, where

$$\begin{aligned} P_{VARsC} &= P_{VA} \text{ x } \sin\theta \\ &= 615.1VA \text{ x } \sin 38° \\ &= 615.1VA \text{ x } .6157 \\ &= 378.72VARs \end{aligned}$$

As you can see, the results are approximately the same as earlier calculated. By using these results, they can be substituted into the Pythagorean theorem, and the apparent power can be verified.

$$\begin{aligned} P_{VA} &= \sqrt{(484.70)^2 + (378.72)^2} \\ &= \sqrt{234{,}934.1 + 143{,}428.84} \\ &= \sqrt{378{,}362.94} \\ &= 615.11VA \end{aligned}$$

Again, with alternate methods, calculations can be either cross-checked or verified to ensure correct answers.

The given series resistance–capacitance circuit has now been completely analyzed, and the objective has been satisfied.

The following display is an overall list of formulas that can be used for solving Series-RC circuits.

Related Series Formulas (Resistance–Capacitance Circuits—Overall)

[Only one source of current (I); therefore, $I = I_R = I_C$.]

Resistance (R) {Series-RC}

$R = E_R/I$ (1) $R = E_R^2/P_W$ (2) $R = P_W/I^2$ (3)

$R = Z \times PF (\cos\theta)$ (4) $R = \sqrt{Z^2 - X_C^2}$ (5) $R = X_C/\tan\theta$ (6)

Capacitance (C) {Series-RC}

$C = 1/2\pi f X_C$

Capacitance Reactance (X_C) {Series-RC}

$X_C = 1/2\pi fC$ (1) $X_C = E_C/I$ (2) $X_C = E_C^2/P_{VARsC}$ (3)

$X_C = P_{VARsC}/I^2$ (4) $X_C = \sqrt{Z^2 - R^2}$ (5) $X_C = Z \times \sin\theta$ (6)

$X_C = R \times \tan\theta$ (7)

Impedance (Z) {Series-RC}

$Z = \sqrt{R^2 + X_C^2}$ (1) $Z = E_T/I$ (2) $Z = E_T^2/P_{VA}$ (3)

$Z = P_{VA}/I^2$ (4) $Z = R/PF (\cos\theta)$ (5) $Z = X_C/\sin\theta$ (6)

Current (I) {Series-RC}

$I = E_R/R$ (1) $I = P_W/E_R$ (2) $I = E_C/X_C$ (3)

$I = P_{VARsC}/E_C$ (4) $I = \sqrt{P_W/R}$ (5) $I = \sqrt{P_{VARsC}/X_C}$ (6)

$I = E_T/Z$ (7) $I = \sqrt{P_{VA}/Z}$ (8) $I = P_{VA}/E_T$ (9)

Voltage Drops (E_R and E_C) {Series-RC}

$E_R = I \times R$ (1) $E_R = P_W/I$ (2) $E_R = \sqrt{P_W \times R}$ (3)

$E_R = \sqrt{E_T^2 - E_C^2}$ (4) $E_R = E_T \times PF (\cos\theta)$ (5) $E_R = E_C/\tan\theta$ (6)

$E_C = I \times X_C$ (1) $E_C = P_{VARsC}/I$ (2) $E_C = \sqrt{P_{VARsC} \times X_C}$ (3)

$E_C = \sqrt{E_T^2 - E_R^2}$ (4) $E_C = E_T \times \sin\theta$ (5) $E_C = E_R \times \tan\theta$ (6)

Total Voltage (E_T) {Series-RC}

$E_T = P_{VA}/I$ (1) $E_T = \sqrt{P_{VA} \times Z}$ (2) $E_T = I \times Z$ (3)

$E_T = \sqrt{E_R^2 + E_C^2}$ (4) $E_T = E_R/\cos\theta$ (PF) (5) $E_T = E_C/\sin\theta$ (6)

The Apparent Power (VA) {Series-RC}

$P_{VA} = E_T \times I$ (1) $P_{VA} = I^2 \times Z$ (2) $P_{VA} = E_T^2/Z$ (3)

$P_{VA} = \sqrt{P_W^2 + P_{VARsC}^2}$ (4) $P_{VA} = P_W/PF\ (\cos\theta)$ (5) $P_{VA} = P_{VARsC}^2/\sin\theta$ (6)

The True Power (W) {Series-RC}

$P_W = I^2 \times R$ (1) $P_W = E_R^2/R$ (2) $P_W = E_R \times I$ (3)

$P_W = \sqrt{P_{VA}^2 - P_{VARsC}^2}$ (4) $P_W = P_{VA} \times PF\ (\cos\theta)$ (5) $P_W = P_{VARsC}/\tan\theta$ (6)

The Reactance Power (VARs) {Series-RC}

$P_{VARsC} = E_C \times I$ (1) $P_{VARsC} = I^2 \times X_C$ (2) $P_{VARsC} = E_C^2/X_C$ (3)

$P_{VARsC} = \sqrt{P_{VA}^2 - P_W^2}$ (4) $P_{VARsC} = P_{VA} \times \sin\theta$ (5) $P_{VARsC} = P_W \times \tan\theta$ (6)

The Power Factor (PF) {Series-RC}

$PF = R/Z\ (\cos\theta)$ (1) $PF = E_R/E_T\ (\cos\theta)$ (5) $PF = P_W/P_{VA}\ (\cos\theta)$ (3)

The Phase Angle (θ) {Series-RC}

$\theta = \cos^{-1}\frac{E_R\,(PF)}{E_T}$ (1) $\theta = \sin^{-1}\frac{E_C}{E_T}$ (2) $\theta = \tan^{-1}\frac{E_C}{E_R}$ (3)

$\theta = \cos^{-1}\frac{R\,(PF)}{Z}$ (4) $\theta = \sin^{-1}\frac{X_C}{Z}$ (5) $\theta = \tan^{-1}\frac{X_C}{R}$ (6)

$\theta = \cos^{-1}\frac{P_W\,(PF)}{P_{VA}}$ (7) $\theta = \sin^{-1}\frac{P_{VARsC}}{P_{VA}}$ (8) $\theta = \tan^{-1}\frac{P_{VARsC}}{P_W}$ (9)

Parallel-RC

Again, in any parallel circuit, regardless of whether being supplied from a *dc* or *ac* source, one thing is for certain: the voltage drop across each parallel component is the same as the applied voltage. Observe figure 26.

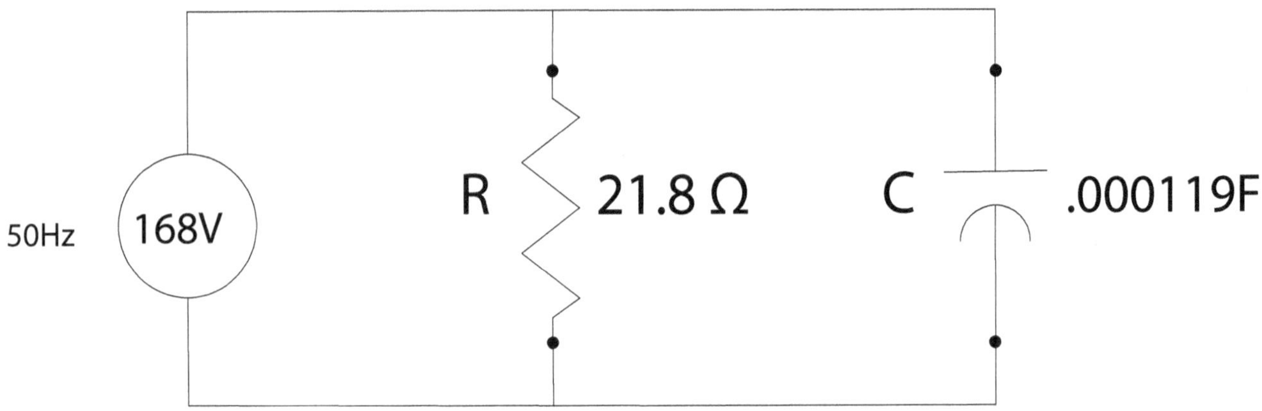

Fig. 26. Parallel-RC circuit.

Based on the parallel resistance–capacitance circuit's operating conditions and the given components, the following quantities will be calculated: the capacitance reactance (X_C), the impedance (Z), the total current (I_T), the individual current across each component (I_R and I_C), the apparent power (VA), the true power (W), the reactance power (VARs), the power factor (PF), and the phase angle (θ).

Because a variation of like quantities will exist while analyzing this circuit, certain calculated quantities will be randomly selected to perform adjoining calculations.

Capacitance Reactance (X_C) {Parallel-RC}

Based on the .000119F capacitor and an operating frequency of 50Hz, the capacitance reactance is

$$X_C = \frac{1}{2 \text{ x } 3.14 \text{ x } 50\text{Hz x } .000119\text{F}} = 26.76\Omega$$

Impedance (Z) {Parallel-RC}

Being in parallel will require the impedance to be calculated differently from that of a series-RC circuit. But the fact still remains, the voltage and current are not in phase, and the impedance has to be calculated vectorially. Therefore, to calculated the impedance for this circuit the vector related formula for a parallel-RC circuit must be used:

$$Z = \frac{R \text{ x } X_C}{\sqrt{R^2 + X_C^2}} \quad \text{or} \quad Z = \frac{1}{\sqrt{1/R^2 + 1/X_C^2}}$$

Applying both formulas,

$$Z = \frac{21.8 \times 26.76}{\sqrt{(21.8)^2 + (26.76)^2}} \quad \text{or} \quad Z = \frac{1}{\sqrt{1/(21.8)^2 + 1/(26.76)^2}}$$

$$Z = \frac{583.37}{\sqrt{475.24 + 716.01}} \quad \text{or} \quad Z = \frac{1}{\sqrt{1/475.24 + 1/716.01}}$$

$$Z = \frac{583.37}{34.51} \quad \text{or} \quad Z = \frac{1}{\sqrt{.0021042 + .0013966}}$$

$$Z = 16.9\Omega \quad \text{or} \quad Z = \frac{1}{.0592}$$

$$Z = 16.892\Omega$$

For this problem, the results derived by use of the product of the sum formula will be used.

Total Current (I_T) {Parallel-RC}

Since E is shown to be 168V, the total current can also be calculated using the formula $I_T = E/Z$:

$$I_T = 168V/16.9\Omega$$
$$= 9.94A$$

Individual Current (I_R and I_C) {Parallel-RC}

Because this is a parallel circuit, more than one path for current flow will exist. Since the ohmic values of R and X_C are not the same, both components will experience different amounts of current flow: I_R and I_C. Therefore,

$$I_R = E/R = 168V/21.8\Omega = 7.71A \quad \text{and} \quad I_C = E/X_C = 168V/26.76\Omega = 6.28A$$

Again, it should be clear that the concept of adding I_R and I_C will not work for this type circuit because the voltage across the capacitor is not in phase with the current flowing about the capacitor. As a result, the total current must also be determined vectorially by means of the vector-related formula, $I_T = \sqrt{I_R^2 + I_C^2}$. Therefore,

$$I_T = \sqrt{(7.71)^2 + (6.28)^2}$$
$$= \sqrt{98.88}$$
$$= 9.94A$$

As expected, when calculated vectorially, the total current should be the same (or approximately the same) as that previously calculated (I_T = 168V/16.9Ω = 9.94A).

The Apparent Power (VA) {Parallel-RC}

Using that which was given as the circuit's voltage (E), calculated current (I), and impedance (Z), the circuit's apparent power can be determined by use of the formulas $P_{VA} = E \times I_T$, $P_{VA} = I_T^2 \times Z$ and $P_{VA} = E^2/Z$.

$$P_{VA} = 168V \times 9.94A = 1669.92VA$$

$$P_{VA} = (9.94A)^2 \times 16.9\Omega = 98.8A \times 16.9\Omega = 1669.72VA$$

$$P_{VA} = (168V)^2/16.9\Omega = 28{,}224V/16.9\Omega = 1670.01VA$$

The True Power (W) {Parallel-RC}

The true power is calculated using the formulas $P_W = I_R^2 \times R$ and $P_W = E^2/R$, where I_R is the current flowing through R.

$$P_W = I_R^2 \times R = (7.71A)^2 \times 21.8\Omega = 1295.88W$$

and

$$P_W = E^2/R = (168V)^2/21.8\Omega = 1294.68W$$

The Reactance Power (VARs) {Parallel-RC}

The reactance power is calculated using the formulas $P_{VARsC} = I_C^2 \times X_C$ and $P_{VARsC} = E^2/X_C$, where I_C is the current flowing about C.

$$P_{VARsC} = I_C^2 \times X_C = (6.28A)^2 \times 26.76\Omega = 1055.37VARs$$

and

$$P_{VARsC} = E^2/X_C = (168V)^2/26.76\Omega = 1054.71VARs$$

The Power Factor (PF) {Parallel-RC}

When calculating the power factor in the parallel-RC circuit, the formula PF = Z/R was used. With a parallel-RC circuit, the formula is the complete opposite:

$$PF = \frac{Z}{R}$$

Using the given and calculated quantities along with the above and previously given power factor formula,

$$PF = \frac{1295.88W}{1669.92VA} = .776\ (77.6\%)$$

and

$$PF = \frac{16.9\Omega}{21.8\Omega} = .775\ (77.5\%)$$

As you can see, when calculated, both sets of quantities yield approximate results.

The Phase Angle (θ) {Parallel-RC}

In a parallel RC circuit, formulas for the phase angle can be derived using any of the three trigonometric formulas:

$$\tan\theta = \frac{I_C}{I_R} \text{ or } \frac{R}{X_C} \quad \text{and} \quad \cos\theta = \frac{Z}{R}$$

From the three formulas, the following derived formulas—along with a previously derived formula—will be used to determine the phase angle:

$$\theta = \tan^{-1}\frac{I_C}{I_R} \text{ or } \frac{R}{X_C} \quad \text{and} \quad \theta = \cos^{-1}\frac{Z}{R}$$

$$\theta = \cos^{-1} PF \text{ (previous)}$$

Based on the derived formulas the circuit's phase angle is

$$\theta = \tan^{-1}\frac{6.28A}{7.71A} \text{ or } \frac{21.8\Omega}{26.76\Omega} \quad \text{and} \quad \theta = \cos^{-1}\frac{16.9\Omega}{21.8\Omega}$$

$$= \tan^{-1} .8145 \qquad = .8146 \qquad = \cos^{-1} .7752$$

$$= 39.16° \qquad = 39.17° \qquad = 39.18°$$

$$\theta = \cos^{-1} .775 \text{ (previous)}$$
$$= 39.195°$$

Using all four formulas, the results are all approximately the same.

Power {Parallel-RC}

Previously the three power quantities were calculated using given or calculated values. The quantities will now be determined using an alternate approach. By using the formula $\cos\theta = P_W/P_{VA}$, the true power can now be determined by use of the derived formula $P_W = P_{VA} \times \cos\theta$, where

$$\begin{aligned} P_W &= P_{VA} \times \cos\theta \\ &= 1669.92VA \times \cos 39.17° \\ &= 1669.92VA \times .7753 \\ &= 1294.69W \end{aligned}$$

By using the formula $\sin\theta = P_{VARsC}/P_{VA}$, the reactance power can now be determined by use of the derived formula $P_{VARsC} = P_{VA} \times \sin\theta$, where

$$\begin{aligned} P_{VARsC} &= P_{VA} \times \sin\theta \\ &= 1669.92VA \times \sin 39.17° \\ &= 1669.92VA \times .6316 \\ &= 1054.72VARs \end{aligned}$$

As you can see, the results are approximately the same as earlier calculated. By using these results, they can be substituted into the Pythagorean theorem, and the apparent power can be verified.

$$P_{VA} = \sqrt{(1294.69)^2 + (1054.72)^2}$$
$$= \sqrt{1,676,222.2 + 1,112,434.28}$$
$$= \sqrt{2,788,656.48}$$
$$= 1669.93\text{VA}$$

The given parallel resistance–capacitance circuit has now been completely analyzed, and the objective has been satisfied.

The following display is an overall list of formulas that can be used for solving Parallel-RC circuits.

Related Parallel Formulas (Resistance–Capacitance Circuits—Overall)

[Only one source of voltage (E); therefore, $E = E_R = E_C$.]

Resistance (R) {Parallel-RC}

$R = E/I_R$ (1) $R = E^2/P_W$ (2) $R = P_W/I_R{}^2$ (3)
$R = Z/\cos\theta$ (PF) (4) $R = X_C \text{ x } \tan\theta$ (5) $R = \dfrac{1}{\sqrt{(1/Z)^2 - (1/X_C)^2}}$ (6)

Capacitance (C) {Parallel-RC}

$C = 1/2\pi f X_C$

Capacitance Reactance (X_C) {Parallel-RC}

$X_C = 1/2\pi fC$ (1) $X_C = E/I_C$ (2) $X_C = E^2/P_{VARsC}$ (3)
$X_C = P_{VARsC}/I_C{}^2$ (4) $X_C = R/\tan\theta$ (5) $X_C = Z/\sin\theta$ (6)
$X_C = \dfrac{1}{\sqrt{(1/Z)^2 - (1/R)^2}}$ (7)

Impedance (Z) {Parallel-RC}

$Z = \dfrac{1}{\sqrt{(1/R)^2 + (1/X_C)^2}} = \dfrac{R \text{ x } X_C}{\sqrt{(R)^2 + (X_C)^2}}$ (1) $Z = E/I_T$ (2)
$Z = E^2/P_{VA}$ (3) $Z = P_{VA}/I_T{}^2$ (4) $Z = R \text{ x } \cos\theta$ (PF) (5)
$Z = X_C \text{ x } \sin\theta$ (6)

Individual Current (I_R and I_C) {Parallel-RC}

$I_R = E/R$ (1) $I_R = P_W/E$ (2) $I_R = \sqrt{P_W/R}$ (3)
$I_R = \sqrt{I_T{}^2 - I_C{}^2}$ (4) $I_R = I_T \text{ x } \cos\theta$ (PF) (5) $I_R = I_C/\tan\theta$ (6)
$I_C = E/X_C$ (1) $I_C = P_{VARsC}/E$ (2) $I_C = \sqrt{P_{VARsC}/X_C}$ (3)
$I_C = \sqrt{I_T{}^2 - I_R{}^2}$ (4) $I_C = I_T \text{ x } \sin\theta$ (5) $I_C = I_R \text{ x } \tan\theta$ (6)

Total Current (I_T) {Parallel-RC}

$I_T = E/Z$ (1) $I_T = P_{VA}/E$ (2) $I_T = \sqrt{P_{VA}/Z}$ (3)

$I_T = \sqrt{I_R^2 + I_C^2}$ (4) $I_T = I_R/\cos\theta$ (PF) (5) $I_T = I_C/\sin\theta$ (6)

Voltage (E) {Parallel-RC}

$E = I_R \times R$ (1) $E = P_W/I_R$ (2) $E = I_C \times X_C$ (3)

$E = P_{VARsC}/I_C$ (4) $E = \sqrt{P_W \times R}$ (5) $E = \sqrt{P_{VARsC} \times X_C}$ (6)

$E = I_T \times Z$ (7) $E = P_{VA}/I_T$ (8) $E = \sqrt{P_{VA} \times Z}$ (9)

The Apparent Power (VA) {Parallel-RC}

$P_{VA} = E \times I_T$ (1) $P_{VA} = I_T^2 \times Z$ (2) $P_{VA} = E^2/Z$ (3)

$P_{VA} = P_W/\cos\theta$ (PF) (4) $P_{VA} = P_{VARsC}/\sin\theta$ (5) $P_{VA} = \sqrt{P_W^2 + P_{VARsC}^2}$ (6)

The True Power (W) {Parallel-RC}

$P_W = E \times I_R$ (1) $P_W = E^2/R$ (2) $P_W = I_R^2 \times R$ (3)

$P_W = P_{VA} \times \cos\theta$ (PF) (4) $P_W = P_{VARsC}/\tan\theta$ (5) $P_W = \sqrt{P_{VA}^2 - P_{VARsC}^2}$ (6)

The Reactance Power (VARs) {Parallel-RC}

$P_{VARsC} = E \times I_C$ (1) $P_{VARsC} = I_C^2 \times X_C$ (2) $P_{VARsC} = E^2/X_C$ (3)

$P_{VARsC} = \sqrt{P_{VA}^2 - P_W^2}$ (4) $P_{VARsC} = P_{VA} \times \sin\theta$ (5) $P_{VARsC} = P_W \times \tan\theta$ (6)

The Power Factor (PF) {Parallel-RC}

PF ($\cos\theta$) = Z/R (1) PF ($\cos\theta$) = I_R/I_T (2) PF ($\cos\theta$) = P_W/P_{VA} (3)

The Phase Angle (θ) {Parallel-RC}

$\theta = \cos^{-1}\frac{I_R}{I_T}$ (PF) (1) $\theta = \sin^{-1}\frac{I_C}{I_T}$ (2) $\theta = \tan^{-1}\frac{I_C}{I_R}$ (3)

$\theta = \cos^{-1}\frac{Z}{R}$ (PF) (4) $\theta = \sin^{-1}\frac{Z}{X_C}$ (5) $\theta = \tan^{-1}\frac{R}{X_C}$ (6)

$\theta = \cos^{-1}\frac{P_W}{P_{VA}}$ (PF) (7) $\theta = \sin^{-1}\frac{P_{VARsC}}{P_{VA}}$ (8) $\theta = \tan^{-1}\frac{P_{VARsC}}{P_W}$ (9)

Resistance (R)-Inductance (L) or Capacitance (C) Circuits

Multiple Components

To shorten this topic, where multiple components (RL or RC) are involved, only one type of reactant components will be analyzed along with resistance components per series and parallel circuits.

Series-RL

When a series resistance–inductance or resistance–capacitance circuit consists of more than one component, be it resistance, inductors, or capacitors, additional steps are required before the circuit can be analyzed as previous circuits. Observe figure 27.

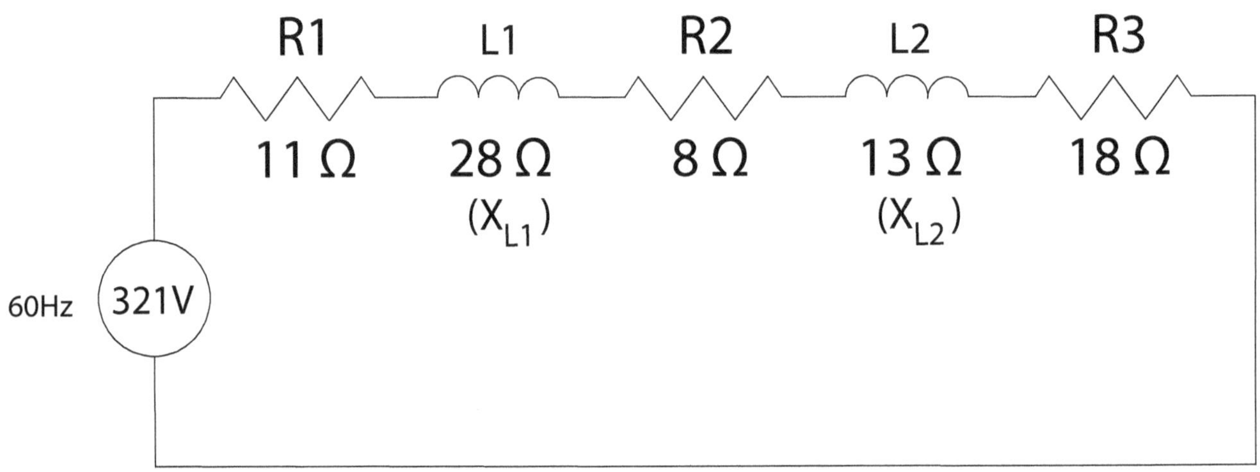

Fig. 27. Series-RL circuit (multiple components).

As you can see, the circuit in figure 27 consists of three resistance and two inductor components. Again, to shorten the process of analyzing this circuit, the inductance reactance for each inductor has been assigned. If the circuit consisted of capacitors instead of inductors, the same approach in analyzing such a circuit would be taken.

Because this is a series circuit, only one source of current will exist. Since the voltage and current about the resistance are in phase, the individual resistance values can be totaled to produce a total resistance value, R_T. Although the voltage and current about the inductors are not in phase, the occurring inductance reactance about each inductor is because each inductor's magnetic field will experience the same expanding and collapsing effects at the same rate. As a result, the individual inductance reactance values can also be totaled to produce a total inductance reactance value, X_{LT}. To sum things up, because all like components share the same in-phase similarities, related quantities ($E_{R1,2...}$, $E_{L1,2...}$, $P_{W1,2...}$, $P_{VARsL1,2...}$) can be totaled to produce a total value. Therefore,

$$R_T = 11\Omega + 8\Omega + 18\Omega = 37\Omega$$

and

$$X_{LT} = 28\Omega + 13\Omega = 41\Omega$$

With these values determined, the circuit can now be analyzed similarly to the previous circuits.

Impedance (Z) {Series-RL (Multiple Components)}
Again, because the current flowing through the resistors and the inductors are not in phase with the voltage across each component, they must be added together vectorially. To calculate the impedance for this circuit the vector related formula, $Z = \sqrt{R_T^2 + X_{LT}^2}$ must be used. Therefore,

$$\begin{aligned} Z &= \sqrt{(37)^2 + (41)^2} \\ &= \sqrt{1369 + 1681} \\ &= \sqrt{3050} \\ &= 55.23\Omega \end{aligned}$$

Now that the circuit's impedance has been determined, other quantities can be determined also.

Current (I) {Series-RL (Multiple Components)}
Since E is shown to be 321V,

$$\begin{aligned} I &= 321V/55.23\Omega \\ &= 5.81A \end{aligned}$$

Voltage Drops ($E_{R1,2,3}$ and $E_{L1,2}$) {Series-RL (Multiple Components)}
Because current will flow through all components in the same amount, a voltage drop will exist across each component. Therefore,

$$\begin{aligned} E_{R1} &= I \times R_1 \\ &= 5.81A \times 11\Omega \\ &= 63.91V \end{aligned}$$

$$\begin{aligned} E_{R2} &= I \times R_2 \\ &= 5.81A \times 8\Omega \\ &= 46.48V \end{aligned}$$

$$\begin{aligned} E_{R3} &= I \times R_3 \\ &= 5.81A \times 18\Omega \\ &= 104.58V \end{aligned}$$

$$\begin{aligned} E_{L1} &= I \times X_{L1} \\ &= 5.81A \times 28\Omega \\ &= 162.68V \end{aligned}$$

$$\begin{aligned} E_{L2} &= I \times X_{L2} \\ &= 5.81A \times 13\Omega \\ &= 75.53V \end{aligned}$$

Total Voltage (E_T) {Series-RL (Multiple Components)}
To determine the circuit's total voltage, the voltage drop across each like component must be totaled together first. Once this is done, the vector formula, $E_T = \sqrt{E_{RT}^2 + E_{LT}^2}$ can then be applied. Therefore,

$$E_{RT} = 63.91V + 46.48V + 104.58V = 214.97V \qquad E_{LT} = 162.68V + 75.53V = 238.21V$$

results in

$$\begin{aligned} E_T &= \sqrt{(214.97)^2 + (238.21)^2} \\ &= \sqrt{46{,}212.1 + 56{,}744} \\ &= \sqrt{102{,}956.1} \\ &= 320.87V \end{aligned}$$

As expected, when calculated vectorially, the calculated total voltage should be the same (or approximately the same) as the given total voltage: 321V.

The Apparent Power (VA) {Series-RL (Multiple Components)}
Using that which was given as the circuit's total voltage (E) and the calculated current (I), the circuit's apparent power can be determined by use of the formula $P_{VA} = E_T \text{ x } I$ in addition to the formulas $P_{VA} = I^2 \text{ x } Z$ and $P_{VA} = E_T^2/Z$.

$$P_{VA} = 321V \text{ x } 5.81A = 1865.01VA$$

$$P_{VA} = (5.81A)^2 \text{ x } 55.23\Omega = 33.76A \text{ x } 55.23\Omega = 1864.56VA$$

$$P_{VA} = (321V)^2/55.23\Omega = 103{,}041V/\,55.23\Omega = 1865.67VA$$

The True Power (W) {Series-RL (Multiple Components)}
Because each resistance component will contribute to the circuit's total true power consumption, the true power consumed by each component must be determined before the total true power is realized. To simplify these calculations, only one formula will be used, $P_W = I^2 \text{ x } R$, to calculate the true power consumption for each component. Therefore,

$$P_{W1} = I^2 \text{ x } R_1 = (5.81A)^2 \text{ x } 11\Omega = 371.32W$$

$$P_{W2} = I^2 \text{ x } R_2 = (5.81A)^2 \text{ x } 8\Omega = 270.05W$$

$$P_{W3} = I^2 \text{ x } R_3 = (5.81A)^2 \text{x } 18\Omega = 607.61W$$

results in

$$P_{W(T)} = 371.32W + 270.05W + 607.61W = 1248.98W$$

The Reactance Power (VARs) {Series-RL (Multiple Components)}
To determine the circuit's total reactance power, the reactance power about each inductor must first be determined before the total reactance power is realized. Again, to simplify these calculations, only one formula will be used, $P_{VARsL} = E_L^2/X_L$ to calculate the reactance power for each component.

Therefore,

$$P_{VARsL1} = E_{L1}^2/X_{L1} = (162.68V)^2/28\Omega = 945.17VARs \qquad P_{VARsL2} = E_{L2}^2/X_{L2} = (75.53V)^2/13\Omega = 438.83VARs$$

results in

$$P_{VARsL(T)} = 945.17VARs + 438.83VARs = 1384VARs$$

The Power Factor (PF) {Series-RL (Multiple Components)}
Using the given and calculated quantities, along with the formulas PF = $P_{W(T)}/P_{VA}$ and PF = R_T/Z, the circuit's power factor is

$$PF = \frac{1248.98W}{1865.01VA} \quad \text{and} \quad PF = \frac{37\Omega}{55.23\Omega}$$

$$= .6697\ (66.97\%) \qquad = .6699\ (66.99\%)$$

Take note that the quantities used to calculate both results were randomly selected.

The Phase Angle (θ) {Series-RL (Multiple Components)}
In a series-RL circuit, formulas for the phase angle can be derived using any of the three trigonometric formulas:

$$\tan\theta = \frac{E_{LT}}{E_{RT}} \quad \text{or} \quad \frac{X_{LT}}{R_T} \quad \text{and} \quad \cos\theta = \frac{R_T}{Z}$$

From the three formulas, the following derived formulas—along with the previously used formula—will be used to determine the phase angle:

$$\theta = \tan^{-1}\frac{E_{LT}}{E_{RT}} \quad \text{or} \quad \frac{X_{LT}}{R_T} \quad \text{and} \quad \theta = \cos^{-1}\frac{R_T}{Z}$$

$$\theta = \cos^{-1} PF \text{ (previous)}$$

Based on the derived formulas, the circuit's phase angle is

$$\theta = \tan^{-1}\frac{238.21V}{214.97V} \text{ or } \frac{41\Omega}{37\Omega} = \tan^{-1} 1.11 = 47.98° \quad \text{and} \quad \theta = \cos^{-1}\frac{37\Omega}{55.23\Omega} = \cos^{-1} .6699 = 47.94°$$

$$\theta = \cos^{-1} .6697 \text{ (previous)} = 47.96°$$

Using all four formulas, the results are all approximately the same.

Power {Series-RL (Multiple Components)}

Previously, the three power quantities were calculated using given or calculated values. Again, the quantities will be determined using an alternate approach. By using the formula $\cos \theta = P_W/P_{VA}$, the true power can now be determined by use of the derived formula $P_W = P_{VA} \times \cos \theta$, where

$$\begin{aligned} P_W &= P_{VA} \times \cos \theta \\ &= 1865.01\text{VA} \times \cos 47.98° \\ &= 1865.01\text{VA} \times .6694 \\ &= 1248.44\text{W} \end{aligned}$$

By using the formula $\sin \theta = P_{VARsL}/P_{VA}$, the reactance power can now be determined by use of the derived formula $P_{VARsL} = P_{VA} \times \sin \theta$, where

$$\begin{aligned} P_{VARsL} &= P_{VA} \times \sin \theta \\ &= 1865.01\text{VA} \times \sin 47.98° \\ &= 1865.01\text{VA} \times .743 \\ &= 1385.7\text{VARs} \end{aligned}$$

As you can see, the results are approximately the same as earlier calculated. By using these results, they can be substituted into the Pythagorean theorem, and the apparent power can be verified.

$$\begin{aligned} P_{VA} &= \sqrt{(1248.44)^2 + (1385.7)^2} \\ &= \sqrt{1,558,602.43 + 1,920,164.49} \\ &= \sqrt{3,478,766.92} \\ &= 1865.15\text{VA} \end{aligned}$$

The following display is an overall list of formulas that can be used for solving Series-RL Circuits—Multiple Components.

Related Series Formulas
(Resistance–Inductance Multiple Components Circuits—Overall)

[Only one source of current (I); therefore, $I = I_R = I_L$.]

Resistance (R) {Series-RL (Multiple Components)}

$R = E_R/I$ (1) $\quad$ $R = E_R{}^2/P_W$ (2) $\quad$ $R = P_W/I^2$ (3)

$R_T = R_1 + R_2 + \ldots\ldots$ (1) $\quad$ $R_T = E_{R(T)}/I$ (2) $\quad$ $R_T = E_{R(T)}{}^2/P_{W(T)}$ (3)

$R_T = P_{W(T)}/I^2$ (4) $\quad$ $R_T = \sqrt{Z^2 - X_{LT}{}^2}$ (5) $\quad$ $R_T = Z \times \cos \theta$ (PF) (6)

$R_T = X_{LT}/\tan \theta$ (7)

Inductance (L) {Series-RL (Multiple Components)}

$L = X_L / 2\pi f$

$L_T = L_1 + L_2 + \ldots\ldots$ (1)
$L_T = X_{LT} / 2\pi f$ (2)

Inductance Reactance (X_L) {Series-RL (Multiple Components)}

$X_L = 2\pi f L$ (1)
$X_L = E_L / I$ (2)
$X_L = E_L^2 / P_{VARsL}$ (3)
$X_L = P_{VARsL} / I^2$ (4)

$X_{LT} = X_{L1} + X_{L2} + \ldots\ldots$ (1)
$X_{LT} = 2\pi f L_{(T)}$ (2)
$X_{LT} = E_{L(T)} / I$ (3)
$X_{LT} = E_{L(T)}^2 / P_{VARsL(T)}$ (4)
$X_{LT} = P_{VARsL(T)} / I^2$ (5)
$X_{LT} = \sqrt{Z^2 - R_T^2}$ (6)
$X_{LT} = Z \times \sin\theta$ (7)
$X_{LT} = R_T \times \tan\theta$ (8)

Impedance (Z) {Series-RL (Multiple Components)}

$Z = \sqrt{R_T^2 + X_{LT}^2}$ (1)
$Z = E_T / I$ (2)
$Z = E_T^2 / P_{VA}$ (3)
$Z = P_{VA} / I^2$ (4)
$Z = R_T / PF (\cos\theta)$ (5)
$Z = X_{LT} / \sin\theta$ (6)

Current (I) {Series-RL (Multiple Components)}

$I = E_R / R$ (1)
$I = E_L / X_L$ (2)
$I = P_W / E_R$ (3)
$I = P_{VARsL} / E_L$ (4)
$I = \sqrt{P_W / R}$ (5)
$I = \sqrt{P_{VARsL} / X_L}$ (6)

$I = E_{R(T)} / R_T$ (1)
$I = E_{L(T)} / X_{LT}$ (2)
$I = P_{W(T)} / E_{R(T)}$ (3)
$I = P_{VARsL(T)} / E_{L(T)}$ (4)
$I = \sqrt{P_{W(T)} / R_T}$ (5)
$I = \sqrt{P_{VARsL(T)} / X_{LT}}$ (6)
$I = P_{VA} / E_T$ (7)
$I = \sqrt{P_{VA} / Z}$ (8)
$I = E_T / Z$ (9)

Voltage Drops (E_R and E_L) {Series-RL (Multiple Components)}

$E_R = I \times R$ (1)
$E_R = P_W / I$ (2)
$E_R = \sqrt{P_W \times R}$ (3)
$E_{R(T)} = E_{R1} + E_{R2} + \ldots.$ (1)
$E_{R(T)} = I \times R_T$ (2)
$E_{R(T)} = P_{W(T)} / I$ (3)
$E_{R(T)} = \sqrt{P_{W(T)} \times R_T}$ (4)
$E_{R(T)} = \sqrt{E_T^2 - E_{L(T)}^2}$ (5)
$E_{R(T)} = E_T \times \cos\theta$ (PF) (6)
$E_{R(T)} = E_{L(T)} / \tan\theta$ (7)

$E_L = I \times X_L$ (1)
$E_L = P_{VARsL} / I$ (2)
$E_L = \sqrt{P_{VARsL} \times X_L}$ (3)

$E_{L(T)} = E_{L1} + E_{L2} + \ldots.$ (1)
$E_{L(T)} = I \times X_{LT}$ (2)
$E_{L(T)} = P_{VARsL(T)} / I$ (3)
$E_{L(T)} = \sqrt{P_{VARsL(T)} \times X_{LT}}$ (4)
$E_{L(T)} = \sqrt{E_T^2 - E_{R(T)}^2}$ (5)
$E_{L(T)} = E_T \times \sin\theta$ (6)
$E_{L(T)} = E_{RT} \times \tan\theta$ (7)

Total Voltage (E_T) {Series-RL (Multiple Components)}

$E_T = P_{VA} / I$ (1) $E_T = I \times Z$ (2) $E_T = \sqrt{P_{VA} / Z}$ (3)

$E_T = \sqrt{E_{R(T)}^2 + E_{L(T)}^2}$ (4) $E_T = E_{R(T)} / PF (\cos\theta)$ (5) $E_T = E_{L(T)} / \sin\theta$ (6)

The Apparent Power (VA) {Series-RL (Multiple Components)}

$P_{VA} = E_T \times I$ (1) $P_{VA} = I^2 \times Z$ (2) $P_{VA} = E_T^2/Z$ (3)

$P_{VA} = P_{W(T)}/PF (\cos\theta)$ (4) $P_{VA} = P_{VARsL(T)}/\sin\theta$ (5) $P_{VA} = \sqrt{P_{W(T)}^2 + P_{VARsL(T)}^2}$ (6)

The True Power (W) {Series-RL (Multiple Components)}

$P_W = I^2 \times R$ (1) $P_W = E_R^2/R$ (2) $P_W = E_R \times I$ (3)

$P_{W(T)} = P_{W1} + P_{W2} + \ldots$ (1) $P_{W(T)} = I^2 \times R_T$ (2) $P_{W(T)} = E_{R(T)}^2 / R_T$ (3)

$P_{W(T)} = E_{R(T)} \times I$ (4) $P_{W(T)} = P_{VA} \times PF (\cos\theta)$ (5) $P_{W(T)} = P_{VARsL(T)}/\tan\theta$ (6)

$P_{W(T)} = \sqrt{P_{VA}^2 - P_{VARsL(T)}^2}$ (7)

The Reactance Power (VARs) {Series-RL (Multiple Components)}

$P_{VARsL} = I^2 \times X_L$ (1) $P_{VARsL} = E_L^2 / X_L$ (2) $P_{VARsL} = E_L \times I$ (3)

$P_{VARsL(T)} = P_{VARsL\,1} + P_{VARsL\,1} + \ldots$ (1) $P_{VARsL(T)} = I^2 \times X_{LT}$ (2) $P_{VARsL(T)} = E_{L(T)}^2/X_{LT}$ (3)

$P_{VARsL(T)} = E_{L(T)} \times I$ (4) $P_{VARsL(T)} = P_{VA} \times \sin\theta$ (5) $P_{VARsL(T)} = P_{W(T)} \times \tan\theta$ (6)

$P_{VARsL(T)} = \sqrt{P_{VA}^2 - P_{W(T)}^2}$ (7)

The Power Factor (PF) {Series-RL (Multiple Components)}

$PF = R_T/Z (\cos\theta)$ (1) $PF = E_{R(T)}/E_T (\cos\theta)$ (2) $PF = P_{W(T)} / P_{VA} (\cos\theta)$ (3)

The Phase Angle (θ) {Series-RL (Multiple Components)}

$\theta = \cos^{-1} \frac{E_{R(T)}}{E_T}$ (PF) (1) $\theta = \sin^{-1} \frac{E_{L(T)}}{E_T}$ (2) $\theta = \tan^{-1} \frac{E_{L(T)}}{E_{R(T)}}$ (3)

$\theta = \cos^{-1} \frac{R_T}{Z}$ (PF) (4) $\theta = \sin^{-1} \frac{X_{LT}}{Z}$ (5) $\theta = \tan^{-1} \frac{X_{LT}}{R_T}$ (6)

$\theta = \cos^{-1} \frac{P_{W(T)}}{P_{VA}}$ (PF) (7) $\theta = \sin^{-1} \frac{P_{VARsL(T)}}{P_{VA}}$ (8) $\theta = \tan^{-1} \frac{P_{VARsL(T)}}{P_{W(T)}}$ (9)

Series-RC

The following display is an overall list of formulas that can be used for solving Series-RC Circuits—Multiple Components.

Related Series Formulas
(Resistance–Capacitance Multiple Components Circuits—Overall)

[Only one source of current (I); therefore, $I = I_R = I_C$.]

Resistance (R) {Series-RC (Multiple Components)}

$R = E_R/I$ (1) $R = E_R^2/P_W$ (2) $R = P_W/I^2$ (3)

$R_T = R_1 + R_2 + \ldots\ldots$ (1) $R_T = E_{R(T)}/I$ (2) $R_T = E_{R(T)}^2/P_{W(T)}$ (3)

$R_T = P_{W(T)}/I^2$ (4) $R_T = \sqrt{Z^2 - X_{CT}^2}$ (5) $R_T = Z \times \cos\theta$ (PF) (6)

$R_T = X_{CT}/\tan\theta$ (7)

Capacitance (C) {Series-RC (Multiple Components)}

$C = 1/2\pi fX_C$

$C_T = \frac{C}{N}$ (1) (see page 67) $C_T = \frac{C_1 \times C_2}{C_1 + C_2}$ (2)

$C_T = \frac{1}{1/C_1 + 1/C_2 + \ldots\ldots}$ (3) $C_T = 1/2\pi fX_{CT}$ (4)

Capacitance Reactance (X_C) {Series-RC (Multiple Components)}

$X_C = 1/2\pi fC$ (1) $X_C = E_C/I$ (2) $X_C = E_C^2/P_{VARsC}$ (3)

$X_C = P_{VARsC}/I^2$ (4)

$X_{CT} = X_{C1} + X_{C2} + \ldots\ldots$ (1) $X_{CT} = 1/2\pi fC_{(T)}$ (2) $X_{CT} = E_{C(T)}/I$ (3)

$X_{CT} = E_{C(T)}^2/P_{VARsC(T)}$ (4) $X_{CT} = P_{VARsC(T)}/I^2$ (5) $X_{CT} = \sqrt{Z^2 - R_T^2}$ (6)

$X_{CT} = Z \times \sin\theta$ (7) $X_{CT} = R_T \times \tan\theta$ (8)

Impedance (Z) {Series-RC (Multiple Components)}

$Z = \sqrt{R_T^2 + X_{CT}^2}$ (1) $Z = E_T/I$ (2) $Z = E_T^2/P_{VA}$ (3)

$Z = P_{VA}/I^2$ (4) $Z = R_T/PF(\cos\theta)$ (5) $Z = X_{CT}/\sin\theta$ (6)

Current (I) {Series-RC (Multiple Components)}

$I = E_R/R$ (1) $I = E_C/X_C$ (2) $I = P_W/E_R$ (3)

$I = P_{VARsC}/E_C$ (4) $I = \sqrt{P_W/R}$ (5) $I = \sqrt{P_{VARsC}/X_C}$ (6)

$I = E_{R(T)}/R_T$ (1) $I = E_{C(T)}/X_{CT}$ (2) $I = P_{W(T)}/E_{R(T)}$ (3)

$I = P_{VARsC(T)}/E_{C(T)}$ (4) $I = \sqrt{P_{W(T)}/R_T}$ (5) $I = \sqrt{P_{VARsC(T)}/X_{CT}}$ (6)

$I = P_{VA}/E_T$ (7) $I = \sqrt{P_{VA}/Z}$ (8) $I = E_T/Z$ (9)

Voltage Drops (E_R and E_C) {Series-R_C (Multiple Components)}

$E_R = I \times R$ (1) $E_R = P_W / I$ (2) $E_R = \sqrt{P_W \times R}$ (3)

$E_{R(T)} = E_{R1} + E_{R2} + \ldots$ (1) $E_{R(T)} = I \times R_T$ (2) $E_{R(T)} = P_{W(T)} / I$ (3)

$E_{R(T)} = \sqrt{P_{W(T)} \times R_T}$ (4) $E_{R(T)} = \sqrt{E_T^2 - E_{C(T)}^2}$ (5) $E_{R(T)} = E_T \times \cos\theta$ (PF) (6)

$E_{R(T)} = E_{C(T)} / \tan\theta$ (7)

$E_C = I \times X_C$ (1) $E_C = P_{VARsC} / I$ (2) $E_C = \sqrt{P_{VARsC} \times X_C}$ (3)

$E_{C(T)} = E_{C1} + E_{C2} + \ldots$ (1) $E_{C(T)} = I \times X_{CT}$ (2) $E_{C(T)} = P_{VARsC(T)} / I$ (3)

$E_{C(T)} = \sqrt{P_{VARsC(T)} \times X_{CT}}$ (4) $E_{C(T)} = \sqrt{E_T^2 - E_{R(T)}^2}$ (5) $E_{C(T)} = E_T \times \sin\theta$ (6)

$E_{C(T)} = E_{RT} \times \tan\theta$ (7)

Total Voltage (E_T) {Series-RC (Multiple Components)}

$E_T = P_{VA} / I$ (1) $E_T = I \times Z$ (2) $E_T = \sqrt{P_{VA} / Z}$ (3)

$E_T = \sqrt{E_{R(T)}^2 + E_{C(T)}^2}$ (4) $E_T = E_{R(T)} /$ PF ($\cos\theta$) (5) $E_T = E_{C(T)} / \sin\theta$ (6)

The Apparent Power (VA) {Series-RC (Multiple Components)}

$P_{VA} = E_T \times I$ (1) $P_{VA} = I^2 \times Z$ (2) $P_{VA} = E_T^2 / Z$ (3)

$P_{VA} = P_{W(T)} /$ PF ($\cos\theta$) (4) $P_{VA} = P_{VARsC(T)} / \sin\theta$ (5) $P_{VA} = \sqrt{P_{W(T)}^2 + P_{VARsC(T)}^2}$ (6)

The True Power (W) {Series-RC (Multiple Components)}

$P_W = I^2 \times R$ (1) $P_W = E_R^2 / R$ (2) $P_W = E_R \times I$ (3)

$P_{W(T)} = P_{W1} + P_{W2} + \ldots$ (1) $P_{W(T)} = I^2 \times R_T$ (2) $P_{W(T)} = E_{R(T)}^2 / R_T$ (3)

$P_{W(T)} = E_{R(T)} \times I$ (4) $P_{W(T)} = P_{VA} \times$ PF ($\cos\theta$) (5) $P_{W(T)} = P_{VARsC(T)} / \tan\theta$ (6)

$P_{W(T)} = \sqrt{P_{VA}^2 - P_{VARsC(T)}^2}$ (7)

The Reactance Power (VARs) {Series-RC (Multiple Components)}

$P_{VARsC} = I^2 \times X_C$ (1) $P_{VARsC} = E_C^2 / X_C$ (2) $P_{VARsC} = E_C \times I$ (3)

$P_{VARsC(T)} = P_{VARsC\,1} + P_{VARsC\,1} + \ldots$ (1) $P_{VARsC(T)} = I^2 \times X_{CT}$ (2) $P_{VARsC(T)} = E_{C(T)}^2 / X_{CT}$ (3)

$P_{VARsC(T)} = E_{C(T)} \times I$ (4) $P_{VARsC(T)} = P_{VA} \times \sin\theta$ (5) $P_{VARsC(T)} = P_{W(T)} \times \tan\theta$ (6)

$P_{VARsC(T)} = \sqrt{P_{VA}^2 - P_{W(T)}^2}$ (7)

The Power Factor (PF) {Series-RC (Multiple Components)}

PF $= R_T / Z$ ($\cos\theta$) (1) PF $= E_{R(T)} / E_T$ ($\cos\theta$) (2) PF $= P_{W(T)} / P_{VA}$ ($\cos\theta$) (3)

$I_{C1} = E/X_{C1}$
$= 74V/13\Omega$
$= 5.69A$

$I_{C2} = E/X_{C2}$
$= 74V/26\Omega$
$= 2.85A$

$I_{C3} = E/X_{C3}$
$= 74V/14\Omega$
$= 5.29A$

Similar to calculating the total voltage for a series circuit based on individual voltage drops, the same has to be done for calculating the total current for a parallel circuit based on individual current paths. To determine the circuit's total current, the current paths about each like component must be totaled together first. Once this is done, the vector formula $I_T = \sqrt{I_{RT}^2 + I_{CT}^2}$ can then be applied. Therefore,

$I_{RT} = 4.93A + 3.36A$
$= 8.29A$

$I_{CT} = 5.69A + 2.85A + 5.29A$
$= 13.83A$

results in

$$I_T = \sqrt{(8.29)^2 + (13.83)^2}$$
$$= \sqrt{68.72 + 191.27}$$
$$= \sqrt{259.99}$$
$$= 16.12A$$

As expected, when calculated vectorially, the total current should be the same (or approximately the same) as that previously calculated ($I_T = 74V/4.59\Omega = 16.12A$).

The Apparent Power (VA) {Parallel-RC (Multiple Components)}
Using that which was given as the circuit's voltage (E), the calculated current (I_T), and impedance (Z), the circuit's apparent power can be determined by use of the formulas $P_{VA} = E \times I_T$, $P_{VA} = I_T^2 \times Z$, and $P_{VA} = E^2/Z$.

$P_{VA} = 74V \times 16.12A$
$= 1192.88VA$

$P_{VA} = (16.12A)^2 \times 4.59\Omega$
$= 259.85A \times 4.59\Omega$
$= 1192.71VA$

$P_{VA} = (74V)^2 / 4.59\Omega$
$= 5476V / 4.59\Omega$
$= 1193.03VA$

The True Power (W) {Parallel-RC (Multiple Components)}
Because each resistance component will contribute to the circuit's total true power consumption, the true power consumed by each component must be determined before the total true power is realized. To simplify these calculations, only one formula will be used, $P_W = E^2/R$ to calculate the true power consumption for each component. Therefore,

$P_{W1} = E^2/R_1$
$= (74V)^2/15\Omega$
$= 365.07W$

$P_{W2} = E^2/R_2$
$= (74V)^2/22\Omega$
$= 248.91W$

results in

$P_{WT} = 365.07W + 248.91W$
$= 613.98W$

The Reactance Power (VARs) {Parallel-RC (Multiple Components)}

To determine the circuit's total reactance power, the reactance power about each capacitor must first be determined before the total reactance power is realized. Again, to simplify these calculations, only one formula will be used, $P_{VARsC} = I_C^2 \times X_C$, to calculate the reactance power for each component. Therefore,

$$\begin{aligned} P_{VARsC1} &= I_{C1}^2 \times X_{C1} \\ &= (5.69A)^2 \times 13\Omega \\ &= 420.89VARs \end{aligned} \qquad \begin{aligned} P_{VARsC2} &= I_{C2}^2 \times X_{C2} \\ &= (2.85A)^2 \times 26\Omega \\ &= 211.19VARs \end{aligned} \qquad \begin{aligned} P_{VARsC3} &= I_{C3}^2 \times X_{C3} \\ &= (5.29A)^2 \times 14\Omega \\ &= 391.78VARs \end{aligned}$$

results in

$$\begin{aligned} P_{VARsC(T)} &= 420.89VARs + 211.19VARs + 391.78VARs \\ &= 1023.86VARs \end{aligned}$$

The Power Factor (PF) {Parallel-RC (Multiple Components)}

When calculating the power factor in the series-RL circuit, the formula PF = R_T/Z was used. Just remember that a series-RL circuit was used in that situation instead of a series-RC circuit, but the same approach would have been taken if the circuit was a series-RC type. Again, with a parallel-RC (or parallel-RL) circuit, the formula is the complete opposite:

$$PF = \frac{Z}{R_T}$$

Using the given and calculated quantities, along with the above and the previously given power factor formula,

$$PF = \frac{613.98W}{1192.71VA} \quad \text{and} \quad PF = \frac{4.59\Omega}{8.92\Omega}$$

$$= .5148\ (51.48\%) \qquad\qquad = .5146\ (51.46\%)$$

As you can see, when calculated, both sets of quantities yield approximate results.

The Phase Angle (θ) {Parallel-RC (Multiple Components)}

In a parallel-RC circuit formulas for the phase angle can be derived using one of the three trigonometric formulas,

$$\tan\theta = \frac{I_{CT}}{I_{RT}} \quad \text{or} \quad \frac{R_T}{X_{CT}} \quad \text{and} \quad \cos\theta = \frac{Z}{R_T}$$

From the three formulas, the following derived formulas—along with a previously derived formula—will be used to determine the phase angle:

$$\theta = \tan^{-1}\frac{I_{CT}}{I_{RT}} \quad \text{or} \quad \frac{R_T}{X_{CT}} \quad \text{and} \quad \theta = \cos^{-1}\frac{Z}{R_T}$$

$$\theta = \cos^{-1} \text{PF (previous)}$$

Based on the derived formulas, the circuit's phase angle is

$$\begin{aligned} \theta &= \tan^{-1}\frac{13.83A}{8.29A} & \text{or} \quad &\frac{8.92\Omega}{5.35\Omega} & \text{and} \quad \theta &= \cos^{-1}\frac{4.59\Omega}{8.92\Omega} \\ &= \tan^{-1} 1.668 & &= 1.667 & &= \cos^{-1} .5146 \\ &= 59.06° & &= 59.04° & &= 59.03° \end{aligned}$$

$$\begin{aligned} \theta &= \cos^{-1} .5146 \text{ (previous)} \\ &= 59.03° \end{aligned}$$

Using all four formulas, the results are all approximately the same.

Power {Parallel-RC (Multiple Components)}

The three power quantities will now be determined using an alternate approach. By using the formula $\cos\theta = P_W/P_{VA}$, the true power can now be determined by use of the derived formula, $P_W = P_{VA} \times \cos\theta$, where

$$\begin{aligned} P_W &= P_{VA} \times \cos\theta \\ &= 1193.03VA \times \cos 59.03° \\ &= 1193.03VA \times .5146 \\ &= 613.93W \end{aligned}$$

By using the formula $\sin\theta = P_{VARsC}/P_{VA}$, the reactance power can now be determined by use of the derived formula, $P_{VARsC} = P_{VA} \times \sin\theta$, where

$$\begin{aligned} P_{VARsC} &= P_{VA} \times \sin\theta \\ &= 1193.03VA \times \sin 59.03° \\ &= 1193.03VA \times .857 \\ &= 1022.43VARs \end{aligned}$$

As you can see, the results are approximately the same as earlier calculated. By using these results, they can be substituted into the Pythagorean theorem, and the apparent power can be verified.

$$\begin{aligned} P_{VA} &= \sqrt{(613.93)^2 + (1022.43)^2} \\ &= \sqrt{376{,}910.04 + 1{,}045{,}363.11} \\ &= \sqrt{1{,}422{,}273.15} \\ &= 1192.59VA \end{aligned}$$

Again, with alternate methods calculations can be either cross-checked or verified to ensure correct answers.

The following display is an overall list of formulas that can be used for solving Parallel-RC Circuits—Multiple Components.

Related Parallel Formulas
(Resistance–Capacitance Multiple Components Circuits—Overall)

[Only one source of voltage (E) – therefore, $E = E_R = E_C$.]

Resistance (R) {Parallel-RC (Multiple Components)}

$R = E/I_R$ (1) $R = E^2/P_W$ (2) $R = P_W/I_R^2$ (3)

$R_T = \dfrac{1}{1/R_1 + 1/R_2 \ldots}$ $R_T = \dfrac{R_1 \times R_2}{R_1 + R_2}$ (2) $R_T = E_{R(T)}/I_{R(T)}$ (3)

$R_T = E^2/Pw_{(T)}$ (4) $R_T = P_{W(T)}/I_{R(T)}^2$ (5) $R_T = Z/\cos\theta$ (PF) (6)

$R_T = X_{CT} \times \tan\theta$ (7) $R_T = \dfrac{1}{\sqrt{(1/Z)^2 - (1/X_{CT})^2}}$ (8)

Capacitance (C) {Parallel-RC (Multiple Components)}

$C = 1/2\pi f X_C$

$C_T = C_1 + C_2 + \ldots\ldots$ (1) $C_T = 1/2\pi f X_{CT}$ (2)

Capacitance Reactance (X_C) {Parallel-RC (Multiple Components)}

$X_C = 1/2\pi fC$ (1) $X_C = E/I_C$ (2) $X_C = E^2/P_{VARsC}$ (3)

$X_C = P_{VARsC}/I_C^2$ (4) $X_C = R/\tan\theta$ (5) $X_C = \dfrac{1}{\sqrt{(1/Z)^2 - (1/R)^2}}$ (6)

$X_{CT} = \dfrac{1}{1/X_{C1} + 1/X_{C2}}$ (1) $X_{CT} = \dfrac{X_{C1} \times X_{C2}}{X_{C1} + X_{C2}}$ (2) $X_{CT} = 1/2\pi fC_T$ (3)

$X_{CT} = E/I_{C(T)}$ (4) $X_{CT} = E^2/P_{VARsC(T)}$ (5) $X_{CT} = P_{VARsC(T)}/I_{C(T)}^2$ (6)

$X_{CT} = R_T/\tan\theta$ (7) $X_{CT} = Z/\sin\theta$ (8) $X_{CT} = \dfrac{1}{\sqrt{(1/Z)^2 - (1/R_T)^2}}$ (9)

Impedance (Z) {Parallel-RC (Multiple Components)}

$Z = E/I_T$ (1) $Z = E^2/P_{VA}$ (2) $Z = P_{VA}/I_T^2$ (3)

$Z = R_T \times \cos\theta$ (PF) (4) $Z = X_{CT} \times \sin\theta$ (5)

$$Z = \frac{1}{\sqrt{(1/R_T)^2 + (1/X_{CT})^2}} = \frac{R_T \times X_{CT}}{\sqrt{(R_T)^2 + (X_{CT})^2}} \quad (6)$$

Individual Current (I_R and I_C) {Parallel-RC (Multiple Components)}

$I_R = E/R$ (1) $I_R = P_W/E$ (2) $I_R = \sqrt{P_W/R}$ (3)
$I_C = E/X_C$ (1) $I_C = P_{VARsC}/E$ (2) $I_C = \sqrt{P_{VARsC}/X_C}$ (3)

Total Individual Currents ($I_{R(T)}$ and $I_{C(T)}$) {Parallel-RC (Multiple Components)}

$I_{R(T)} = I_{R1} + I_{R2} + \ldots$ (1) $I_{R(T)} = E/R_T$ (2) $I_{R(T)} = P_{W(T)}/E_{R(T)}$ (3)
$I_R = \sqrt{P_{W(T)}/R_T}$ (4) $I_{R(T)} = \sqrt{I_T^2 - I_{C(T)}^2}$ (5) $I_{R(T)} = I_T \text{ x } \cos\theta$ (PF) (6)
$I_{R(T)} = I_{C(T)}/\tan\theta$ (7)

$I_{C(T)} = I_{C1} + I_{C2} + \ldots$ (1) $I_{C(T)} = E/X_{CT}$ (2) $I_{C(T)} = P_{VARsC(T)}/E$ (3)
$I_{C(T)} = \sqrt{P_{VARs\,C(T)}/X_{CT}}$ (4) $I_{C(T)} = \sqrt{I_T^2 - I_{R(T)}^2}$ (5) $I_{C(T)} = I_T \text{ x } \sin\theta$ (6)
$I_{C(T)} = I_{R(T)} \text{ x } \tan\theta$ (7)

Total Current (I_T) {Parallel-RC (Multiple Components)}

$I_T = E/Z$ (1) $I_T = P_{VA}/E$ (2) $I_T = \sqrt{P_{VA}/Z}$ (3)
$I_T = \sqrt{I_{R(T)}^2 + I_{C(T)}^2}$ (4) $I_T = I_{R(T)}/\cos\theta$ (PF) (5) $I_T = I_{C(T)}/\sin\theta$ (6)

Voltage (E) {Parallel-RC (Multiple Components)}

$E = I_R \text{ x } R$ (1) $E = I_{R(T)} \text{ x } R_T$ (2) $E = P_{W(T)}/I_{R(T)}$ (3)
$E = I_C \text{ x } X_C$ (4) $E = I_{C(T)} \text{ x } X_{C(T)}$ (5) $E = \sqrt{P_{VARsC} \text{ x } I_C}$ (6)
$E = P_{VARsC(T)}/I_{C(T)}$ (7) $E = \sqrt{P_W \text{ x } R}$ (8) $E = \sqrt{P_{W(T)} \text{ x } R_T}$ (9)
$E = \sqrt{P_{VARsC} \text{ x } X_C}$ (10) $E = \sqrt{P_{VARsC(T)} \text{ x } X_{C(T)}}$ (11) $E = I_T \text{ x } Z$ (12)
$E = \sqrt{P_{VA}/I_T}$ (13) $E = \sqrt{P_{VA} \text{ x } Z}$ (14)

The Apparent Power (VA) {Parallel-RC (Multiple Components)}

$P_{VA} = E \text{ x } I_T$ (1) $P_{VA} = I_T^2 \text{ x } Z$ (2) $P_{VA} = E^2/Z$ (3)
$P_{VA} = P_{W(T)}/\cos\theta$ (PF) (4) $P_{VA} = P_{VARsC(T)}/\sin\theta$ (5)
$P_{VA} = \sqrt{P_{W(T)}^2 + P_{VARsC(T)}^2}$ (6)

The True Power (W) {Parallel-RC (Multiple Components)}

$P_W = E \text{ x } I_R$ (1) $P_W = E^2/R$ (2) $P_W = I_R^2 \text{ x } R$ (3)

$P_{W(T)} = P_{W1} + P_{W2} + \ldots$ (1) $P_{W(T)} = E \text{ x } I_{R(T)}$ (2) $P_{W(T)} = E^2/R_T$ (3)
$P_{W(T)} = I_{R(T)}^2 \text{ x } R_T$ (4) $P_{W(T)} = P_{VA} \text{ x } \cos\theta$ (PF) (5) $P_{W(T)} = P_{VARsC(T)}/\tan\theta$ (6)
$P_{W(T)} = \sqrt{P_{VA}^2 - P_{VARsC(T)}^2}$ (7)

The Reactance Power (VARs) {Parallel-RC (Multiple Components)}

$P_{VARsC} = E \text{ x } I_C$ (1) $P_{VARsC} = I_C^2 \text{ x } X_C$ (2) $P_{VARsC} = E_C^2/X_C$ (3)

$P_{VARsC(T)} = P_{VARsC1} + P_{VARsC2} + \ldots.$ (1) $P_{VARsC(T)} = E \times I_{C(T)}$ (2) $P_{VARsC(T)} = I_{C(T)}^2 \times X_{CT}$ (3)

$P_{VARsC(T)} = E_{C(T)}^2/X_{CT}$ (4) $P_{VARsC(T)} = \sqrt{P_{VA}^2 - P_{W(T)}^2}$ (5) $P_{VARsC(T)} = P_{VA} \times \sin\theta$ (6)

$P_{VARsC(T)} = P_{W(T)} \times \tan\theta$ (7)

The Power Factor (PF) {Parallel-RC (Multiple Components)}

$PF = Z/R_T (\cos\theta)$ (1) $PF = I_{R(T)}/I_T (\cos\theta)$ (2) $PF = P_{W(T)}/P_{VA} (\cos\theta)$ (3)

The Phase Angle (θ){Parallel-RC (Multiple Components)}

$\theta = \cos^{-1}\dfrac{I_{R(T)}}{I_T}$ (PF) (1) $\theta = \sin^{-1}\dfrac{I_{C(T)}}{I_T}$ (2) $\theta = \tan^{-1}\dfrac{I_{C(T)}}{I_{R(T)}}$ (3)

$\theta = \cos^{-1}\dfrac{Z}{R_T}$ (PF) (4) $\theta = \sin^{-1}\dfrac{Z}{X_{CT}}$ (5) $\theta = \tan^{-1}\dfrac{R_T}{X_{CT}}$ (6)

$\theta = \cos^{-1}\dfrac{P_{W(T)}}{P_{VA}}$ (PF) (7) $\theta = \sin^{-1}\dfrac{P_{VARsC(T)}}{P_{VA}}$ (8) $\theta = \tan^{-1}\dfrac{P_{VARsC(T)}}{P_{W(T)}}$ (9)

The following display is an overall list of formulas that can be used for solving Parallel-RL circuits—Multiple Components.

Related Parallel Formulas
(Resistance–Inductance Multiple Components Circuits—Overall)

[Only one source of voltage (E); – therefore, $E = E_R = E_L$.]

Resistance (R) {Parallel-RL (Multiple Components)}

$R = E/I_R$ (1) $R = E^2/P_W$ (2) $R = P_W/I_R^2$ (3)

$R_T = \dfrac{1}{1/R_1 + 1/R_2 \ldots}$ (1) $R_T = \dfrac{R_1 \times R_2}{R_1 + R_2}$ (2) $R_T = E_{R(T)}/I$ (3)

$R_T = E^2/Pw_{(T)}$ (4) $R_T = P_{W(T)}/I_{R(T)}^2$ (5) $R_T = Z/\cos\theta$ (PF) (6)

$R_T = X_{LT} \times \tan\theta$ (7) $R_T = \dfrac{1}{\sqrt{(1/Z)^2 - (1/X_{LT})^2}}$ (8)

Inductance (L) {Parallel-RL (Multiple Components)}

$L = X_L/2\pi f$ $L_T = \dfrac{L}{N}$ (1) (see page 57) $L_T = \dfrac{L_1 \times L_2}{L_1 + L_2}$ (2)

$L_T = \dfrac{1}{1/L_1 + 1/L_2 + \ldots\ldots.}$ (3) $L_T = X_{LT}/2\pi f$ (4)

Inductance Reactance (X_L) {Paralle- RL (Multiple Components)}

$X_L = 2\pi fL$ (1) $X_L = E/I_L$ (2) $X_L = E^2/P_{VARsL}$ (3)

$X_L = P_{VARsL}/I_L^2$ (4) $X_L = R/\tan\theta$ (5) $X_L = \dfrac{1}{\sqrt{(1/Z)^2 - (1/R)^2}}$ (6)

$X_{LT} = \dfrac{1}{1/X_{L1} + 1/X_{L2}}$ (1) $X_{LT} = \dfrac{X_{L1} \times X_{L2}}{X_{L1} + X_{L2}}$ (2) $X_{LT} = L_T/2\pi f$ (3)

$X_{LT} = E/I_{L(T)}$ (4) $X_{LT} = E^2/P_{VARsL(T)}$ (5) $X_{LT} = P_{VARsL(T)}/I_{L(T)}^2$ (6)

$X_{LT} = R_T/\tan\theta$ (7) $X_{LT} = Z/\sin\theta$ (8) $X_{LT} = \dfrac{1}{\sqrt{(1/Z)^2 - (1/R_T)^2}}$ (9)

Impedance (Z) {Parallel-RL (Multiple Components)}

$Z = E/I_T$ (1) $Z = E^2/P_{VA}$ (2) $Z = P_{VA}/I_T^2$ (3)

$Z = R_T \times \cos\theta$ (PF) (4) $Z = X_{LT} \times \sin\theta$ (5)

$$Z = \frac{1}{\sqrt{(1/R_T)^2 + (1/X_{LT})^2}} = \frac{R_T \times X_{LT}}{\sqrt{(R_T)^2 + (X_{LT})^2}} \quad (6)$$

Individual Current (I_R and I_L) {Parallel RL (Multiple Components)}

$I_R = E/R$ (1) $I_R = P_W/E$ (2) $I_R = \sqrt{P_W/R}$ (3)

$I_L = E/X_L$ (1) $I_L = P_{VARsL}/E$ (2) $I_L = \sqrt{P_{VARsL}/X_L}$ (3)

Total Individual Currents ($I_{R(T)}$ and $I_{L(T)}$) {Parallel-RL (Multiple Components)}

$I_{R(T)} = I_{R1} + I_{R2} + \ldots$ (1) $I_{R(T)} = E/R_T$ (2) $I_{R(T)} = P_{W(T)}/E_{R(T)}$ (3)

$I_R = \sqrt{P_{W(T)}/R_T}$ (4) $I_{R(T)} = \sqrt{I_T^2 - I_{L(T)}^2}$ (5) $I_{R(T)} = I_T \times \cos\theta$ (PF) (6)

$I_{R(T)} = I_{L(T)}/\tan\theta$ (7)

$I_{L(T)} = I_{L1} + I_{L2} + \ldots$ (1) $I_{L(T)} = E/X_{LT}$ (2) $I_{L(T)} = P_{VARsL(T)}/E$ (3)

$I_{L(T)} = \sqrt{P_{VARs\,L(T)}/X_{LT}}$ (4) $I_{L(T)} = \sqrt{I_T^2 - I_{R(T)}^2}$ (5) $I_{L(T)} = I_T \times \sin\theta$ (6)

$I_{L(T)} = I_{R(T)} \times \tan\theta$ (7)

Total Current (I_T) {Parallel-RL (Multiple Components)}

$I_T = E/Z$ (1) $I_T = P_{VA}/E$ (2) $I_T = \sqrt{P_{VA}/Z}$ (3)

$I_T = \sqrt{I_{R(T)}^2 + I_{L(T)}^2}$ (4) $I_T = I_{R(T)}/\cos\theta$ (PF) (5) $I_T = I_{L(T)}/\sin\theta$ (6)

Voltage (E) {Parallel-RL (Multiple Components)}

$E = I_R \times R$ (1) $E = I_{R(T)} \times R_T$ (2) $E = P_{W(T)}/I_{R(T)}$ (3)

$E = I_L \times X_L$ (4) $E = I_{L(T)} \times X_{L(T)}$ (5) $E = \sqrt{P_{VARsL} \times I_L}$ (6)

$E = P_{VARsL(T)} / I_{L(T)}$ (7) $E = \sqrt{P_W \times R}$ (8) $E = \sqrt{P_{W(T)} \times R_T}$ (9)

$E = \sqrt{P_{VARsL} \times X_L}$ (10) $E = \sqrt{P_{VARsL(T)} \times X_{L(T)}}$ (11) $E = I_T \times Z$ (12)

$E = \sqrt{P_{VA} / I_T}$ (13) $E = \sqrt{P_{VA} \times Z}$ (14)

The Apparent Power (VA) {Parallel-RL (Multiple Components)}

$P_{VA} = E \times I_T$ (1) $P_{VA} = I_T^2 \times Z$ (2) $P_{VA} = E^2/Z$ (3)

$P_{VA} = P_{W(T)}/\cos\theta$ (PF) (4) $P_{VA} = P_{VARsL(T)} / \sin\theta$ (5)

$P_{VA} = \sqrt{P_{W(T)}^2 + P_{VARsL(T)}^2}$ (6)

The True Power (W) {Parallel-RL (Multiple Components)}

$P_W = E \times I_R$ (1) $P_W = E^2/R$ (2) $P_W = I_R^2 \times R$ (3)

$P_{W(T)} = P_{W1} + P_{W2} + \ldots.$ (1) $P_{W(T)} = E \times I_{R(T)}$ (2) $P_{W(T)} = E^2/R_T$ (3)

$P_{W(T)} = I_{R(T)}^2 \times R_T$ (4) $P_{W(T)} = P_{VA} \times \cos\theta$ (PF) (5) $P_{W(T)} = P_{VARsL(T)}/\tan\theta$ (6)

$P_{W(T)} = \sqrt{P_{VA}^2 - P_{VARsL(T)}^2}$ (7)

The Reactance Power (VARs) {Parallel-RL (Multiple Components)}

$P_{VARsL} = E \times I_L$ (1) $P_{VARsL} = I_L^2 \times X_L$ (2) $P_{VARsL} = E_L^2/X_L$ (3)

$P_{VARsL(T)} = P_{VARsL1} + P_{VARsL2} + \ldots.$ (1) $P_{VARsL(T)} = E \times I_{L(T)}$ (2)

$P_{VARsL(T)} = I_{L(T)}^2 \times X_{LT}$ (3) $P_{VARsL(T)} = E_{L(T)}^2/X_{LT}$ (4)

$P_{VARsL(T)} = \sqrt{P_{VA}^2 - P_{W(T)}^2}$ (5) $P_{VARsL(T)} = P_{VA} \times \sin\theta$ (6)

$P_{VARsL(T)} = P_{W(T)} \times \tan\theta$ (7)

The Power Factor (PF) {Parallel-RL (Multiple Components)}

$PF = Z/R_T$ ($\cos\theta$) (1) $PF = I_{R(T)}/I_T$ ($\cos\theta$) (2) $PF = P_{W(T)}/P_{VA}$ ($\cos\theta$) (3)

The Phase Angle (θ){Parallel-RL (Multiple Components)}

$\theta = \cos^{-1} \frac{I_{R(T)}}{I_T}$ (PF) (1) $\theta = \sin^{-1} \frac{I_{L(T)}}{I_T}$ (2) $\theta = \tan^{-1} \frac{I_{L(T)}}{I_{R(T)}}$ (3)

$\theta = \cos^{-1} \frac{Z}{R_T}$ (PF) (4) $\theta = \sin^{-1} \frac{Z}{X_{LT}}$ (5) $\theta = \tan^{-1} \frac{R_T}{X_{LT}}$ (6)

$\theta = \cos^{-1} \frac{P_{W(T)}}{P_{VA}}$ (PF) (7) $\theta = \sin^{-1} \frac{P_{VARsL(T)}}{P_{VA}}$ (8) $\theta = \tan^{-1} \frac{P_{VARsL(T)}}{P_{W(T)}}$ (9)

Resistance (R)–Inductance (L)-Capacitance (C) Circuits

Up until now, only circuits with resistance and inductors or resistance and capacitors components have been dealt with. Now we will look at circuits consisting of resistance, inductors, and capacitors components that are both series and parallel connected.

Series-RLC
Consider the Series-RLC circuit in figure 29.

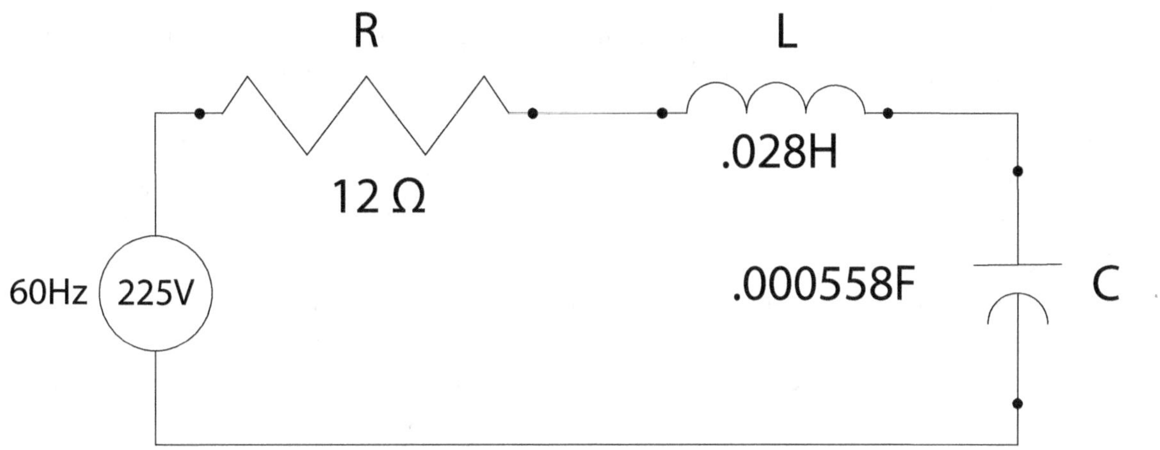

Fig. 29. Series-RLC circuit.

Inductance Reactance (X_L) and Capacitance Reactance (X_C) {Series-RLC}
Based on the given inductor, capacitor and operating frequency the inductance reactance is

$$X_L = 2 \text{ x } 3.14 \text{ x } 60\text{Hz x } .028\text{H} = 10.55\Omega$$

and

the capacitance reactance is

$$X_C = \frac{1}{2 \text{ x } 3.14 \text{ x } 60\text{Hz x } .000558\text{F}} = 4.76\Omega$$

Impedance (Z) {Series-RLC}
In order to calculate the impedance for this circuit the vector related formula for a series-RLC circuit must be used; that is, $Z = \sqrt{R^2 + (X_L - X_C)^2}$, where X_L is larger than X_C. Therefore,

$$\begin{aligned} Z &= \sqrt{(12)^2 + (10.55 - 4.76)^2} \\ &= \sqrt{144 + 33.52} \\ &= \sqrt{177.52} \\ &= 13.32\Omega \end{aligned}$$

Now that the circuit's impedance has been determined, other quantities can be determined as well.

Current (I) {Series-RLC}

Because this is a series circuit, only one source of current I will exist. Since E_T is shown to be 225V and Z was calculated at 13.32Ω,

$$I = 225V/13.32\Omega$$
$$= 16.89A$$

Voltage Drops (E_R, E_L and E_C) {Series-RLC}

As with all series circuits, the component having the largest ohmic value will experience the largest voltage drop and so forth being that the circuit's current is common to all components.

$$E_R = I \times R = 16.89A \times \underline{12\Omega} = 202.68V$$

$$E_L = I \times X_L = 16.89A \times \underline{10.55\Omega} = 178.19V$$

$$E_C = I \times X_C = 16.89A \times \underline{4.76\Omega} = 80.40V$$

Since the voltage drops and current flow about the inductors and capacitors will not be in phase with each other, these two components, along with the resistance, have to be analyzed vectorially for this series circuit. Referring to figure 30, you will notice that the vector relationship between each voltage drop E_R, E_L, and E_C and current I as applied to a purely resistance, inductance, and capacitance circuit, is a combination of figures 12(a)–(c).

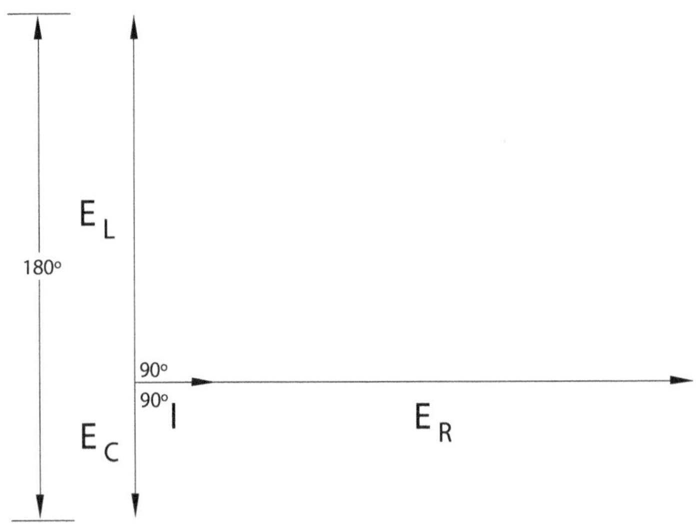

Fig. 30. E_{RLC}—I vector relationships.

Using current as the reference vector, figure 30 is a three-part representation of all three series components in terms of their vector relationships with respect to voltage and current. As you can see, the vector representation reflects the voltage drop E_R across the resistor R being in phase with the circuit's current I. Also, the voltage drop E_L across the inductor L leads the current I by 90°, and the voltage drop E_C across the capacitor C lags the current I by 90°. Because the vectors E_L and E_C are 180° out of phase, the larger vector will be reduced in magnitude while the smaller vector is eliminated. In a series circuit, when two vectors are 180° out of phase, the only mathematical operation required in determining their difference is the method of subtraction. Take notice that the unequal magnitudes of vectors E_L and E_C in figure 31(a) represents the voltage drop across E_L being of a larger magnitude than the voltage drop across E_C. It can also be said that these unequal vectors will serve to represent the differences in magnitudes of the reactance quantities relative to these vectors when calculated.

To evaluate the given series-RLC circuit vectorially, only two vectors can be analyzed at a time. Earlier we discussed in detail the head and tail method, which identified the 90° displacement of two adjoining vectors forming a right triangle, and which allowed the two vectors to be added together to yield a resulting vector. Well, the same approach has been taken in figures 31(a), (b), and (c), where the two opposing vectors E_L and E_C, which are 180° out of phase, are converted into a common vector E_{LC}. This common vector was derived by subtracting the smaller vector from the larger vector (E_L-E_C).

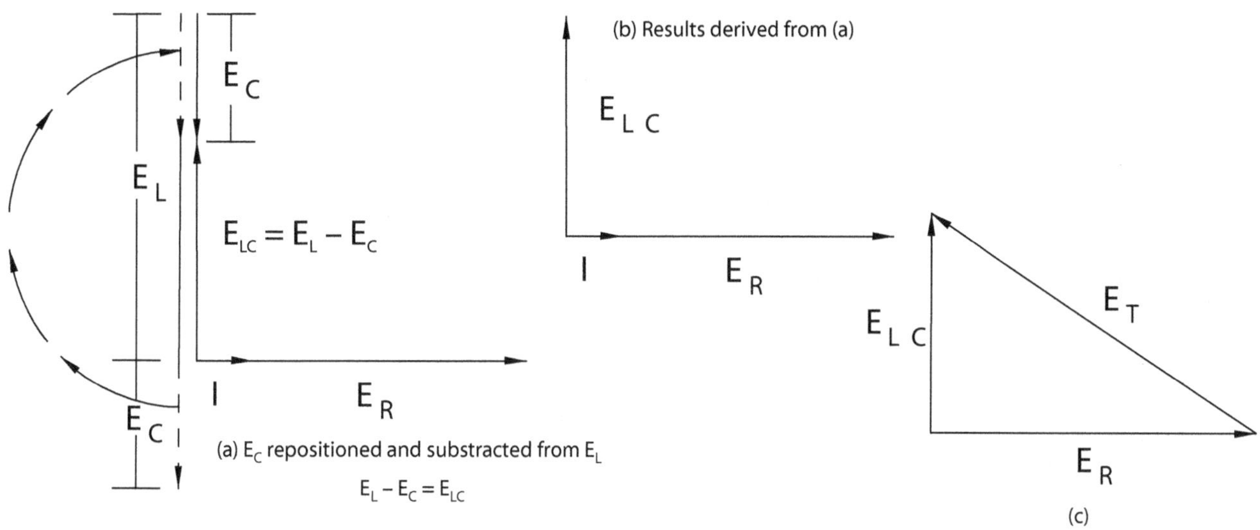

Fig. 31. Common vector $E_{(LC)}$.

As with all situations, the common vector will always be positioned in the same direction as the larger vector. If E_C was larger than E_L the common vector would then be labeled E_{CL} and positioned in the opposite direction. As a result, this conclusion as shown in figure 31(c) leads to a similar vector diagram and right triangle as displayed in figure 15(a). From here we can then began to evaluate the given Series-RLC circuit.

Total Voltage (E_T) {Series-RLC}

The calculated total voltage for this circuits is the realization of figures 31(a), (b), and (c). Because E_C experiences the smallest voltage drop compared to E_L, E_C is subtracted from E_L to derive a common voltage drop E_{LC}; that is,

$$E_{LC} = E_L - E_C$$

Because the total voltage must be determined based on all three components and must be calculated vectorially, we'll start with the vector formula, $E_T = \sqrt{E_R^2 + E_L^2}$. Since this formula only contains a resistance and inductance component it must be extended to include the third component E_C. Using the derived common voltage drop E_{LC} the formula can now be inclusive of the third component where $E_T = \sqrt{E_R^2 + (E_L - E_C)^2}$ or $E_T = \sqrt{E_R^2 + (E_{LC})^2}$. By applying this formula the circuit's total voltage can now be calculated. Therefore,

$$\begin{aligned} E_T &= \sqrt{(202.68)^2 + (178.19 - 80.4)^2} \\ &= \sqrt{41{,}079.18 + (97.79)^2} \\ &= \sqrt{41{,}079.18 + 9562.88} \\ &= \sqrt{50{,}642.1} \\ &= 225.04\text{V} \end{aligned}$$

Did you notice in the formula that E_{LC} is equal to 97.79V?

Even when three components are included the calculated total voltage should still be the same (or approximately the same) as the given total voltage; that is, 225V.

The Apparent Power (VA) {Series-RLC}

Again the same formulas used with previous series resistance–reactance circuits can be used with this three component series circuit for calculating the apparent power.

$$\begin{aligned} P_{VA} &= 225\text{V} \times 16.89\text{A} \\ &= 3800.25\text{VA} \end{aligned}$$

$$\begin{aligned} P_{VA} &= (16.89\text{A})^2 \times 13.32\Omega \\ &= 285.27\text{A} \times 13.32\Omega \\ &= 3799.8\text{VA} \end{aligned}$$

$$\begin{aligned} P_{VA} &= (225\text{V})^2/13.32\Omega \\ &= 50{,}625\text{V}/13.32\Omega \\ &= 3800.68\text{VA} \end{aligned}$$

The True Power (W) {Series-RLC}

In calculating the true power for this circuit the same formulas are used as before.

$$\begin{aligned} P_W &= I^2 \times R \\ &= (16.89\text{A})^2 \times 12\Omega \\ &= 3423.27\text{W} \end{aligned}$$

and

$$\begin{aligned} P_W &= E_R^2/R \\ &= (202.68\text{V})^2/12\Omega \\ &= 3423.27\text{W} \end{aligned}$$

The Reactance Powers (VARs) {Series-RLC}

The reactance power is always associated with the reactance part of an *ac* circuit; that is, the inductance or capacitance reactance. In a circuit involving both inductance and capacitance two likewise reactance powers will exist, inductance reactance power, P_{VARsL} and capacitance reactance power, P_{VARsC}. To determine each reactance power for this circuit, the formulas $P = I^2 \times R$ and $P = (E_{L\text{ or }C})^2/R$ will be used where R is substituted with X_L and X_C to derive the formulas $P_{VARsL} = I^2 \times X_L$ and E^2/X_L and $P_{VARsC} = I^2 \times X_C$ and E^2/X_C. Applying both formulas,

$$P_{VARsL} = (16.89A)^2 \text{ x } 10.55\Omega = 3009.62\text{VARs} \quad \text{or} \quad P_{VARsL} = (178.19V)^2/10.55\Omega = 3009.64\text{VARs}$$

and

$$P_{VARsC} = (16.89A)^2 \text{ x } 4.76\Omega = 1357.90\text{VARs} \quad \text{or} \quad P_{VARsC} = (80.4V)^2/4.76\Omega = 1358.02\text{VARs}$$

Of the two sets of results, we will later see which combination is more accurate.

The Power Factor (PF) {Series-RLC}

With the circuit's impedance and resistance alone, the circuit's power factor could be determined solely with those quantities. However, the same could be said about the circuit's apparent power and true power. With each set of quantities, the power factor can be calculated, and best of all, they will both yield the same results or an approximate. Because the power factor is defined as the ratio between the true power and the apparent power, and since the true power is relative to the circuit's resistance and the apparent power is relative to the impedance, this makes both quantities available for use in determining the power factor. Take a look:

$$PF = \frac{P_W}{P_{VA}} \quad \text{and} \quad PF = \frac{R}{Z}$$

Using the given and calculated quantities,

$$PF = \frac{3423.27W}{3800.25VA} = .9008\ (90.08\%) \quad \text{and} \quad PF = \frac{12\Omega}{13.32\Omega} = .9009\ (90.09\%)$$

The Phase Angle (θ) {Series-RLC}

In a Series-RLC circuit, formulas for the phase angle can be derived using any one of the four trigonometric formulas,

$$\tan\theta = \frac{E_L - E_C{}^*}{E_R} \quad \text{or} \quad \frac{X_L - X_C}{R} \quad \text{and} \quad \cos\theta = \frac{R}{Z} \quad \text{or} \quad \frac{E_R{}^*}{E_T}$$

* See figure 31(c).

From the four formulas, the following derived formulas along with a previously derived formula will be used to determine the phase angle:

$$\theta = \tan^{-1}\frac{E_L - E_C}{E_R} \quad \text{or} \quad \frac{X_L - X_C}{R} \quad \text{and} \quad \theta = \cos^{-1}\frac{R}{Z} \quad \text{or} \quad \frac{E_R}{E_T}$$

$$\theta = \cos^{-1} PF \text{ (previous)}$$

Based on the derived formulas, the circuit's phase angle is,

$$\theta = \tan^{-1}\frac{97.79V}{202.68V} = \tan^{-1} .482 = 25.73° \quad \text{or} \quad \frac{5.79\Omega}{12\Omega} = .483 = 25.78° \quad \text{and} \quad \theta = \cos^{-1}\frac{12\Omega}{13.32\Omega} = \cos^{-1} .9009 = 25.72° \quad \text{or} \quad \frac{202.68V}{225V} = .9008 = 25.74°$$

$$\theta = \cos^{-1} .9008 \text{ (previous)} = 25.74°$$

Using all five formulas, the results are all approximate.

Power {Series-RLC}

Previously, the three power quantities were calculated using given or calculated values. The quantities will now be determined using an alternate approach. By using the formula cos $\theta = P_W/P_{VA}$, the true power can now be determined by use of the derived formula, $P_W = P_{VA}$ x cos θ, where

$$P_W = P_{VA} \text{ x } \cos\theta = 3800.25VA \text{ x } \cos 25.72° = 3800.25VA \text{ x } .9009 = 3423.65W$$

By using the formula sin $\theta = P_{VARs}/P_{VA}$, the circuit's reactance power can now be determined by use of the derived formula, $P_{VARs} = P_{VA}$ x sin θ, where

$$P_{VARs} = P_{VA} \text{ x } \sin\theta = 3800.25VA \text{ x } \sin 25.72° = 3800.25VA \text{ x } .434 = 1649.31VARs$$

Earlier the reactance power per reactance was determined. With these calculated quantities the circuit's reactance power can also be determined by subtracting the smaller reactance power from the larger reactance power being that both quantities are 180° out of phase, observe.

$$P_{VARsL} = (16.89A)^2 \text{ x } 10.55\Omega = 3009.62VARs \quad \text{or} \quad P_{VARsL} = (178.19V)^2/10.55\Omega = 3009.64VARs$$

and

$$P_{VARsC} = (16.89A)^2 \text{ x } 4.76\Omega = 1357.90VARs \quad \text{or} \quad P_{VARsC} = (80.4V)^2/4.76\Omega = 1358.02VARs$$

$$3009.62VARs - 1357.90VARs = 1651.72VARs$$

and

$$3009.64VARs - 1358.02VARs = 1651.62VARs$$

The circuit's reactance power quantity (1649.31VARs) calculated earlier proves to be more in line with both quantities derived by subtraction (1651.72VARs, 1651.62VARs). At times regardless of quantities used one formula may prove to provide more accuracy than another. Just remember the objective is to derive quantities that are approximate and with approximate quantities results can be cross-checked to verify comparable calculated results.

Now, by using the calculated results of the circuit's true power and the reactance power per trigonometric functions, these results can be substituted into the Pythagorean theorem and the apparent power can be verified.

$$P_{VA} = \sqrt{(3423.65)^2 + (1649.31)^2}$$
$$= \sqrt{11{,}721{,}379.3 + 2{,}720{,}223.5}$$
$$= \sqrt{14{,}441{,}602.8}$$
$$= 3800.21\text{VA}$$

The following display is an overall list of formulas that can be used for solving Series-RLC circuits.

Related Series Formulas (Resistance–Inductance–Capacitance Circuits—Overall)

[Only one source of current (I)—therefore, $I = I_R = I_L = I_C$.]

(IMPORTANT: Regardless of how shown in relative formulas, where the differences between L and C components are required, the smaller component is always subtracted from the larger component.)

Resistance (R) {Series-RLC}

$R = E_R/I$ (1) $R = E_R^2/P_W$ (2) $R = P_W/I^2$ (3)

$R = Z \times PF (\cos\theta)$ (4) $R = \sqrt{Z^2 - (X_L - X_C)^2}$ (5) $R = X_L - X_C/\tan\theta$ (6)

Inductance (L) and Capacitance (C) {Series-RLC}

$L = X_L/2\pi f$ $C = 1/2\pi f X_C$

Reactances (X_L, X_C and $X_L - X_C$) {Series-RLC}

$X_L = 2\pi fL$ (1) $X_L = E_L/I$ (2)
$X_L = E_L^2/P_{VARsL}$ (3) $X_L = P_{VARsL}/I^2$ (4)

$X_C = 1/2\pi fC$ (1) $X_C = E_C/I$ (2)
$X_C = E_C^2/P_{VARsC}$ (3) $X_C = P_{VARsC}/I^2$ (4)

$X_L - X_C = \sqrt{Z^2 - R^2}$ (1) $X_L - X_C = Z \times \sin\theta$ (2) $X_L - X_C = R \times \tan\theta$ (3)

Impedance (Z) {Series-RLC}

$Z = \sqrt{R^2 + (X_L - X_C)^2}$ (1) $Z = E_T/I$ (2) $Z = E_T^2/P_{VA}$ (3)
$Z = P_{VA}/I^2$ (4) $Z = R/PF (\cos\theta)$ (5) $Z = (X_L - X_C)/\sin\theta$ (6)

Current (I) {Series-RLC}

$I = E_R/R$ (1) $I = P_W/E_R$ (2) $I = \sqrt{P_W/R}$ (3)
$I = E_T/Z$ (4) $I = \sqrt{P_{VA}/Z}$ (5) $I = P_{VA}/E_T$ (6)

$I = E_L/X_L$ (1) $I = P_{VARsL}/E_L$ (2) $I = \sqrt{P_{VARsL}/X_L}$ (3)
$I = E_C/X_C$ (1) $I = P_{VARsC}/E_C$ (2) $I = \sqrt{P_{VARsC}/X_C}$ (3)

Voltage Drops (E_R, E_L, E_C and E_L–E_C) {Series-RLC}

$E_R = I \times R$ (1) $E_R = P_W/I$ (2)
$E_R = \sqrt{P_W \times R}$ (3) $E_R = E_T \times PF\ (\cos\theta)$ (4)

$E_R = \sqrt{E_T^2 - (E_L - E_C)^2}$ (1) $E_R = (E_L - E_C)/\tan\theta$ (2)

$E_L = I \times X_L$ (1) $E_L = P_{VARsL}/I$ (2) $E_L = \sqrt{P_{VARsL} \times X_L}$ (3)

$E_C = I \times X_C$ (1) $E_C = P_{VARsC}/I$ (2) $E_C = \sqrt{P_{VARsC} \times X_C}$ (3)

$E_L - E_C = \sqrt{E_T^2 - E_R^2}$ (1) $E_L - E_C = E_T \times \sin\theta$ (2) $E_L - E_C = E_R \times \tan\theta$ (3)

Total Voltage (E_T) {Series-RLC}

$E_T = P_{VA}/I$ (1) $E_T = \sqrt{P_{VA} \times Z}$(2) $E_T = I \times Z$ (3)
$E_T = \sqrt{E_R^2 + (E_L - E_C)^2}$ (4) $E_T = E_R/\cos\theta\ (PF)$ (5) $E_T = E_L - E_C/\sin\theta$ (6)

The Apparent Power (VA) {Series-RLC}

$P_{VA} = E_T \times I$ (1) $P_{VA} = I^2 \times Z$ (2) $P_{VA} = E_T^2/Z$ (3)
$P_{VA} = \sqrt{P_W^2 + P_{VARs}^2}$ (4) $P_{VA} = P_W/PF\ (\cos\theta)$ (5) $P_{VA} = P_{VARs}/\sin\theta$ (6)

The True Power (W) {Series-RLC}

$P_W = I^2 \times R$ (1) $P_W = E_R^2/R$ (2) $P_W = E_R \times I$ (3)
$P_W = \sqrt{P_{VA}^2 - P_{VARs}^2}$ (4) $P_W = P_{VA} \times PF\ (\cos\theta)$ (5)
$P_W = P_{VARs}/\tan\theta$ (6)

The Reactance Powers (VARs) {Series-RLC}

$P_{VARsL} = E_L \times I$ (1) $P_{VARsL} = I^2 \times X_L$ (2) $P_{VARsL} = E_L^2/X_L$ (3)
$P_{VARsL} = P_{VARs} + P_{VARsC}$ (4)

$P_{VARsC} = E_C \times I$ (1) $P_{VARsC} = I^2 \times X_C$ (2) $P_{VARsC} = E_C^2/X_C$ (3)
$P_{VARsC} = P_{VARsL} - P_{VARs}$ (4)

The Total Reactance Power (VARs) {Series-RLC}

$P_{VARs} = \sqrt{P_{VA}^2 - P_W^2}$ (1) $P_{VARs} = P_{VA} \times \sin\theta$ (2) $P_{VARs} = P_W \times \tan\theta$ (3)

$P_{VARs} = I^2 \times (X_L - X_C)$ (4) $P_{VARs} = (E_L - E_C)^2/(X_L - X_C)$ (5)

$P_{VARs} = P_{VARsL} - P_{VARsC}$ (6)

The Power Factor (PF) {Series-RLC}

$PF = R/Z\ (\cos\theta)$ (1) $PF = E_R/E_T\ (\cos\theta)$ (2) $PF = P_W/P_{VA}\ (\cos\theta)$ (3)

The Phase Angle (θ) {Series-RLC}

$\theta = \cos^{-1}\frac{E_R}{E_T}$ (PF) (1) $\theta = \sin^{-1}\frac{E_L - E_C}{E_T}$ (2) $\theta = \tan^{-1}\frac{E_L - E_C}{E_R}$ (3)

$\theta = \cos^{-1}\frac{R}{Z}$ (PF) (4) $\theta = \sin^{-1}\frac{X_L - X_C}{Z}$ (5) $\theta = \tan^{-1}\frac{X_L - X_C}{R}$ (6)

$\theta = \cos^{-1}\frac{P_W}{P_{VA}}$ (PF) (7) $\theta = \sin^{-1}\frac{P_{VARs}}{P_{VA}}$ (8) $\theta = \tan^{-1}\frac{P_{VARs}}{P_W}$ (9)

Parallel-RLC

Consider the parallel RLC circuit in figure 32. Just as with other circuits, we'll begin by determining needed quantities to get started.

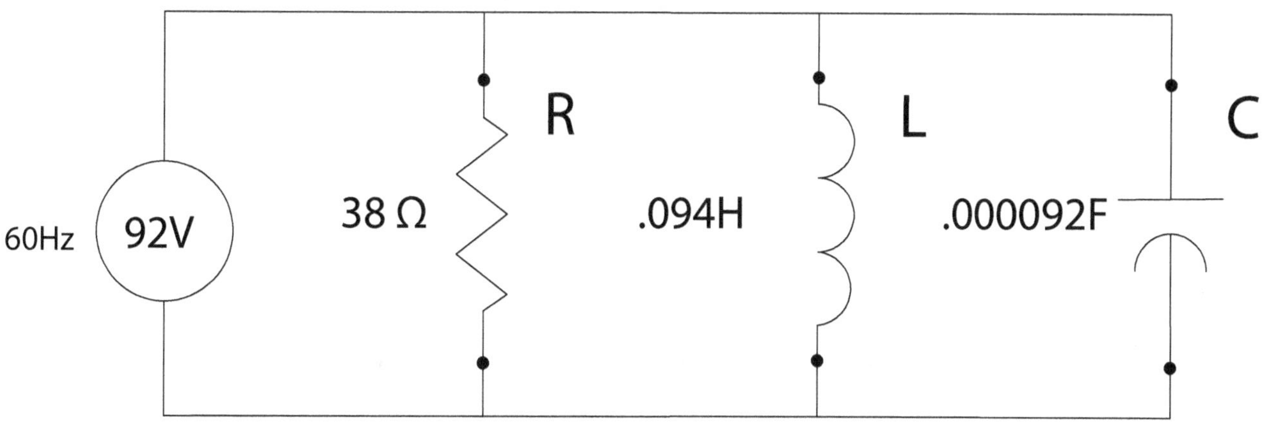

Fig. 32. Parallel-RLC circuit.

Inductance Reactance (X_L) and Capacitance Reactance (X_C) {Parallel-RLC}

Based on the given inductor, capacitor and operating frequency the inductance reactance is,

$$X_L = 2 \text{ x } 3.14 \text{ x } 60\text{Hz x } .094\text{H} = 35.42\Omega$$

and

the capacitance reactance is

$$X_C = \frac{1}{2 \text{ x } 3.14 \text{ x } 60\text{Hz x } .000092\text{F}} = 28.85\Omega$$

Impedance (Z) {Parallel-RLC}

In order to calculate the impedance for this circuit the vector related formula for a parallel-RLC circuit must be used; that is, $Z = 1/\sqrt{(1/R)^2 + (1/X_L - 1/X_C)^2}$, where X_L is larger than X_C. However, when substituted into the formula as a reciprocal regardless of order the results will always yield a positive value. Therefore,

$$\begin{aligned} Z &= \frac{1}{\sqrt{(1/38)^2 + (1/35.42 - 1/28.85)^2}} \\ &= \frac{1}{\sqrt{(1/38)^2 + (.0282326 - .0346620)^2}} \\ &= \frac{1}{\sqrt{.000693 + (-.0064294)^2}} \\ &= \frac{1}{\sqrt{.000693 + .0000413}} \\ &= \frac{1}{\sqrt{.0007343}} \\ &= \frac{1}{.0270980} \\ &= 36.9\Omega \end{aligned}$$

Now that the circuit's impedance has been determined, other vital quantities can be determined.

Total Current (I_T) {Parallel-RLC}

Since E is shown to be 92V, the total current can also be calculated using the formula $I_T = E/Z$.

$$\begin{aligned} I_T &= 92\text{V}/36.9\Omega \\ &= 2.49\text{A} \end{aligned}$$

Individual Current (I_R, I_L and I_C) {Parallel-RLC}

Because this is a parallel circuit, only one source of voltage and multiple paths of current will exist. Since the ohmic values of R, X_L, and X_C are not the same each component will experience different amounts of current flow; that is, I_R, I_L, and I_C. Therefore,

$I_R = E/R$
$= 92V/38\Omega$
$= 2.42A$

$I_L = E/X_L$
$= 92V/35.42\Omega$
$= 2.6A$

$I_C = E/X_C$
$= 92V/28.85\Omega$
$= 3.19A$

Again, it should be clear that the concept of adding the three current quantities together will not work for this type circuit because the voltage across the inductor and capacitor is not in phase with the current flowing through the inductor and about the capacitor. In reference to figure 30, which displays each component's voltage drop (E_R,E_L,E_C) with respect to the series circuit's current I, figure 33 is a illustration of each component's current (I_R,I_L,I_C) with respect to the parallel circuit's voltage source E.

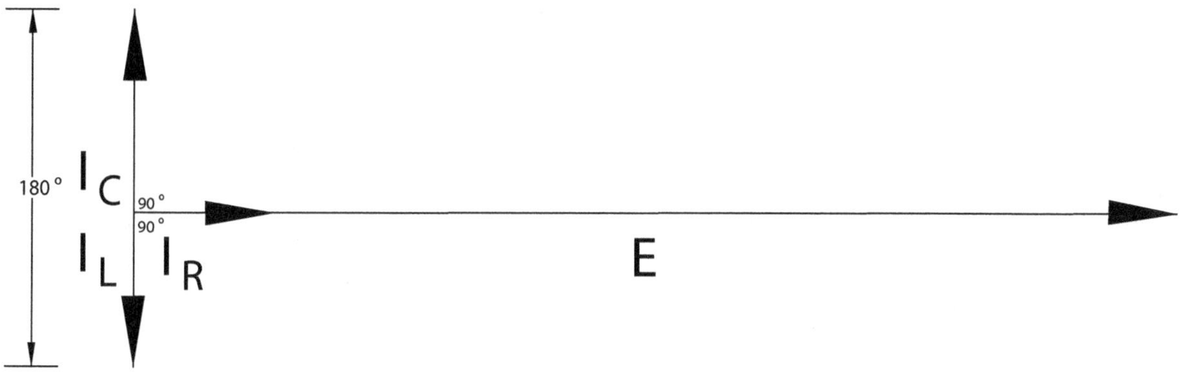

Fig. 33. I_{RLC}—E vector relationships.

As you can see, the vector representation reflects the current I_R through the resistor R being in phase with the circuit's voltage E. Also, the current I_C about the capacitor C leads the voltage E by 90°, and the current I_L through the inductor L lags the voltage E by 90°. Take notice that the unequal magnitudes of vectors I_C and I_L in figure 34(a), although close, represent the current about I_C being of a larger magnitude than the current through I_L. Because the vectors I_C and I_L are 180° out of phase, they will undoubtedly alter each other, thus causing the larger vector to be reduced while eliminating the smaller vector.

To evaluate the given parallel RLC circuit vectorially, only two vectors can be analyzed at a time. Just as with figures 31(a), (b), and (c), the same approach has been taken in figures 34(a), (b), and (c). Because of their out-of-phase difference, the vector currents can simply be resolved by means of subtraction ($I_C - I_L$) and converted into a common vector I_{CL}, where vector I_C is the larger of the two vectors.

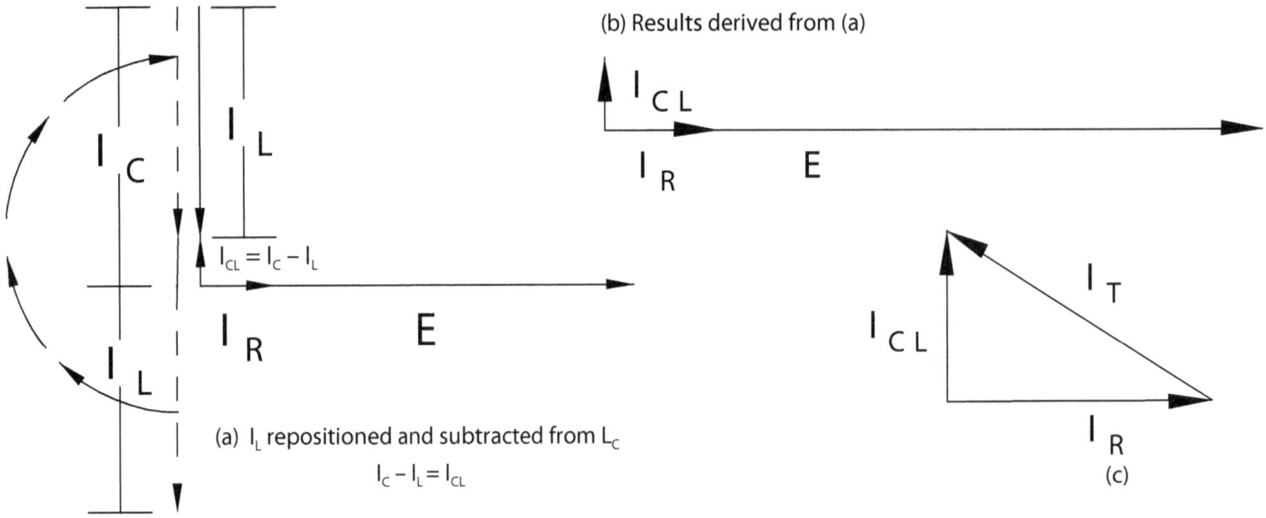

Fig. 34. Common vector $I_{(CL)}$.

As with all situations, the common vector will always be positioned in the same direction as the larger vector. If I_L was larger than I_C, the common vector would then be labeled I_{LC} and positioned in the opposite direction. As a result, this conclusion, also [as shown in figure 34(c)], leads to a similar vector diagram and right triangle as displayed in figure 15(a). From here we can then began to evaluate the given parallel RLC circuit.

Because the total current must be determined based on all three components and must be calculated vectorially, we'll start with the vector formula, $I_T = \sqrt{I_R^2 + I_L^2}$. Since this formula only contains a resistance and inductance component, it must be extended to include the third component I_C. Using the derived common vector I_{CL}, the formula can now be inclusive of the third component, where $I_T = \sqrt{I_R^2 + (I_C - I_L)^2}$ or $I_T = \sqrt{I_R^2 + (I_{CL})^2}$. By applying this formula, the circuit's total current can now be calculated. Therefore,

$$\begin{aligned} I_T &= \sqrt{(2.42)^2 + (3.19 - 2.6)^2} \\ &= \sqrt{5.856 + (.59)^2} \\ &= \sqrt{5.856 + .348} \\ &= \sqrt{6.204} \\ &= 2.49\text{A} \end{aligned}$$

Did you notice in the formula that I_{CL} is equal to .59A?

Even when three components are included, the calculated total current should still be the same (or approximately the same) as previously calculated; that is, $I_T = 92\text{V}/36.9\Omega = 2.49\text{A}$.

The Apparent Power (VA) {Parallel-RLC}

Using that which was given as the circuit's voltage (E), the calculated current (I), and impedance (Z), the circuit's apparent power can be determined by use of the formulas $P_{VA} = E \times I_T$, $P_{VA} = I_T^2 \times Z$ and $P_{VA} = E^2/Z$.

$$P_{VA} = 92V \times 2.49A = 229.08VA$$

$$P_{VA} = (2.49A)^2 \times 36.9\Omega = 6.20A \times 36.9\Omega = 228.78VA$$

$$P_{VA} = (92V)^2 / 36.9\Omega = 8464V / 36.9\Omega = 229.38VA$$

Because the results of the first and last calculated quantities (229.08VA and 229.38VA) included a given quantity (92V) in each formula the results in each case would prove to be more accurate than the results of the second calculated quantity (228.78VA) because this quantity was determined based on the use of two rounded-off calculated quantities (2.49A and 36.9Ω).

The True Power (W) {Parallel-RLC}

The true power is calculated using the formulas $P_W = I_R^2 \times R$ and $P_W = E^2/R$, where I_R is the current flowing through R.

$$P_W = I_R^2 \times R = (2.42A)^2 \times 38\Omega = 222.54W$$

and

$$P_W = E^2/R = (92V)^2/38\Omega = 222.74W$$

Because the voltage and resistance quantities were given opposed to being calculated, the formula, which includes both quantities (92V and 38Ω), would prove to provide the most accurate answer although both formulas produced approximate results.

The Reactance Power (VARs) {Parallel-RLC}

Considering the fact that this circuit consists of two combinations of reactance components (X_L and I_L and X_C and I_C) the previously used formulas $P_{VARs} = I_{(L\ or\ C)}^2 \times X_{(L\ or\ C)}$ and $P_{VARs} = E^2/X_{(L\ or\ C)}$ cannot be used as is.

Since this is a parallel circuit, the difference between X_L and X_C ($X_L - X_C$) can be neither derived nor evaluated similarly to that of a series circuit. In other words, the common reactance (X_{LC}) between the two reactances cannot be determined as previously done. By modifying one of the parallel resistance formulas we all are familiar with, and that being the product of the sum method, the common reactance can be determined. In this case, since X_L is larger than X_C, the denominator of this formula is modified to reflect the difference between the two reactances. As a result of both reactance currents I_C and I_L (figure 33) being 180° out of phase, so are both reactance (X_L and X_C). The common reactance is a reflection of the 180° difference or opposites. With this modified method, the common reactance can be determined by the following formula:

$$X_{LC} = \frac{X_L \times X_C}{X_L - X_C}$$

As previously stated and illustrated in figure 33, the vectors I_C and I_L can be subtracted from each other to derive the common vector I_{CL} without modifications.

Using the modified formula, the common reactance results in

$$X_{LC} = \frac{35.42\Omega \times 28.85\Omega}{35.42\Omega - 28.85\Omega}$$

$$= \frac{1021.87\Omega}{6.57\Omega}$$

$$= 155.54\Omega$$

and the common vector $I_{CL} = 3.19A - 2.6A = .59A$. With both common quantities in place the previously used reactance power formulas with modifications can now be used, where

$$P_{VARsLC} = I_{CL}^{\ 2} \times X_{LC} = (.59A)^2 \times 155.54\Omega = 54.14VARs$$

and

$$P_{VARsLC} = E^2/X_{LC} = (92V)^2/155.54\Omega = 54.42VARs$$

The consistency of these results will be realized when an alternate approach is applied.

The Power Factor (PF) {Parallel-RLC}

When calculating the power factor in a parallel-RLC circuit, the formula PF = Z/R is used.

Using the given and calculated quantities along with the above and the previously given power factor formula,

$$PF = \frac{222.74W}{229.08VA} = .9723\ (97.23\%)$$

and

$$PF = \frac{36.9\Omega}{38\Omega} = .9711\ (97.11\%)$$

As you can see, when calculated, both sets of quantities yield approximate results.

The Phase Angle (θ) {Parallel-RLC}

In a parallel-RLC circuit, formulas for the phase angle can be derived using any one of the four trigonometric formulas,

$$\tan\theta = \frac{I_C - I_L}{I_R} \text{ or } \frac{R}{X_{LC}} \quad \text{and} \quad \cos\theta = \frac{Z}{R} \text{ or } \frac{I_R{}^*}{I_T}$$

* See figure 34(c).

From the four formulas the following derived formulas along with a previously derived formula will be used to determine the phase angle,

$$\theta = \tan^{-1}\frac{I_C - I_L}{I_R} \quad \text{or} \quad \frac{R}{X_{LC}} \quad \text{and} \quad \theta = \cos^{-1}\frac{Z}{R} \quad \text{or} \quad \frac{I_R}{I_T}$$

$$\theta = \cos^{-1} \text{PF (previous)}$$

Based on the derived formulas, the circuit's phase angle is

$$\theta = \tan^{-1}\frac{.59\text{A}}{2.42\text{A}} = \tan^{-1} .2438 = 13.70° \quad \text{or} \quad \frac{38\Omega}{155.54\Omega} = .2443 = 13.73° \quad \text{and} \quad \theta = \cos^{-1}\frac{36.9\Omega}{38\Omega} = \cos^{-1} .9710 = 13.83° \quad \text{or} \quad \frac{2.42\text{A}}{2.49\text{A}} = .9719 = 13.61°$$

$$\theta = \cos^{-1} .9722 \text{ (previous)} = 13.54°$$

Again, using all five formulas, the results are all approximate.

Power {Parallel-RLC}

Previously the three power quantities were calculated using given or calculated values. The quantities will now be determined using an alternate approach. By using the formula $\cos\theta = P_W/P_{VA}$, the true power can now be determined by use of the derived formula, $P_W = P_{VA} \times \cos\theta$, where

$$\begin{aligned} P_W &= P_{VA} \text{ x } \cos\theta \\ &= 229.08\text{VA x } \cos 13.73° \\ &= 229.08\text{VA x } .9714 \\ &= 222.53\text{W} \end{aligned}$$

The result of this calculation is aligned with the level of accuracy discussed earlier for the true power quantity.

By using the formula $\sin\theta = P_{VARs}/P_{VA}$, the reactance power can now be determined by use of the derived formula, $P_{VARs} = P_{VA} \times \sin\theta$, where

$$\begin{aligned} P_{VARs} &= P_{VA} \text{ x } \sin\theta \\ &= 229.08\text{VA x } \sin 13.73° \\ &= 229.08\text{VA x } .2373 \\ &= 54.36\text{VARs} \end{aligned}$$

As you can see, the results are approximately the same as earlier calculated. These results can be substituted into the Pythagorean theorem, and the apparent power can be verified.

$$P_{VA} = \sqrt{(222.53)^2 + (54.36)^2}$$
$$= \sqrt{49{,}519.6 + 2955}$$
$$= \sqrt{52{,}474.6}$$
$$= 229.07\text{VA}$$

The following display is an overall list of formulas that can be used for solving Parallel-RLC circuits.

Related Parallel Formulas (Resistance–Inductance–Capacitance Circuits—Overall)

[Only one source of voltage (E) – therefore, $E = E_R = E_L = E_C$.]

(IMPORTANT: Regardless of how shown in relative formulas, where the differences between L and C components are required, the smaller component is always subtracted from the larger component.)

Resistance (R) {Parallel-RLC}

$R = E/I_R$ (1) $\quad R = E^2/P_W$ (2) $\quad R = P_W/I_R^{\,2}$ (3)

$R = Z/\cos\theta$ (PF) (4) $\quad R = (X_C \times X_L / X_C - X_L) \times \tan\theta$ (5)

$$R = \frac{1}{\sqrt{(1/Z)^2 - (1/X_L - 1/X_C)^2}} \quad (6)$$

Inductance (L) and Capacitance (C) {Parallel-RLC}

$L = X_L/2\pi f \quad C = 1/2\pi f X_C$

Reactances (X_C, X_L and $X_C \times X_L / X_C - X_L$) {Parallel-RLC}

$X_C = 1/2\pi fC$ (1) $\quad X_C = E/I_C$ (2)

$X_C = E^2/P_{VARsC}$ (3) $\quad X_C = P_{VARsC}/I_C^{\,2}$ (4)

$X_L = 2\pi fL$ (1) $\quad X_L = E/I_L$ (2)

$X_L = E^2/P_{VARsL}$ (3) $\quad X_L = P_{VARsL}/I_L^{\,2}$ (4)

$(X_C \times X_L / X_C - X_L) = R/\tan\theta$ (1) $\quad (X_C \times X_L / X_C - X_L) = Z/\sin\theta$ (2)

$$(X_C \times X_L / X_C - X_L) = \frac{1}{\sqrt{(1/Z)^2 - (1/R)^2}} \quad (3)$$

Impedance (Z) {Parallel-RLC}

$$Z = \frac{1}{\sqrt{(1/R)^2 + (1/X_C - X_L)^2}} \quad (1)$$

$Z = E/I_T$ (2)

$Z = E^2/P_{VA}$ (3) $\quad Z = P_{VA}/I_T^{\,2}$ (4) $\quad Z = R \times \cos\theta$ (PF) (5)

$Z = (X_C \times X_L / X_C - X_L) \times \sin\theta$ (6)

Individual Currents (I_R, I_C, I_L and I- I_L) {Parallel-RLC}

$I_R = E/R$ (1) $I_R = P_W/E$ (2) $I_R = \sqrt{P_W/R}$ (3)

$I_R = \sqrt{I_T^2 - (I_C - I_L)^2}$ (4) $I_R = I_T \times \cos\theta$ (PF) (5) $I_R = I_C - I_L/\tan\theta$ (6)

$I_C = E/X_C$ (1) $I_C = P_{VARsC}/E$ (2) $I_C = \sqrt{P_{VARsC}/X_C}$ (3)

$I_L = E/X_L$ (1) $I_L = P_{VARsL}/E$ (2) $I_L = \sqrt{P_{VARsL}/X_L}$ (3)

$I_C - I_L = \sqrt{I_T^2 - I_R^2}$ (1) $I_C - I_L = I_T \times \sin\theta$ (2) $I_C - I_L = I_R \times \tan\theta$ (3)

Total Current (I_T) {Parallel-RLC}

$I_T = E/Z$ (1) $I_T = P_{VA}/E$ (2) $I_T = \sqrt{P_{VA}/Z}$ (3)

$I_T = \sqrt{I_R^2 + (I_C - I_L)^2}$ (4) $I_T = I_R/\cos\theta$ (PF) (5) $I_T = I_C - I_L/\sin\theta$ (6)

Voltage (E) {Parallel-RLC}

$E = I_R \times R$ (1) $E = P_W/I_R$ (2) $E = I_C \times X_C$ (3)
$E = I_L \times X_L$ (4) $E = P_{VARsC}/I_C$ (5) $E = P_{VARsL}/I_L$ (6)

$E = \sqrt{P_W \times R}$ (7) $E = \sqrt{P_{VARsC} \times X_C}$ (8) $E = \sqrt{P_{VARsL} \times X_L}$ (9)

$E = I_T \times Z$ (10) $E = P_{VA}/I_T$ (11) $E = \sqrt{P_{VA} \times Z}$ (12)

The Apparent Power (VA) {Parallel-RLC}

$P_{VA} = E \times I_T$ (1) $P_{VA} = I_T^2 \times Z$ (2) $P_{VA} = E^2/Z$ (3)

$P_{VA} = P_W/\cos\theta$ (PF) (4) $P_{VA} = P_{VARs}/\sin\theta$ (5) $P_{VA} = \sqrt{P_W^2 + P_{VARs}^2}$ (6)

The True Power (W) {Parallel-RLC}

$P_W = E \times I_R$ (1) $P_W = E^2/R$ (2) $P_W = I_R^2 \times R$ (3)

$P_W = P_{VA} \times \cos\theta$ (PF) (4) $P_W = P_{VARs}/\tan\theta$ (5) $P_W = \sqrt{P_{VA}^2 - P_{VARs}^2}$ (6)

The Reactance Power (VARs) {Parallel-RLC}

$P_{VARsC} = E \times I_C$ (1) $P_{VARsC} = I_C^2 \times X_C$ (2) $P_{VARsC} = E^2/X_C$ (3)
$P_{VARsC} = P_{VARs} + P_{VARsL}$ (4)

$P_{VARsL} = E \times I_L$ (1) $P_{VARsL} = I_L^2 \times X_L$ (2) $P_{VARsL} = E^2/X_L$ (3)
$P_{VARsL} = P_{VARsC} - P_{VARs}$ (4)

The Total Reactance Power (VARs) {Parallel-RLC}

$P_{VARs} = \sqrt{P_{VA}^2 - P_W^2}$ (1) $P_{VARs} = P_{VA} \times \sin\theta$ (2)
$P_{VARs} = P_W \times \tan\theta$ (3) $P_{VARs} = P_{VARsC} - P_{VARsL}$ (4)

The Power Factor (PF) {Parallel-RLC}

PF $(\cos\theta) = Z/R$ (1) PF $(\cos\theta) = I_R/I_T$ (2) PF $(\cos\theta) = P_W/P_{VA}$ (3)

The Phase Angle (θ) {Parallel-RLC}

$$\theta = \cos^{-1}\frac{I_R}{I_T}\ (PF)\ (1) \qquad \theta = \sin^{-1}\frac{I_C - I_L}{I_T}\ (2) \qquad \theta = \tan^{-1}\frac{I_C - I_L}{I_R}\ (3)$$

$$\theta = \cos^{-1}\frac{Z}{R}\ (PF)\ (4)$$

$$\theta = \sin^{-1}\frac{Z}{X_C \times X_L / X_C - X_L}\ (5) \qquad \theta = \tan^{-1}\frac{R}{X_C \times X_L / X_C - X_L}\ (6)$$

$$\theta = \cos^{-1}\frac{P_W}{P_{VA}}\ (PF)\ (7) \qquad \theta = \sin^{-1}\frac{P_{VARs}}{P_{VA}}\ (8) \qquad \theta = \tan^{-1}\frac{P_{VARs}}{P_W}\ (9)$$

Resistance (R)–Inductance (L)–Capacitance (C) Circuits

Multiple Components

Earlier we dealt with multiple component circuits that consisted of either a combination of RL or RC components. Combined with these type circuits and the RLC circuits just discussed, we are now ready to use such knowledge to focus on RLC circuits consisting of multiple components.

Series-RLC Multiple Components

Let's begin by observing the multiple component RLC series circuit in figure 35.

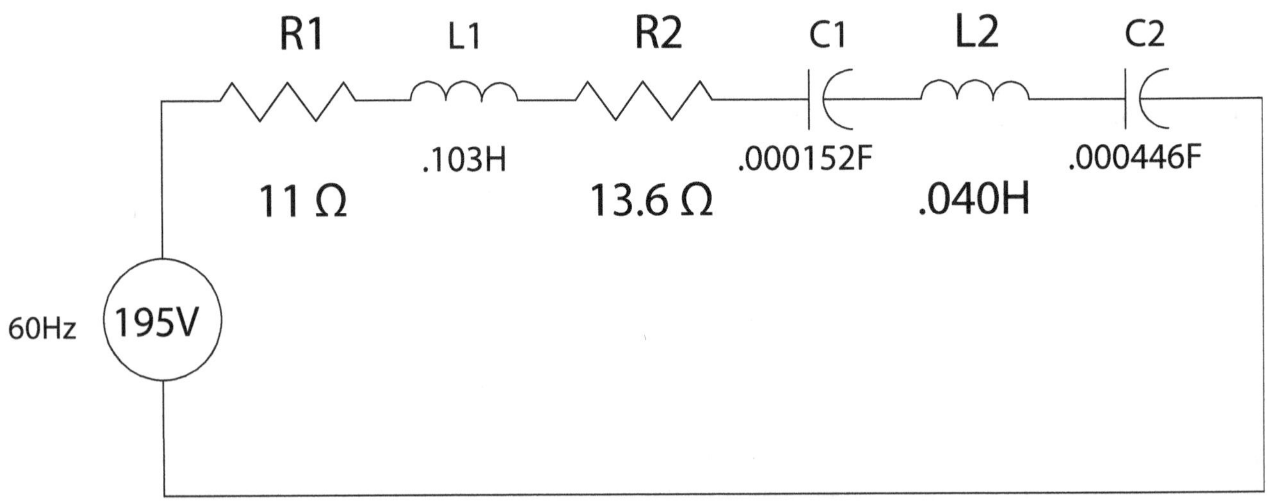

Fig. 35. Series-RLC circuit (multiple components).

Resistance (R), Inductance Reactance (X_L) and Capacitance Reactance (X_C) {Series-RLC (Multiple Components)}

Our first approach in analyzing this circuit is to determine the individual reactance of each component followed by the total resistance and reactance of all like components. Based on the given inductors, capacitors, and operating frequency, the inductance reactance of each inductor is

$$X_{L1} = 2 \times 3.14 \times 60Hz \times .103H = 38.81\Omega$$

$$X_{L2} = 2 \times 3.14 \times 60Hz \times .040H = 15.07\Omega$$

and

the capacitance reactance of each capacitor is

$$X_{C1} = \frac{1}{2 \times 3.14 \times 60Hz \times .000152F} = 17.46\Omega$$

$$X_{C2} = \frac{1}{2 \times 3.14 \times 60Hz \times .000446F} = 5.95\Omega$$

The total of all like components is

$$R_T = 11\Omega + 13.6\Omega = 24.6\Omega$$
$$X_{LT} = 38.81\Omega + 15.07\Omega = 53.88\Omega$$
$$X_{CT} = 17.46\Omega + 5.95\Omega = 23.41\Omega$$

Impedance (Z) {Series-RLC (Multiple Components)}

In order to calculate the impedance for this circuit the vector related formula for a Series-RLC circuit must be used where the totals of each combination of like components are used; that is, $Z = \sqrt{R_T^{\,2} + (X_{LT} - X_{CT})^2}$ where X_{LT} is larger than X_{CT}. Therefore,

$$Z = \sqrt{(24.6)^2 + (53.88 - 23.41)^2}$$
$$= \sqrt{605.16 + (30.47)^2}$$
$$= \sqrt{605.16 + 928.42}$$
$$= \sqrt{1533.58}$$
$$= 39.16\Omega$$

Just as other circuits with the impedance being determined other quantities can be determined also.

Current (I) {Series-RLC (Multiple Components)}

Because this is a series circuit, only one source of current I will exist. Since E_T is shown to be 195V and Z being calculated at 39.16Ω,

$$I = 195V/39.16\Omega$$
$$= 4.98A$$

Voltage Drops (E_R, E_L and E_C) {Series-RLC (Multiple Components)}
Because this is a series circuit, each component will share the same current but will vary in voltage drop based solely on the component's ohmic value.

$$E_{R1} = I \times R_1 = 4.98A \times \underline{11\Omega} = 54.78V$$

$$E_{L1} = I \times X_{L1} = 4.98A \times \underline{38.81\Omega} = 193.27V$$

$$E_{R2} = I \times R_2 = 4.98A \times \underline{13.6\Omega} = 67.73V$$

$$E_{C1} = I \times X_{C1} = 4.98A \times \underline{17.46\Omega} = 86.95V$$

$$E_{L2} = I \times X_{L2} = 4.98A \times \underline{15.07\Omega} = 75.05V$$

$$E_{C2} = I \times X_{C2} = 4.98A \times \underline{5.95\Omega} = 29.63V$$

Total Voltage (E_T) {Series-RLC (Multiple Components)}
Again, to determine the circuit's total voltage the voltage drop across each like component must be totaled together first. Once this is done the vector formula $E_T = \sqrt{E_{RT}^2 + (E_{LT} - E_{CT})^2}$ can then be applied (see figure 31). Therefore,

$$E_{RT} = 54.78V + 67.73V = 122.51V$$

$$E_{LT} = 193.27V + 75.05V = 268.32V$$

$$E_{CT} = 86.95V + 29.63V = 116.58V$$

results in

$$E_T = \sqrt{(122.51)^2 + (268.32 - 116.58)^2}$$
$$= \sqrt{15{,}008.7 + (151.74)^2}$$
$$= \sqrt{15{,}008.7 + 23{,}025.03}$$
$$= \sqrt{38{,}033.73}$$
$$= 195.02V$$

Just as other series resistance–reactance circuits the same holds true for this series multiple component RLC circuit. When calculated vectorially, the calculated total voltage turned out to be approximately the same as the given total voltage (195V).

Did you notice in the formula that $E_{LC(T)}$ is equal to 151.74V?

The Apparent Power (VA) {Series-RLC (Multiple Components)}
Using that which was given as the circuit's total voltage (E) and the calculated current (I), the circuit's apparent power can be determined by use of the formula $P_{VA} = E_T \times I$ in addition to the formulas $P_{VA} = I^2 \times Z$ and $P_{VA} = E_T^2/Z$.

$$P_{VA} = 195V \times 4.98A = 971.1VA$$

$$P_{VA} = (4.98A)^2 \times 39.16\Omega = 24.8A \times 39.16\Omega = 971.17VA$$

$$P_{VA} = (195V)^2/39.16\Omega = 38{,}025V/39.16\Omega = 971.02VA$$

The True Power (W) {Series-RLC (Multiple Components)}

Because each resistance component will contribute to the circuit's total true power consumption, the true power consumed by each component must be determined before the total true power is realized. To simplify these calculations, only one formula will be used; that is, $P_W = I^2 \times R$ to calculate the true power consumption for each component. Therefore,

$$P_{W1} = I^2 \times R_1 = (4.98A)^2 \times 11\Omega = 272.8W$$

$$P_{W2} = I^2 \times R_2 = (4.98A)^2 \times 13.6\Omega = 337.29W$$

results in

$$P_{W(T)} = 272.8W + 337.29W = 610.09W$$

To verify the results of this calculation, let's use the formula $P_{W(T)} = I^2 \times R_T$, where the circuit's true power is determined based on the circuit's total resistance (R_T).

$$P_{W(T)} = (4.98A)^2 \times 24.6\Omega = 610.09W$$

Results verified!

The Reactance Power (VARs) {Series-RLC (Multiple Components)}

To determine the circuit's total reactance power the reactance power about each reactance component must first be determined. To realize these reactance quantities, the same formula will be used but with modifications; that is, $P_{VARs(L\ or\ C)} = E_{(L\ or\ C)}{}^2/X_{(L\ or\ C)}$. Therefore,

$$P_{VARsL1} = E_{L1}{}^2/X_{L1} = (193.27V)^2/38.81\Omega = 962.47VARs$$

$$P_{VARsL2} = E_{L2}{}^2/X_{L2} = (75.05V)^2/15.07\Omega = 373.76VARs$$

$$P_{VARsC1} = E_{C1}{}^2/X_{C1} = (86.95V)^2/17.46\Omega = 433.01VARs$$

$$P_{VARsC2} = E_{C2}{}^2/X_{C2} = (29.63V)^2/5.95\Omega = 147.55VARs$$

results in

$$P_{VARsL(T)} = 962.47VARs + 373.76VARs = 1336.23VARs$$

$$P_{VARsC(T)} = 433.01VARs + 147.55VARs = 580.56VARs$$

Just as the voltage drops across the inductors and capacitors differ by 180° and the common voltage drop between them results in $E_{LC} = E_L - E_C$, where E_L is larger than E_C, the same procedure can be followed to determine the circuit's reactance power. Because the reactance of each component is a product of the voltage drops across each, the total reactance power of each group of reactance will also differ by 180°. As a result, the total reactance power for this circuit can be determined as follows:

$$P_{VARs(T)} = P_{VARsL(T)} - P_{VARsC(T)}$$

where $P_{VARsL(T)}$ is larger than $P_{VARsC(T)}$. Therefore, this circuit's total reactance power is

$$\begin{aligned} P_{VARs(T)} &= 1336.23\text{VARs} - 580.56\text{VARs} \\ &= 755.67\text{VARs} \end{aligned}$$

If the total reactance power gathered from this procedure is incorrect it will be identified when an alternate approach is used to determine comparable results.

The Power Factor (PF) {Series-RLC (Multiple Components)}

To calculate the power factor for this RLC multiple component series circuit, the formulas $PF = P_{W(T)}/P_{VA}$ and $PF = R_T/Z$ are used. Therefore,

$$PF = \frac{610.09\text{W}}{971.1\text{VA}} \qquad \text{and} \qquad PF = \frac{24.6\Omega}{39.16\Omega}$$

$$= .6282\ (62.82\%) \qquad\qquad = .6282\ (62.82\%)$$

The Phase Angle (θ) {Series-RLC (Multiple Components)}

In a Series-RLC circuit, formulas for the phase angle can be derived using one of the three trigonometric formulas,

$$\tan\theta = \frac{E_{LT} - E_{CT}}{E_{RT}} \quad \text{or} \quad \frac{X_{LT} - X_{CT}}{R_T} \qquad \text{and} \qquad \cos\theta = \frac{R_T}{Z}$$

From the three formulas, the following derived formulas—along with the previously used formula—will be used to determine the phase angle:

$$\theta = \tan^{-1}\frac{E_{LT} - E_{CT}}{E_{RT}} \quad \text{or} \quad \frac{X_{LT} - X_{CT}}{R_T} \qquad \text{and} \qquad \theta = \cos^{-1}\frac{R_T}{Z}$$

$$\theta = \cos^{-1} PF \text{ (previous)}$$

Based on the derived formulas, the circuit's phase angle is

$$\theta = \tan^{-1} \frac{151.74V}{122.51V} \quad \text{or} \quad \frac{30.47\Omega}{24.6\Omega} \qquad \text{and} \qquad \theta = \cos^{-1} \frac{24.6\Omega}{39.16\Omega}$$

$$= \tan^{-1} 1.239 \qquad\qquad = \cos^{-1} .6282$$

$$= 51.09^\circ \qquad\qquad = 51.08^\circ$$

$$\theta = \cos^{-1} .6282 \text{ (previous)}$$
$$= 51.08^\circ$$

Using all four formulas, the results are all approximately the same.

Power {Series-RLC (Multiple Components)}

Previously, the three power quantities were calculated using given or calculated values. Again, the quantities will be determined using an alternate approach. By using the formula $\cos\theta = P_{W(T)}/P_{VA}$, the true power can now be determined by use of the derived formula $P_{W(T)} = P_{VA} \text{ x } \cos\theta$, where

$$P_{W(T)} = P_{VA} \text{ x } \cos\theta$$
$$= 971.1VA \text{ x } \cos 51.08^\circ$$
$$= 971.1VA \text{ x } .6282$$
$$= 610.05W$$

By using the formula $\sin\theta = P_{VARs(T)}/P_{VA}$, the reactance power can now be determined by use of the derived formula $P_{VARs(T)} = P_{VA} \text{ x } \sin\theta$, where

$$P_{VARs(T)} = P_{VA} \text{ x } \sin\theta$$
$$= 971.1VA \text{ x } \sin 51.08^\circ$$
$$= 971.1VA \text{ x } .778$$
$$= 755.52VARs$$

As you can see, the results are approximately the same as earlier calculated. By using these results, they can be substituted into the Pythagorean theorem, and the apparent power can be verified.

$$P_{VA} = \sqrt{(610.05)^2 + (755.52)^2}$$
$$= \sqrt{372{,}161.0 + 570{,}810.5}$$
$$= \sqrt{942{,}971.5}$$
$$= 971.07VA$$

The following display is an overall list of formulas that can be used for solving Series-RLC circuits—Multiple Components.

Related Series Formulas
(Resistance–Inductance–Capacitance Multiple Component Circuits—Overall)

[Only one source of current (I); therefore, $I = I_R = I_L = I_C$.]

(IMPORTANT: Regardless of how shown in relative formulas, where the differences between L and C components are required, the smaller component is always subtracted from the larger component.)

Resistance (R) {Series-RLC (Multiple Components)}

$R = E_R / I$ (1) $R = E_R^2 / P_W$ (2) $R = P_W / I^2$ (3)

$R_T = R_1 + R_2 + \ldots\ldots$ (1) $R_T = E_{R(T)} / I$ (2) $R_T = E_{R(T)}^2 / P_{W(T)}$ (3)

$R_T = P_{W(T)} / I^2$ (4) $R_T = \sqrt{Z^2 - (X_{LT} - X_{CT})^2}$ (5) $R_T = Z \times \cos\theta$ (PF) (6)

$R_T = X_{LT} - X_{CT} / \tan\theta$ (7)

Inductance (L) and Capacitance (C) {Series-RLC (Multiple Components)}

$L = X_L / 2\pi f$ $C = 1/2\pi f X_C$

$L_T = L_1 + L_2 + \ldots\ldots$ (1) $C_T = \frac{C}{N}$ (1) (see page 67) $C_T = \frac{C_1 \times C_2}{C_1 + C_2}$ (2)

$L_T = X_{LT} / 2\pi f$ (2) $C_T = \frac{1}{1/C_1 + 1/C_2 + \ldots\ldots}$ (3) $C_T = 1/2\pi f X_{CT}$ (4)

Reactances (X_L, X_C and X_L–X_C) {Series-RLC (Multiple Components)}

$X_L = 2\pi f L$ (1) $X_L = E_L / I$ (2)

$X_L = E_L^2 / P_{VARsL}$ (3) $X_L = P_{VARsL} / I^2$ (4)

$X_{LT} = X_{L1} + X_{L2} + \ldots.$ (1) $X_{LT} = 2\pi f L_{(T)}$ (2) $X_{LT} = E_{L(T)} / I$ (3)

$X_{LT} = E_{L(T)}^2 / P_{VARsL(T)}$ (4) $X_{LT} = P_{VARsL(T)} / I^2$ (5)

$X_C = 1/2\pi f C$ (1) $X_C = E_C / I$ (2)

$X_C = E_C^2 / P_{VARsC}$ (3) $X_C = P_{VARsC} / I^2$ (4)

$X_{CT} = X_{C1} + X_{C2} + \ldots.$ (1) $X_{CT} = 1/2\pi f C_{(T)}$ (2) $X_{CT} = E_{C(T)} / I$ (3)

$X_{CT} = E_C^2 / P_{VARsC(T)}$ (4) $X_C = P_{VARsC(T)} / I^2$ (5)

$X_{LT} - X_{CT} = \sqrt{Z^2 - R_T^2}$ (1) $X_{LT} - X_{CT} = E_{L(T)} - E_{C(T)} / I$ (2)

$X_{LT} - X_{CT} = Z \times \sin\theta$ (3) $X_{LT} - X_{CT} = R_T \times \tan\theta$ (3)

Impedance (Z) {Series-RLC (Multiple Components)}

$Z = \sqrt{R_T^2 + (X_{LT} - X_{CT})^2}$ (1) $Z = E_T / I$ (2) $Z = E_T^2 / P_{VA}$ (3)

$Z = P_{VA} / I^2$ (4) $Z = R_T / PF (\cos\theta)$ (5) $Z = (X_{LT} - X_{CT}) / \sin\theta$ (6)

Current (I) {Series-RLC (Multiple Components)}

$I = E_R/R$ (1) $I = P_W/E_R$ (2) $I = \sqrt{P_W/R}$ (3)

$I = E_{R(T)}/R_T$ (4) $I = P_W/E_{R(T)}$ (5) $I = \sqrt{P_W/R_T}$ (6)

$I = E_L/X_L$ (1) $I = P_{VARsL}/E_L$ (2) $I = \sqrt{P_{VARsL}/X_L}$ (3)

$I = E_{L(T)}/X_{LT}$ (4) $I = P_{VARsL(T)}/E_{L(T)}$ (5) $I = \sqrt{P_{VARsL(T)}/X_{LT}}$ (6)

$I = E_C/X_C$ (1) $I = P_{VARsC}/E_C$ (2) $I = \sqrt{P_{VARsC}/X_C}$ (3)

$I = E_{C(T)}/X_{CT}$ (4) $I = P_{VARsC(T)}/E_{C(T)}$ (5) $I = \sqrt{P_{VARsC(T)}/X_{CT}}$ (6)

$I = E_T/Z$ (1) $I = \sqrt{P_{VA}/Z}$ (2) $I = P_{VA}/E_T$ (3)

$I = E_{L(T)} - E_{C(T)}/X_{LT} - X_{CT}$ (4) $I = P_{VARs(T)}/E_{L(T)} - E_{C(T)}$ (5) $I = \sqrt{P_{VARs(T)}/X_{LT} - X_{CT}}$ (6)

Voltage Drops (E_R, E_L, E_C, and E_L-E_C) {Series-RLC (Multiple Components)}

$E_R = I \text{ x } R$ (1) $E_R = P_W/I$ (2) $E_R = \sqrt{P_W \text{ x } R}$ (3)

$E_{R(T)} = E_{R1} + E_{R2} + \ldots$ (1) $E_{R(T)} = I \text{ x } R_T$ (2) $E_{R(T)} = P_{W(T)}/I$ (3)

$E_{R(T)} = \sqrt{P_{W(T)} \text{ x } R_T}$ (4) $E_{R(T)} = E_T \text{ x PF } (\cos\theta)$ (5) $E_{R(T)} = \sqrt{E_T^2 - (E_{L(T)} - E_{C(T)})^2}$ (6)

$E_{R(T)} = (E_{L(T)} - E_{C(T)})/\tan\theta$ (7)

$E_L = I \text{ x } X_L$ (1) $E_L = P_{VARsL}/I$ (2) $E_L = \sqrt{P_{VARsL} \text{ x } X_L}$ (3)

$E_{L(T)} = I \text{ x } X_{LT}$ (1) $E_{L(T)} = P_{VARsL(T)}/I$ (2) $E_{L(T)} = \sqrt{P_{VARsL(T)} \text{ x } X_{LT}}$ (3)

$E_{L(T)} = E_{L1} + E_{L2} + \ldots$ (4)

$E_C = I \text{ x } X_C$ (1) $E_C = P_{VARsC}/I$ (2) $E_C = \sqrt{P_{VARsC} \text{ x } X_C}$ (3)

$E_{C(T)} = I \text{ x } X_{CT}$ (1) $E_{C(T)} = P_{VARsC(T)}/I$ (2) $E_{C(T)} = \sqrt{P_{VARsC(T)} \text{ x } X_{C(T)}}$ (3)

$E_{C(T)} = E_{C1} + E_{C2} + \ldots$ (4)

$E_{L(T)} - E_{C(T)} = I \text{ x } (X_{LT} - X_{CT})$ (1) $E_{L(T)} - E_{C(T)} = P_{VARs(T)}/I$ (2) $E_{L(T)} - E_{C(T)} = \sqrt{E_T^2 - E_{R(T)}^2}$ (3)

$E_{L(T)} - E_{C(T)} = E_T \text{ x } \sin\theta$ (4) $E_{L(T)} - E_{C(T)} = E_{R(T)} \text{ x } \tan\theta$ (5)

Total Voltage (E_T) {Series-RLC (Multiple Components)}

$E_T = P_{VA}/I$ (1) $E_T = \sqrt{P_{VA} \text{ x } Z}$ (2) $E_T = I \text{ x } Z$ (3)

$E_T = \sqrt{E_{R(T)}^2 + (E_{L(T)} - E_{C(T)})^2}$ (4) $E_T = E_{R(T)}/\cos\theta \text{ (PF)}$ (5) $E_T = E_{L(T)} - E_{C(T)}/\sin\theta$ (6)

The Apparent Power (VA) {Series-RLC (Multiple Components)}

$P_{VA} = E_T \times I$ (1) $P_{VA} = I^2 \times Z$ (2) $P_{VA} = E_T^2/Z$ (3)

$P_{VA} = \sqrt{P_{W(T)}^2 + P_{VARs(T)}^2}$ (4) $P_{VA} = P_{W(T)}/PF\ (\cos\theta)$ (5) $P_{VA} = P_{VARs(T)}/\sin\theta$ (6)

The True Power (W) {Series-RLC (Multiple Components)}

$P_W = I^2 \times R$ (1) $P_W = E_R^2/R$ (2) $P_W = E_R \times I$ (3)

$P_{W(T)} = P_{W1} + P_{W2} + \ldots$ (1) $P_{W(T)} = I^2 \times R_T$ (2) $P_{W(T)} = E_{R(T)}^2/R_T$ (3)

$P_{W(T)} = E_{R(T)} \times I$ (4) $P_{W(T)} = \sqrt{P_{VA}^2 - P_{VARs(T)}^2}$ (5) $P_{W(T)} = P_{VA} \times PF\ (\cos\theta)$ (6)

$P_{W(T)} = P_{VARs(T)}/\tan\theta$ (7)

The Reactance Powers (VARs) {Series-RLC (Multiple Components)}

$P_{VARsL} = I^2 \times X_L$ (1) $P_{VARsL} = E_L^2/X_L$ (2) $P_{VARsL} = E_L \times I$ (3)

$P_{VARsL(T)} = P_{VARsL1} + P_{VARsL2} + \ldots$ (1) $P_{VARsL(T)} = I^2 \times X_{LT}$ (2) $P_{VARsL(T)} = E_{L(T)}^2/X_{LT}$ (3)

$P_{VARsL(T)} = E_{L(T)} \times I$ (4) $P_{VARsL(T)} = P_{VARs(T)} + P_{VARsC(T)}$ (5)

$P_{VARsC} = I^2 \times X_C$ (1) $P_{VARsC} = E_C^2/X_C$ (2) $P_{VARsC} = E_C \times I$ (3)

$P_{VARsC(T)} = P_{VARsC1} + P_{VARsC2} + \ldots$ (1) $P_{VARsC(T)} = I^2 \times X_{CT}$ (2) $P_{VARsC(T)} = E_{C(T)}^2/X_{CT}$ (3)

$P_{VARsC(T)} = E_{C(T)} \times I$ (4) $P_{VARsC(T)} = P_{VARsL(T)} - P_{VARs(T)}$ (5)

The Total Reactance Power (VARs) {Series-RLC (Multiple Components)}

$P_{VARs(T)} = \sqrt{P_{VA}^2 - P_{W(T)}^2}$ (1) $P_{VARs(T)} = P_{VA} \times \sin\theta$ (2) $P_{VARs(T)} = P_{W(T)} \times \tan\theta$ (3)

$P_{VARs(T)} = I^2 \times (X_{LT} - X_{CT})$ (4) $P_{VARs(T)} = (E_{L(T)} - E_{C(T)})^2/(X_{LT} - X_{CT})$ (5)

$P_{VARs(T)} = P_{VARsL(T)} - P_{VARsC(T)}$ (6)

The Power Factor (PF) {Series-RLC (Multiple Components)}

$PF = R_T/Z\ (\cos\theta)$ (1) $PF = E_{R(T)}/E_T\ (\cos\theta)$ (2) $PF = P_{W(T)}/P_{VA}\ (\cos\theta)$ (3)

The Phase Angle (θ) {Series-RLC (Multiple Components)}

$\theta = \cos^{-1} \dfrac{E_{R(T)}}{E_T}$ (PF) (1) $\theta = \sin^{-1} \dfrac{E_{L(T)} - E_{C(T)}}{E_T}$ (2) $\theta = \tan^{-1} \dfrac{E_{L(T)} - E_{C(T)}}{E_{R(T)}}$ (3)

$\theta = \cos^{-1} \dfrac{R_T}{Z}$ (PF) (4) $\theta = \sin^{-1} \dfrac{X_{LT} - X_{CT}}{Z}$ (5) $\theta = \tan^{-1} \dfrac{X_{LT} - X_{CT}}{R_T}$ (6)

$\theta = \cos^{-1} \dfrac{P_{W(T)}}{P_{VA}}$ (PF) (7) $\theta = \sin^{-1} \dfrac{P_{VARs(T)}}{P_{VA}}$ (8) $\theta = \tan^{-1} \dfrac{P_{VARs(T)}}{P_{W(T)}}$ (9)

Parallel-RLC Multiple Components

Observe the multiple component RLC parallel circuit in figure 36.

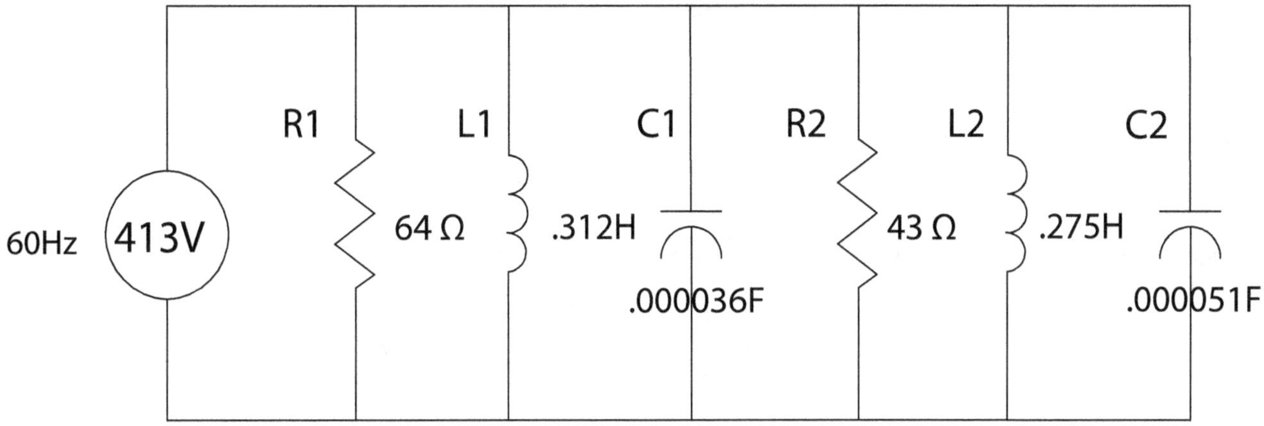

Fig. 36. Parallel RLC circuit (multiple components).

Resistance (R), Inductance Reactance (X_L) and Capacitance Reactance (X_C) {Parallel-RLC (Multiple Components)}

Again, our first approach in analyzing this circuit is to determine the individual reactance of each component followed by the total resistance and reactance of all like components. Based on the given resistance, the total resistance is

$$R_T = \frac{1}{1/64\Omega + 1/43\Omega}$$

$$= \frac{1}{.01563\Omega + .02326\Omega}$$

$$= \frac{1}{.0389\Omega}$$

$$= 25.71\Omega$$

Based on the given inductors and operating frequency, the inductance reactance of each inductor is

$$X_{L1} = 2 \text{ x } 3.14 \text{ x } 60\text{Hz x } .312\text{H} = 117.56\Omega$$
$$X_{L2} = 2 \text{ x } 3.14 \text{ x } 60\text{Hz x } .275\text{H} = 103.62\Omega$$

where the total inductance reactance is

$$X_{LT} = \frac{1}{1/117.56\Omega + 1/103.62\Omega}$$

$$= \frac{1}{.00851\Omega + .00965\Omega}$$

$$= \frac{1}{.0182\Omega}$$

$$= 54.95\Omega$$

Based on the given capacitors and operating frequency, the capacitance reactance of each capacitor is

$$X_{C1} = \frac{1}{2 \text{ x } 3.14 \text{ x } 60\text{Hz x } .000036\text{F}} = 73.72\Omega$$

$$X_{C2} = \frac{1}{2 \text{ x } 3.14 \text{ x } 60\text{Hz x } .000051\text{F}} = 52.04\Omega$$

Where the total capacitance reactance is

$$X_{CT} = \frac{1}{1/73.72\Omega + 1/52.04\Omega}$$

$$= \frac{1}{.01356\Omega + .01922\Omega}$$

$$= \frac{1}{.03278\Omega}$$

$$= 30.51\Omega$$

Impedance {Parallel-RLC (Multiple Components)}
To calculate the impedance for this circuit, the vector-related formula for a parallel-RLC circuit must be used, $Z = 1/\sqrt{(1/R_T)^2 + (1/X_{LT} - 1/X_{CT})^2}$, where X_{LT} is larger than X_{CT}. Therefore,

$$Z = \frac{1}{\sqrt{(1/25.71)^2 + (1/30.51 - 1/54.95)^2}}$$

$$= \frac{1}{\sqrt{(1/25.71)^2 + (.03278 - .01819)^2}}$$

$$= \frac{1}{\sqrt{.001513 + (.01459)^2}}$$

$$= \frac{1}{\sqrt{.001513 + .000213}}$$

$$= \frac{1}{\sqrt{.00173}}$$

$$= \frac{1}{.04159}$$

$$= 24.04\Omega$$

With the circuit's impedance, other quantities can now be determined.

Total Current (I_T) {Parallel-RLC (Multiple Components)}
Since E is shown to be 413V, the total current can also be calculated using the formula $I_T = E/Z$.

$$I_T = 413V/24.04\Omega$$
$$= 17.18A$$

Individual Current (I_R, I_L and I_C) {Parallel-RLC (Multiple Components)}
Because this is a parallel circuit, only one source of voltage and multiple paths of current will exist. Since the ohmic values of $R_{1\text{-}2}$, $X_{L1\text{-}2}$, and $X_{C1\text{-}2}$ are not the same, each component will experience different amounts of current flow; that is: $I_{R1\text{-}2}$, $I_{L1\text{-}2}$, and $I_{C1\text{-}2}$. Therefore,

$$I_{R1} = E/R_1 = 413V/64\Omega = 6.45A$$

$$I_{L1} = E/X_{L1} = 413/117.56\Omega = 3.51A$$

$$I_{C1} = E/X_{C1} = 413V/73.72\Omega = 5.60A$$

$$I_{R2} = E/R_2 = 413V/43\Omega = 9.60A$$

$$I_{L2} = E/X_{L2} = 413/103.62\Omega = 3.99A$$

$$I_{C2} = E/X_{C2} = 413V/52.04\Omega = 7.94A$$

$$I_{R1\&2(T)} = 6.45A + 9.6A = 16.05A$$

$$I_{L1\&2(T)} = 3.51A + 3.99A = 7.50A$$

$$I_{C1\&2(T)} = 5.6A + 7.94A = 13.54A$$

Modifying the previously used formula, where $I_T = \sqrt{I_R^2 + (I_C - I_L)^2}$ to include the total current quantities per component, we get $I_T = \sqrt{(I_{R1\&2(T)})^2 + (I_{C1\&2(T)} - I_{L1\&2(T)})^2}$, where the results of $I_{C1\&2(T)}$ are larger than $I_{L1\&2(T)}$. By applying this formula, the circuit's total current can now be calculated. Therefore,

$$I_T = \sqrt{(16.05)^2 + (13.54 - 7.5)^2}$$
$$= \sqrt{257.6 + (6.04)^2}$$
$$= \sqrt{257.6 + 36.48}$$
$$= \sqrt{294.08}$$
$$= 17.15A$$

Compared to the total current calculated earlier (I_T = 413V/24.04Ω = 17.18A), the results are approximately the same.

The Apparent Power (VA) {Parallel-RLC (Multiple Components)}

Using that which was given as the circuit's voltage (E), the calculated current (I), and impedance (Z), the circuit's apparent power can be determined by use of the formulas $P_{VA} = E \times I_T$, $P_{VA} = I_T^2 \times Z$, and $P_{VA} = E^2/Z$.

$$P_{VA} = 413V \times 17.18A = 7095.34VA$$

$$P_{VA} = (17.18A)^2 \times 24.04\Omega = 295.15A \times 24.04\Omega = 7095.41VA$$

$$P_{VA} = (413V)^2 / 24.04\Omega = 170{,}569V / 24.04\Omega = 7095.22VA$$

The True Power (W) {Parallel-RLC (Multiple Components)}

The true power consumed by each resistance R is calculated using the formulas $P_W = I_R^2 \times R$ and $P_W = E^2/R$, where I_R is the current flowing through each respective resistance R component.

$$P_{W1} = (I_{R1})^2 \times R_1 = (6.45A)^2 \times 64\Omega = 2662.56W$$ and $$P_{W1} = E^2/R_1 = (413V)^2/64\Omega = 2665.14W$$

$$P_{W2} = (I_{R2})^2 \times R_2 = (9.6A)^2 \times 43\Omega = 3962.88W$$ and $$P_{W2} = E^2/R_2 = (413V)^2/43\Omega = 3966.72W$$

With these calculated quantities per formula, the circuit's total true power is

$$P_{W(T)} = 2662.56W + 3962.88W = 6625.44W$$

$$P_{W(T)} = 2665.14W + 3966.72W = 6631.86W$$

To verify the results of these calculations, let's use the formulas $P_{W(T)} = (I_{R1\&2(T)})^2 \times R_T$ and $P_{W(T)} = E^2/R_T$, where the circuit's true power is determined based on the circuit's total resistance (R_T).

$$P_{W(T)} = (16.05A)^2 \times 25.71\Omega = 6622.96W$$

$$P_{W(T)} = (413V)^2/25.71\Omega = 6634.34W$$

As you can see, the results are approximately the same as those calculated by means of summation.

The Reactance Power (VARs) {Parallel-RLC (Multiple Components)}

Relying upon that which was previously learned about determining the reactance power for single component RLC circuits that are parallel connected, the same approach will be taken here, except total quantities will be used. Again, by using the modified product of the sum formula, a totaled common reactance can be determined.

Once more, the denominator of this formula will reflect the difference between the two totaled reactance X_{LT} and X_{CT}, where X_{LT} is larger than X_{CT}. As a result of both totaled reactance currents ($I_{C1\&2(T)}$ and $I_{L1\&2(T)}$) being 180° out of phase, so are both totaled reactances (X_{LT} and X_{CT}). With this modified formula, the totaled common reactance is

$$X_{LC(T)} = \frac{X_{LT} \times X_{CT}}{X_{LT} - X_{CT}}$$

where the totaled common reactance results in

$$X_{LC(T)} = \frac{54.95\Omega \times 30.51\Omega}{54.95\Omega - 30.51\Omega}$$

$$= \frac{1676.52\Omega}{24.44}$$

$$= 68.60\Omega$$

and the difference between the totaled reactance currents is $I_{CL(T)} = 13.54A - 7.5A = 6.04A$. With both common totaled quantities in place, the previously used reactance power formulas with modifications can now be used, where

$$P_{VARs(T)} = I_{CL(T)}^{2} \times X_{LC(T)} = (6.04A)^2 \times 68.6\Omega = 2502.64VARs$$

and

$$P_{VARs(T)} = E^2/X_{LC(T)} = (413V)^2/68.6\Omega = 2486.43VARs$$

The consistency of these results will be realized when an alternate approach is applied.

The Power Factor (PF)) {Parallel-RLC (Multiple Components)}

To calculate the power factor for this RLC multiple component parallel circuit, the formulas $PF = P_{W(T)}/P_{VA}$ and $PF = Z/R_T$ are used. Therefore,

$$PF = \frac{6631.86W}{7095.41VA} = .9347\ (93.47\%)$$

and

$$PF = \frac{24.04\Omega}{25.71\Omega} = .9350\ (93.50\%)$$

As you can see, when calculated, both sets of quantities yield approximate results.

The Phase Angle (θ) {Parallel-RLC (Multiple Components)}
In a Parallel-RLC circuit, formulas for the phase angle can be derived using any one of the four trigonometric formulas:

$$\tan\theta = \frac{I_{CT} - I_{LT}}{I_{RT}} \quad \text{or} \quad \frac{R_T}{X_{L\,C(T)}} \quad \text{and} \quad \cos\theta = \frac{Z}{R_T} \quad \text{or} \quad \frac{I_{RT}}{I_T}$$

From the four formulas, the following derived formulas along with a previously derived formula will be used to determine the phase angle,

$$\theta = \tan^{-1}\frac{I_{CT} - I_{LT}}{I_{RT}} \quad \text{or} \quad \frac{R_T}{X_{L\,C(T)}} \quad \text{and} \quad \theta = \cos^{-1}\frac{Z}{R_T} \quad \text{or} \quad \frac{I_{RT}}{I_T}$$

$$\theta = \cos^{-1} PF \text{ (previous)}$$

Based on the derived formulas, the circuit's phase angle is

$$\theta = \tan^{-1}\frac{6.04A}{16.05A} = \tan^{-1} .3763 = 20.62° \quad \text{or} \quad \frac{25.71\Omega}{68.6\Omega} = .3748 = 20.55°$$

$$\text{and} \quad \theta = \cos^{-1}\frac{24.04\Omega}{25.71\Omega} = \cos^{-1} .9350 = 20.77° \quad \text{or} \quad \frac{16.05A}{17.18A} = .9342 = 20.9°$$

$$\theta = \cos^{-1} .9347 \text{ (previous)} = 20.82°$$

Again, using all five formulas, the results are all approximate.

Power {Parallel-RLC (Multiple Components)}
Previously, the three power quantities were calculated using given or calculated values. The quantities will now be determined using an alternate approach. By using the formula $\cos\theta = P_{W(T)}/P_{VA}$, the true power can now be determined by use of the derived formula $P_{W(T)} = P_{VA} \times \cos\theta$, where

$$\begin{aligned} P_{W(T)} &= P_{VA} \times \cos\theta \\ &= 7095.34VA \times \cos 20.82° \\ &= 7095.34VA \times .9347 \\ &= 6632.01W \end{aligned}$$

The result of this calculation is aligned with the level of accuracy discussed earlier for the true power quantity.

By using the formula $\sin\theta = P_{VARs(T)}/P_{VA}$, the reactance power can now be determined by use of the derived formula $P_{VARs(T)} = P_{VA} \times \sin\theta$, where

$$
\begin{aligned}
P_{VARs(T)} &= P_{VA} \text{ x } \sin\theta \\
&= 7095.34\text{VA x } \sin 20.82° \\
&= 7095.34\text{VA x } .3554 \\
&= 2521.68\text{W}
\end{aligned}
$$

As you can see, the results are approximately the same as earlier calculated. By using these results, they can be substituted into the Pythagorean theorem, and the apparent power can be verified.

$$
\begin{aligned}
P_{VA} &= \sqrt{(6632.01)^2 + (2521.68)^2} \\
&= \sqrt{43{,}983{,}556.64 + 6{,}358{,}870.02} \\
&= \sqrt{50{,}342{,}426.66} \\
&= 7095.24\text{VA}
\end{aligned}
$$

The following display is an overall list of formulas that can be used for solving Parallel-RLC Circuits—Multiple Components.

Related Parallel Formulas
(Resistance–Inductance–Capacitance Multiple Component Circuits—Overall)

[Only one source of voltage (E); therefore, $E = E_R = E_L = E_C$.]

(IMPORTANT: Regardless of how shown in relative formulas, where the differences between L and C components are required, the smaller component is always subtracted from the larger component.)

Resistance (R) {Parallel-RLC (Multiple Components)}

$R = E/I_R$ (1) $\quad$ $R = E^2/P_W$ (2) $\quad$ $R = P_W/I_R{}^2$ (3)

$R_T = \dfrac{1}{1/R_1 + 1/R_2 + \ldots}$ (1) $\quad$ $R_T = \dfrac{R_1 \text{ x } R_2}{R_1 + R_2}$ (2) $\quad$ $R = E/I_{R(T)}$ (3)

$R = E^2/P_{W(T)}$ (4) $\quad$ $R_T = Z/\cos\theta$ (PF) (5)

$R_T = (X_{CT} \text{ x } X_{LT} / X_{CT} - X_{LT}) \text{ x } \tan\theta$ (5)

$R_T = \dfrac{1}{\sqrt{(1/Z)^2 - (1/X_{LT} - 1/X_{CT})^2}}$ (6)

Inductance (L) and Capacitance (L) {Parallel-RLC (Multiple Components)}

$L = X_L/2\pi f$ $\quad$ $L_T = \dfrac{L}{N}$ (1) (seepage 57) $\quad$ $L_T = \dfrac{L_1 \text{ x } L_2}{L_1 + L_2}$ (2)

$L_T = \dfrac{1}{1/L_1 + 1/L_2 + \ldots\ldots}$ (3) $\quad$ $L_T = X_{LT}/2\pi f$ (4)

$C = 1/2\pi f X_C$ $\quad$ $C_T = C_1 + C_2 + \ldots\ldots$ (1) $\quad$ $C_T = 1/2\pi f X_{CT}$ (2)

Reactances (X_C, X_L and X_C x X_L /X_C–X_L) {Parallel-RLC (Multiple Components)}

$X_C = 1/2\pi fC$ (1) $X_C = E/I_C$ (2)

$X_C = E^2/P_{VARsC}$ (3) $X_C = P_{VARsC}/I_C^2$ (4)

$$X_{CT} = \frac{1}{1/X_{C1} + 1/X_{C2} + \ldots} \quad (1) \qquad X_{CT} = \frac{X_{C1} \times X_{C2}}{X_{C1} + X_{C2}} \quad (2)$$

$X_{CT} = 1/2\pi fC_T$ (3) $X_{CT} = E/I_{C(T)}$ (4)

$X_L = 2\pi fL$ (1) $X_L = E/I_L$ (2)

$X_L = E^2/P_{VARsL}$ (3) $X_L = P_{VARsL}/I_L^2$ (4)

$$X_{LT} = \frac{1}{1/X_{L1} + 1/X_{L2} + \ldots} \quad (1) \qquad X_{LT} = \frac{X_{L1} \times X_{L2}}{X_{L1} + X_{L2}} \quad (2)$$

$X_{LT} = 2\pi fL_T$ (3) $X_{LT} = E/I_{L(T)}$ (4)

$(X_{CT} \times X_{LT}/X_{CT} - X_{LT}) = R_T/\tan\theta$ (1) $(X_{CT} \times X_{LT}/X_{CT} - X_{LT}) = Z/\sin\theta$ (2)

$$(X_{CT} \times X_{LT}/X_{CT} - X_{LT}) = \frac{1}{\sqrt{(1/Z)^2 - (1/R_T)^2}} \quad (3)$$

$(X_{CT} \times X_{LT}/X_{CT} - X_{LT}) = P_{VARs(T)}/(I_{C(T)} - I_{L(T)})^2$ (4)

$(X_{CT} \times X_{LT}/X_{CT} - X_{LT}) = E^2/P_{VARs(T)}$ (5)

Impedance (Z) {Parallel-RLC (Multiple Components)}

$$Z = \frac{1}{\sqrt{(1/R_T)^2 + (1/(X_{CT} \times X_{LT}/X_{CT} - X_{LT})^2}} \quad (1)$$

$Z = E/I_T$ (2)

$Z = E^2/P_{VA}$ (3) $Z = P_{VA}/I_T^2$ (4) $Z = R_T \times \cos\theta$ (PF) (5)

$Z = (X_{CT} \times X_{LT}/X_{CT} - X_{LT}) \times \sin\theta$ (6)

Individual Currents (I_R, I_C, I_L and I_C–I_L) {Parallel-RLC (Multiple Components)}

$I_R = E/R$ (1) $I_R = P_W/E$ (2) $I_R = \sqrt{P_W/R}$ (3)

$I_{R(T)} = I_{R1} + I_{R2} + \ldots$ (1) $I_{R(T)} = E/R_T$ (2) $I_{R(T)} = P_{W(T)}/E$ (3)

$I_{R(T)} = \sqrt{P_{W(T)}/R_T}$ (4) $I_{R(T)} = \sqrt{I_T^2 - (I_{C(T)} - I_{L(T)})^2}$ (5)

$I_{R(T)} = I_T \times \cos\theta$ (PF) (6) $I_{R(T)} = I_{C(T)} - I_{L(T)}/\tan\theta$ (7)

$I_C = E/X_C$ (1) $I_C = P_{VARs\,C}/E$ (2) $I_C = \sqrt{P_{VARsC}/X_C}$ (3)

$I_{C(T)} = E/X_{CT}$ (1) $I_{C(T)} = P_{VARs\,C(T)}/E$ (2) $I_{C(T)} = \sqrt{P_{VARs\,C(T)}/X_{CT}}$ (3)

$I_L = E / X_L$ (1) $I_L = P_{VARsL} / E$ (2) $I_L = \sqrt{P_{VARsL} / X_L}$ (3)

$I_{L(T)} = E / X_{LT}$ (1) $I_{L(T)} = P_{VARs\,L(T)} / E$ (2) $I_{L(T)} = \sqrt{P_{VARs\,L(T)} / X_{LT}}$ (3)

$I_{C(T)} - I_{L(T)} = \sqrt{I_T^2 - I_{R(T)}^2}$ (1) $I_{C(T)} - I_{L(T)} = I_T \times \sin\theta$ (2) $I_{C(T)} - I_{L(T)} = I_{R(T)} \times \tan\theta$ (3)

$I_{C(T)} - I_{L(T)} = P_{VARs(T)} / E$ (4) $I_{C(T)} - I_{L(T)} = \sqrt{P_{VARs(T)} / (X_{CT} \times X_{LT} / X_{CT} - X_{LT})}$ (5)

Total Current (I_T) {Parallel-RLC (Multiple Components)}

$I_T = E/Z$ (1) $I_T = P_{VA} / E$ (2) $I_T = \sqrt{P_{VA} / Z}$ (3)

$I_T = \sqrt{I_{R(T)}^2 + (I_{C(T)} - I_{L(T)})^2}$ (4) $I_T = I_{R(T)} / \cos\theta$ (PF) (5) $I_T = I_{C(T)} - I_{L(T)} / \sin\theta$ (6)

Voltage (E) {Parallel-RLC (Multiple Components)}

$E = I_R \times R$ (1) $E = P_W / I_R$ (2) $E = I_C \times X_C$ (3)

$E = I_{R(T)} \times R_T$ (4) $E = P_{W(T)} / I_{R(T)}$ (5) $E = I_{C(T)} \times X_{C(T)}$ (6)

$E = I_L \times X_L$ (1) $E = P_{VARsC} / I_C$ (2) $E = P_{VARsL} / I_L$ (3)

$E = I_{L(T)} \times X_{LT}$ (4) $E = P_{VARsC(T)} / I_{C(T)}$ (5) $E = P_{VARsL(T)} / I_{L(T)}$ (6)

$E = \sqrt{P_W \times R}$ (1) $E = \sqrt{P_{VARsC} \times X_C}$ (2) $E = \sqrt{P_{VARsL} \times X_L}$ (3)

$E = \sqrt{P_{W(T)} \times R_T}$ (4) $E = \sqrt{P_{VARsC(T)} \times X_{C(T)}}$ (5) $E = \sqrt{P_{VARsL(T)} \times X_{LT}}$ (6)

$E = I_T \times Z$ (1) $E = \sqrt{P_{VA} / I_T}$ (2) $E = \sqrt{P_{VA} \times Z}$ (3)

$E = P_{VARs(T)} / I_{C(T)} - I_{L(T)}$ (4) $E = \sqrt{P_{VARs(T)} \times (X_{CT} \times X_{LT} / X_{CT} - X_{LT})}$ (5)

The Apparent Power (VA) {Parallel-RLC (Multiple Components)}

$P_{VA} = E \times I_T$ (1) $P_{VA} = I_T^2 \times Z$ (2) $P_{VA} = E^2 / Z$ (3)

$P_{VA} = P_{W(T)} / \cos\theta$ (PF) (4) $P_{VA} = P_{VARs(T)} / \sin\theta$ (5) $P_{VA} = \sqrt{P_{W(T)}^2 + P_{VARs(T)}^2}$ (6)

The True Power (W) {Parallel-RLC (Multiple Components)}

$P_W = E \times I_R$ (1) $P_W = E^2 / R$ (2) $P_W = I_R^2 \times R$ (3)

$P_{W(T)} = P_{W1} + P_{W2} + \ldots.$ (1) $P_{W(T)} = E \times I_{R(T)}$ (2) $P_{W(T)} = E^2 / R_T$ (3)

$P_{W(T)} = I_{R(T)}^2 \times R_T$ (4) $P_{W(T)} = P_{VA} \times \cos\theta$ (PF) (5) $P_{W(T)} = P_{VARs(T)} / \tan\theta$ (6)

$P_{W(T)} = \sqrt{P_{VA}^2 - P_{VARs(T)}^2}$ (7)

The Reactance Power (VARs) {Parallel-RLC (Multiple Components)}

$P_{VARsC} = E \times I_C$ (1) $P_{VARsC} = I_C{}^2 \times X_C$ (2) $P_{VARsC} = E^2/X_C$ (3)

$P_{VARsC(T)} = P_{VARsC1} + P_{VARsC2} + \ldots$ (1) $P_{VARsC(T)} = E \times I_{C(T)}$ (2) $P_{VARsC(T)} = I_{C(T)}{}^2 \times X_{CT}$ (3)
$P_{VARsC(T)} = E^2/X_{CT}$ (4)

$P_{VARsL} = E \times I_L$ (1) $P_{VARsL} = I_L{}^2 \times X_L$ (2) $P_{VARsL} = E^2/X_L$ (3)

$P_{VARsL(T)} = P_{VARsL1} + P_{VARsL2} + \ldots$ (1) $P_{VARsL(T)} = E \times I_{L(T)}$ (2) $P_{VARsL(T)} = I_{L(T)}{}^2 \times X_{LT}$ (3)
$P_{VARsL(T)} = E^2/X_{LT}$ (4)

The Total Reactance Power (VARs) {Parallel-RLC (Multiple Components)}

$P_{VARs(T)} = \sqrt{P_{VA}{}^2 - P_{W(T)}{}^2}$ (1) $P_{VARs(T)} = P_{VA} \times \sin\theta$ (2) $P_{VARs(T)} = P_{W(T)} \times \tan\theta$ (3)

$P_{VARs(T)} = P_{VARsC(T)} - P_{VARsL(T)}$ (4) $P_{VARs(T)} = E \times I_{C(T)} - I_{L(T)}$ (5)

$P_{VARs(T)} = (X_{CT} \times X_{LT} / X_{CT} - X_{LT}) \times (I_{C(T)} - I_{L(T)})^2$ (6)

$P_{VARs(T)} = E^2 / (X_{CT} \times X_{LT} / X_{CT} - X_{LT})$ (7)

The Power Factor (PF) {Parallel-RLC (Multiple Components)}

$PF = Z/R_T (\cos\theta)$ (1) $PF = I_{R(T)}/I_T (\cos\theta)$ (2) $PF = P_{W(T)}/P_{VA} (\cos\theta)$ (3)

The Phase Angle (θ) {Parallel-RLC (Multiple Components)}

$\theta = \cos^{-1} \frac{I_{R(T)}}{I_T}$ (PF) (1) $\theta = \sin^{-1} \frac{I_{C(T)} - I_{L(T)}}{I_T}$ (2) $\theta = \tan^{-1} \frac{I_{C(T)} - I_{L(T)}}{I_{R(T)}}$ (3)

$\theta = \cos^{-1} \frac{Z}{R_T}$ (PF) (4)

$\theta = \sin^{-1} \frac{Z}{X_{CT} \times X_{LT} / X_{CT} - X_{LT}}$ (5) $\theta = \tan^{-1} \frac{R_T}{X_{CT} \times X_{LT} / X_{CT} - X_{LT}}$ (6)

$\theta = \cos^{-1} \frac{P_{W(T)}}{P_{VA}}$ (PF) (7)

$\theta = \sin^{-1} \frac{P_{VARs(T)}}{P_{VA}}$ (8) $\theta = \tan^{-1} \frac{P_{VARs(T)}}{P_{W(T)}}$ (9)

Resistance (R)–Inductance (L)–Capacitance (C) Combination Circuits

Although they may appear simple and can be easily solved using the same approach as purely resistance combination circuits, parallel-series or series-parallel circuits containing resistance, inductance, and capacitance can be quite challenging while attempting to solve them.

In the final stages of this topic on resistance—reactance circuits—we will conclude by looking at combination circuits that are parallel-series and series-parallel connected. Based on what's been covered and the vast amount of knowledge gained, these types of circuits MAY (again) APPEAR SIMPLE, and CAN BE EASILY SOLVED using the same approach as others; but beware, they carry a twist unparallel to others. To help make these types of circuits a little easier to solve, let's introduce another topic to assist us in these final stages.

Perhaps you may have heard of or may be already familiar with the term "phasor," or maybe you have used phasors at one time or another in solving RLC related circuits. Regardless, this topic will be covered with the same thing in mind as other topics and that's to ensure a clear and thorough understanding of phasors.

Phasors

Whether used in discussing electrical or non-electrical applications, vectors are mostly perceived as being fixed or stationary objects, although that's not always the case. From an electrical point of view, a phasor is recognized as a time-varying combination vector because part of it occurs in phase (at the same time) with an adjacent vector, while the other part occurs out of phase (at a different time) with a distant vector. From a mathematical perspective, a phasor is a complex number, where such number is a combination of a real number and an imaginary number, which results in the format

$$a + jb$$

where **a** is the real number, **b** is the imaginary number, and the small letter **j** is used as a substitute to represent the square root of negative one ($\sqrt{-1}$), therefore resulting in the relationship $\mathbf{j = \sqrt{-1}}$.

Because of the lengthy details that are involved with deriving this relational conclusion and the fact that $j = \sqrt{-1}$ is already an accepted truth, we can move forward with this topic.

Where complex numbers and phasors are concerned, j is known as an operator and is referred to as the operator j. At one time, the small letter i was used instead, but that was changed when i was later used as a symbol to identify instantaneous current (see topic Instantaneous Value).

In dealing with the square root of a positive number, we will always derive a positive number; for example, $\sqrt{16}$ is 4, where 4 x 4 or 4^2 is 16. However, as it pertains to the square root of a negative number, there is no such number that, when multiplied by itself or raised to any even power, will produce a negative result. For example, what purpose or good would $\sqrt{-16}$ (or the square of any negative number) serve when there is no number that when multiplied by itself will yield –16? As a result, the square root of any negative number is recognized as an imaginary number. Looking back at the format a + jb, we can actually say that b is also a real number but is recognized as an imaginary number because of its affiliation with the operator j.

To make use of an imaginary number and the relation $j = \sqrt{-1}$, all imaginary numbers are expressed as the product of a positive number and $\sqrt{-1}$. For example,

$$\begin{aligned}\sqrt{-16} &= \sqrt{(16)(-1)}\\ &= \sqrt{16} \times \sqrt{-1}\\ &= 4 \times \sqrt{-1}\\ &= 4\sqrt{-1}\end{aligned}$$

Again, because $j = \sqrt{-1}$, the product can be expressed as 4j or j4.

Now if the relationship $j = \sqrt{-1}$ is squared on both sides, we get the derived relationship

$$j^2 = -1$$

If the derived relationship is again multiplied on both sides by j, $j^2 \times j = -1 \times j$, where $j^2 \times j = j^3$ and $-1 \times j = -j$, another relationship is derived that results in

$$j^3 = -j$$

Finally since $j^2 = -1$ then $j^2 \times j^2 = -1 \times -1$ and

$$j^4 = 1$$

Without these four relationships, $j = \sqrt{-1}$, $j^2 = -1$, $j^3 = -j$, and $j^4 = 1$, the task of solving problems involving complex numbers or phasors would prove to be as useless as trying to derive the square root of a negative number.

<u>The Rectangular and Polar Forms of a Phasor</u>

From an electrical perspective, **a** is recognized as the **in-phase** component of the format **a** (+/–) **jb**, while **b** is recognized as the **out-of-phase** component of the format, and **j** is used exclusively with the out-of-phase component of a phasor to signify a 90° displacement. The symbol (+/–) means plus or minus. Where *ac* electrical circuits are concerned, the format R +/– jX is a conversion of the format a (+/–) jb, where R can represent resistance or resistance-related quantities , and +/– jX can represent inductance reactance or capacitance reactance or reactance related quantities. From a mathematical perspective, the format a +/– jb and R +/– jX are said to be the **rectangular form** of a phasor (or complex number).

The rectangular form can be converted into the **polar form**. The polar form consists of a magnitude **r** and a phase angle **θ**, and is written in the format **r** $\underline{/\boldsymbol{\theta}}$. Refer to figure 37. Figure 37 is a replica of figure 16(a), with the exception of its being labeled **r**, **a**, and **b**.

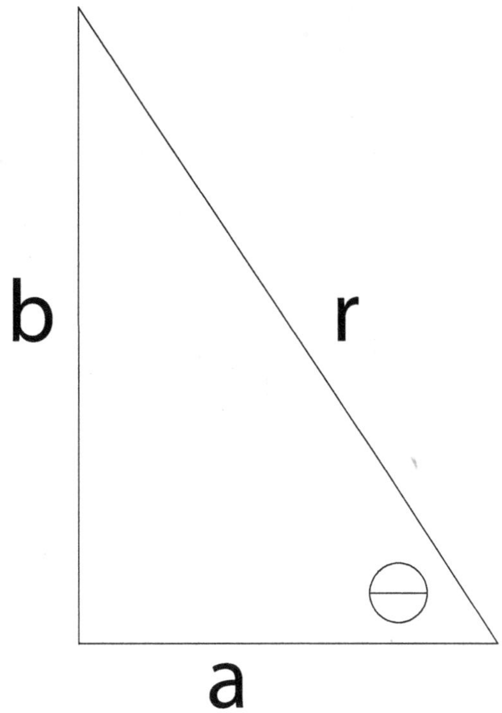

Fig. 37. The polar form.

As a result, a (+/–) jb is equivalent to r $\underline{/(+/-)\theta}$, where both formats are corresponding representations of the relationship between the rectangular and polar forms. Now based on the triangular relationship illustrated in figure 37, as it relates to trigonometric functions, $a = r \times \cos\theta$ and $b = r \times \sin\theta$, while $b/a = \tan\theta$. When converted, $r = a/\cos\theta$ and $b/\sin\theta$ and the phase angle $\theta = \tan^{-1} b/a$. Because all of these formulas have been thoroughly covered in previous topics, understanding each application should now come with ease.

Like any other combination of numbers, phasors can also be added, subtracted, multiplied, and divided. Therefore, let's take a look at each operation by first focusing on phasors in the rectangular form.

<u>Adding and Subtracting Phasors in Rectangular Form</u>

Add

1. 6 + j4 + 3 + j2

Subtract

2. 5 + j4 – 3 + j
3. (–26 – j11) – (–18 – j9)
4. (–15 – j9) – (–26 + j11)

When phasors are added or subtracted, the approach is always to add real numbers together and then the imaginary numbers. For certain operations, phasors can be arranged and solved in both horizontal and column forms. As a result,

1. $6 + j4 + 3 + j2 = (6 + 3) + (j4 + j2)$
$= 9 + j6$

2. $5 + j4 - (3 + j) = 5 + j4 - 3 - j$
$= (5 - 3) + (j4 - j)$
$= 2 + j3$

(in horizontal form)
3. $(-26 - j11) - (-18 - j9) = (-26 - [-18]) + (-j11 - [-j9])$
$= (-26 - [-18]) + (-j11 - [-j9])$
$= (-26 + 18) + (-j11 + j9)$
$= -8 - j11 + j9$
$= -8 - j2$

or

(in column form)
$-26 - j11$
$-(-18 - j9) = 18 + j9$

resulting in
$$\begin{array}{r} -26 - j11 \\ 18 + j9 \\ \hline -8 - j2 \end{array}$$

and

(in horizontal form)
4. $(-15 - j9) - (-26 + j11) = (-15 - [-26]) + (-j9 - j11)$
$= -15 + 26 - j9 - j11$
$= 11 - j20$

(in column form)
$-15 - j9$
$-(-26 + j11) = 26 - j11$

resulting in
$$\begin{array}{r} 26 - j11 \\ -15 - j9 \\ \hline 11 - j20 \end{array}$$

<u>Multiplying and Dividing Phasors in Rectangular Form</u>

Multiply

5. $6 + j4 \text{ x } (-3 + j2)$
6. $-26 - j11 \text{ x } (-18 - j9)$

When phasors are multiplied, the approach is always to multiply the first phasor by the real and imaginary numbers of the second phasor individually. Once this is done, the four relationships involving **j** are applied accordingly, and all like numbers are added together. As a result,

(in horizontal form)

$$
\begin{aligned}
5.\ (6 + j4) \times (-3 + j2) &= [(6 + j4) \times -3] + [(6 + j4) \times j2] \\
&= [-18 - j12] + [j12 + j^2 8] \\
&= -18 - j12 + j12 + j^2 8 \\
&= -18 + j^2 8 \text{ (where } j^2 = -1) \\
&= -18 + (-1) \times 8 \\
&= -18 - 8 \\
&= -26
\end{aligned}
$$

(in column form)

$$
\begin{array}{l}
\quad 6 + j4 \\
\underline{\times\ (-3 + j2)} \\
\quad 12j + j^2 8 \quad \text{(where } j^2 = 1) \\
\underline{-12j + 0 \quad -18} \\
\quad (-1)8 - 18 = -26 \\
\quad -8 - 18 = -26
\end{array}
$$

$$
\begin{aligned}
6.\ (-26 - j11) \times (-18 - j9) &= [(-26 - j11) \times -18] + [(-26 - j11) \times -j9] \\
&= [468 + j198] + [(j234 + j^2 99] \\
&= 468 + j198 + j234 + j^2 99 \text{ (where } j^2 = -1) \\
&= 468 + j432 + (-1)\times 99 \\
&= 468 - 99 + j432 \\
&= 369 + j432
\end{aligned}
$$

Divide

7. $26/(2 - j5)$,
8. $(9 - j7)/(5 + j11)$

When phasors are divided, the approach is always to multiply both phasors in the numerator and denominator by the phasor in the denominator using the opposite sign (+ or –) of the sign used in the original phasor in the denominator. That is, if the phasor in the denominator is "a + jb," then the phasor in the numerator and the denominator are both multiplied by "a – jb." Mathematically, "a + jb" is called the **conjugate** of "a – jb" and vice versa. In this application, the term conjugate means opposite in signs preceding the imaginary number. Observe:

$$7.\ \frac{26}{2 - j5} = \frac{26 \times (2 + j5)}{(2 - j5) \times (2 + j5)}$$

$$= \frac{52 + j130}{(2 - j5) \times 2 + (2 - j5) \times j5}$$

$$= \frac{52 + j130}{4 - j10 + j10 - j^2 25} \quad \text{(where } j^2 = -1\text{)}$$

$$= \frac{52 + j130}{4 - (-1)25}$$

$$= \frac{52 + j130}{29} \quad \text{or} \quad 52/29 + j130/29 \quad \text{or} \quad 1.793 + j4.483$$

8. $\frac{9 - j7}{5 + j11} = \frac{(9 - j7) \text{ x } (5 - j11)}{(5 + j11) \text{ x } (5 - j11)}$

$$= \frac{(9 - j7) \text{ x } 5 + (9 - j7) \text{ x } - j11}{(5 + j11) \text{ x } 5 + (5 + j11) \text{ x } - j11}$$

$$= \frac{45 - j35 - j99 + j^2 77}{25 + j55 - j55 - j^2 121} \quad \text{(where } j^2 = -1\text{)}$$

$$= \frac{45 - j134 + (-1)77}{25 - (-1)121}$$

$$= \frac{-32 - j134}{146} \quad \text{or} \quad -.2192 - j.9178$$

Based on the gathered results, let's now convert each phasor from rectangular to polar form. However, before getting started, observe figure 38. We'll be using it to define whether the correct phase angle is obtained per phasor.

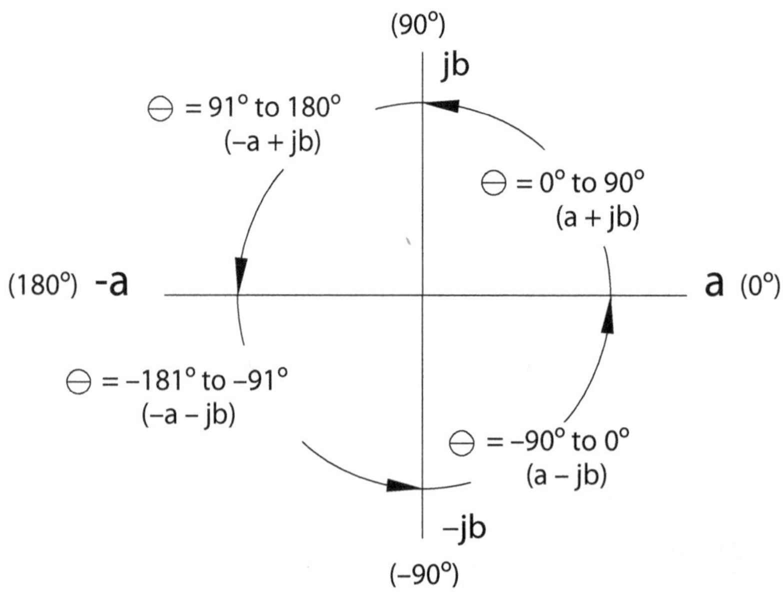

Fig. 38. Defining correct phase angles.

<u>Converting Phasors from Rectangular to Polar Form</u>

1a. $9 + j6$

$\theta = \tan^{-1} 6/9$ $\quad r = 9/\cos 33.69°$
$= \tan^{-1} .6667$ $\quad = 9/.8321$
$= 33.69°$ $\quad = 10.82$
$9 + j6 = 10.82\underline{/33.69°}$

2a. $2 + j3$

$\theta = \tan^{-1} 3/2$ $\quad r = 3/\sin 56.31°$
$= \tan^{-1} 1.5$ $\quad = 3/.8321$
$= 56.31°$ $\quad = 3.61$
$2 + j3 = 3.61\underline{/56.31°}$

Refer to figure 38. Because both values representing a and jb in problems 1 and 2 were positive, the phase angles must fall between 0° and 90°, where in both problems they do.

3a. $-8 - j2$

$\theta = \tan^{-1} -2/-8$
$= \tan^{-1} .25$
$= $ ~~14.04~~°
$= -165.96°$

Refer to figure 38. Because a and jb are negative numbers, that is, –8 and – j2, the phase angle must fall between –181° and –91°. To obtain the correct phase angle, the derived results (14.04°) must be added to –180°. As a result, the correct phase angle is –180° + 14.04° = –165.96°.

$r = -2/\cos -165.96°$
$= -8/-.9701$
$= 8.25$
$-8 - j2 = 8.25\underline{/-165.96°}$

If 14.04° was used as the phase angle when trying to convert from the polar form (8.25/–165.96°) back to the rectangular form the existing results (–8 – j2) could not be obtained.

4a. $11 - j20$

$\theta = \tan^{-1} -20/11$
$= \tan^{-1} -1.8182$
$= -61.19°$

Refer to figure 38. Because 11 is positive and j20 is negative, the phase angle must fall between –90° and 0°, where in this case it does.

$r = 11/\cos -61.19°$ and $r = -20/\sin -61.19°$
$= 11/.4819$ $\quad = -20/-.8762$
$= 22.83$ $\quad = 22.83$
$11 - j20 = 22.83\underline{/-61.19°}$

5a. –26. Although this problem resulted in only an in-phase component, the phasor is written as –26 + j0, which is the same as –26. As it pertains to figure 38, the phase angle for this phasor should fall between 91° and 180°. Now let's proceed.

$\theta = \tan^{-1} 0/{-26}$
$= \tan^{-1} 0$
$= 0$
$= 0 + 180° = 180°$

$r = -26/\cos 180°$
$= -26/{-1}$
$= 26$
-26 (or $-26 + j0$) $= 26\underline{/180°}$

6a. $369 + j432$
$\theta = \tan^{-1} 432/369$ $\quad$ $r = 369/\cos 49.50°$
$= \tan^{-1} 1.171$ $\quad$ $= 369/.6494$
$= 49.50°$ $\quad$ $= 568.22$
$369 + j432 = 568.22\underline{/49.50°}$

7a. $1.793 + j4.483$
$\theta = \tan^{-1} 4.483/1.793$
$= \tan^{-1} 2.5$
$= 68.20°$

$r = 1.793/\cos 68.20°$ and $r = 4.483/\sin 68.20°$
$= 1.793/.3714$ $\quad$ $= 4.483/.9285$
$= 4.83$ $\quad$ $= 4.83$
$1.793 + j4.483 = 4.83\underline{/68.20°}$

8a. $-.2192 - j.9178$
$\theta = \tan^{-1} -.9178/{-.2192}$
$= \tan^{-1} 4.187$
= ~~76.57~~° Refer to figure 38. Again, just like problem 3a, because .2192 and
$= -103.43°$ j.9178 are negative numbers, the phase angle must fall between –181° and –91°. To obtain the correct phase angle, the derived results (76.57°) must be added to –180°. As a result, the correct phase angle is –180° + 76.57° = –103.43°.

$r = -.9178/\sin -103.43°$
$= -.9178/{-.9727}$
$= .9436$
$-.2192 - j.9178 = .9436\underline{/-103.43°}$

Now that we have an understanding of phasors and how they are applied in both rectangular and polar forms, let's take a look at adding, subtracting, multiplying, and dividing phasors exclusively in the polar form.

Adding and Subtracting Phasors in Polar Form

To add or subtract phasors in the polar form, the phasor must first be converted to the rectangular form and then converted back to the polar form.

Add

9. $21\underline{/-73.5^\circ} + 13.4\underline{/66.27^\circ}$

We'll get started by converting the given phasors into rectangular form.

$$\begin{aligned}21\underline{/-73.5^\circ} &= 21(\cos -73.5^\circ + j \sin -73.5^\circ)\\ &= 21(.2840 - j.9588)\\ &= 5.964 - j20.135\end{aligned}$$

$$\begin{aligned}13.4\underline{/66.27^\circ} &= 13.4(\cos 66.27^\circ + j \sin 66.27^\circ)\\ &= 13.4(.4024 + j.9155)\\ &= 5.392 + j12.268\end{aligned}$$

The derived phasors in rectangular form are now added together.

$$\begin{array}{r} 5.964 - j20.135 \\ \underline{5.392 + j12.268} \\ 11.356 - j7.867 \end{array}$$

The sum of the two phasors is then converted into polar form.

$$11.356 - j7.867 = 13.81\underline{/-34.71^\circ}$$

(Magnitude and phase angle rounded to nearest hundredth.)

Subtract

10. $(23.47\underline{/58.8^\circ}) - (11.22\underline{/-42^\circ})$

The same approach is used for subtracting phasors.

$$\begin{aligned}23.47\underline{/58.8^\circ} &= 23.47(\cos 58.8^\circ + j \sin 58.8^\circ)\\ &= 23.47(.5180 + j.8554)\\ &= 12.157 + j20.076\end{aligned}$$

$$\begin{aligned}11.22\underline{/-42^\circ} &= 11.22(\cos -42^\circ + j \sin -42^\circ)\\ &= 11.22(.7431 - j.6691)\\ &= 8.338 - j7.507\end{aligned}$$

The derived phasors in rectangular form are now subtracted from each other.

$$\begin{aligned} &12.157 + j20.076 \\ &-(8.338 - j7.507) = -8.338 + j7.507 \end{aligned}$$

resulting in

$$\begin{array}{r} 12.157 + j20.076 \\ \underline{-8.338 + j7.507} \\ 3.819 + j27.583 \end{array}$$

The difference between the two phasors is then converted into polar form.

$$3.819 + j27.583 = 27.86\underline{/82.12°}$$

(Magnitude and phase angle rounded to nearest hundredth.)

Multiply

11. (–26 – j11) x (–18 – j9)

This is problem 6 being used again, but now using a different approach to derive the same results. Remember the results in rectangular form came out to be 369 + j432 or, if converted to polar form, 568 $\underline{/49.50}$° (rounded off).

When phasors are in polar form, they can be used as is. The magnitudes are multiplied and the phase angles are added together. As a result, let's began this approach by converting the two given phasors into polar form.

$$\begin{aligned} -26 - j11 &= 28.23\underline{/-157.07°} \\ -18 - j9 &= 20.13\underline{/-153.43°} \end{aligned}$$

(Both magnitudes and phase angles are rounded to nearest hundredth.)

With the phasors in polar form, the magnitudes can be multiplied and the phase angles added.

$$\begin{aligned} (28.23\underline{/-157.07°}) \times (20.12\underline{/-153.43°}) &= (28.23 \times 20.12)\underline{/(-157.07° + (-153.43)} \\ &= 567.99\underline{/-310.50°} \end{aligned}$$

Compared to problem 6 (in polar form [568/$\underline{49.50°}$]), the magnitudes are approximately the same, but what about the phase angles? Well there are several ways to come to the conclusion that both phase angles are the same. If we refer to figure 38 and imagine (starting at 0°) rotating –310.50° in a clockwise manner, the results would be 49.50°. This is the same as adding 360° and –310.50°, where the results would also be 49.50°. Another way of looking at it is considering the 567.99$\underline{/-310.50}$° in polar form, and, if converted into rectangular form, the results are approximately the same as that in problem 6, which is 368.88 + j431.90. Because 368.88 and j431.90 are positive numbers, the phase angle according to figure 38 should fall between 0° and 90°, which in this case it does because 49.5° and –310.50° are the same and only differ in displacement. To verify that both phase angles are the same if the cosine, sine, and tangent functions of both phase angles were computed, the paired results per trigonometric function would be same. Therefore, the results of the two polar form phasors is 567.99$\underline{/-310.50°}$ or 567.99$\underline{/49.50°}$.

Divide
12. $(9 - j7)/(5 + j11)$

This is problem 8 being used again but now using a different approach to derive the same results.

Similar to the operation of multiplication, phasors must also be in polar form to divide them. The magnitudes are divided, and the phase angles are subtracted from each other. Observe:

$$9 - j7 = 11.40\underline{/-37.87^\circ}$$
$$5 + j11 = 12.08\underline{/65.56^\circ}$$

(Both magnitudes and phase angles are rounded to nearest hundredth.)

With the phasors in polar form, the magnitudes can be divided and the phase angles subtracted.

$$\frac{11.40\ \underline{/-37.87^\circ}}{12.08\ \underline{/65.56^\circ}} = \frac{11.40}{12.08}\ \underline{/-37.87^\circ - 65.56^\circ}$$

$$= .9437\ \underline{/-103.43^\circ}\ \text{(or } -.2192 - j.9179 \text{ [rounded off])}$$

And there you have it—results that are the same as the results of problem 8.

Now to get back to where we initially started, we'll begin by analyzing RLC components that are parallel-series connected, followed by components that are series-parallel connected.

Parallel-Series RLC
Because we will be applying that which we have learned about phasors after analyzing the given circuit (using the same approach as with other circuits), we will refer to our usual way of analyzing this circuit as Method 1, followed by the application of Method 2. Method 2 will consist of analyzing certain quantities of the given circuit using phasors.

Observe figure 39.

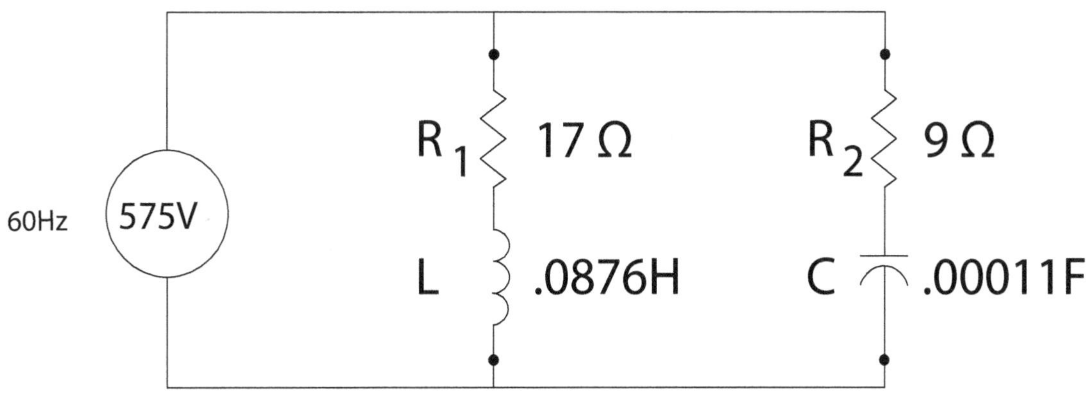

Fig. 39. Parallel-Series RLC combination circuit.

In observing the circuit, the first thing you will notice is that the RL and RC series components will experience different current, impedance, and phase angle quantities. However, what you might not realize is that the circuit will also experience a total current, a total impedance, and a phase angle that are independent of the series components. Because the two sets of parallel components are in series, each individual component of the same set will experience the same current flow, while both sets of components will experience the same voltage drop relative to the circuit's applied voltage.

With respect to the applied voltage, the individual current flow through each set of components will be in phase with the two resistance components and out of phase with the two reactance components. To be specific, by a difference of 90°, the applied voltage will lead the individual current flow through the inductor and lag the individual current flow about the capacitor. Because of such displacements, the individual flow of currents (I_{RL} and I_{RC}) through the RL and RC components cannot be vectorially totaled as in previous parallel-RLC components ($I_T = 1/\sqrt{[1/I_{RL}]^2 + [1/I_{RC}]^2}$) to determine the circuit's total current. As a result, the impedance and phase angle quantities about the RL and RC series components must first be determined to derive the in-phase currents (I_{R1} and I_{R2}) and the out-of-phase reactance currents (I_L and I_C) about each set of components. Once these are determined, the like current quantities can be totaled, and the circuit's total current can then be determined using the formula $I_T = \sqrt{(I_{R1/R2})^2 + ([I_{L-C}]_{or}[I_{C-L}])^2}$. So let's get started by first analyzing each group of series components individually.

Method 1. Applying the usual approach.

Inductance Reactance (X_L) and Capacitance Reactance (X_C)
{Parallel-Series RLC (Combination)}

Our first approach in analyzing this circuit is to determine the individual reactance of each reactance component. Based on the given components and operating frequency, the inductance reactance is

$$X_L = 2 \text{ x } 3.14 \text{ x } 60\text{Hz x } .0876\text{H} = 33.00\Omega$$

and the capacitance reactance is

$$X_C = \frac{1}{2 \text{ x } 3.14 \text{ x } 60\text{Hz x } .00011\text{F}} = 24.13\Omega$$

The Series Component's Impedance {Parallel-Series RLC (Combination)}

Now that both reactances are determined, (the magnitudes of) the impedances for each set of series components can also be determined. Because these components are in series, the formula $Z = \sqrt{(R)^2 + (X_{L} \text{ or } _{C})^2}$ will be used. Beginning with the RL components followed by the RC components,

$$Z_{RL} = \sqrt{(17)^2 + (33)^2} = \sqrt{289 + 1089} = \sqrt{1378} = 37.12\Omega$$

and

$$Z_{RC} = \sqrt{(9)^2 + (24.13)^2} = \sqrt{81 + 582.26} = \sqrt{663.26} = 25.75\Omega$$

Because the impedances for both sets of series components have been determined, you would think that the circuit's impedance (Z_T) could now be determined simply by applying the parallel resistance method, but it can't, and you will see why in the next procedure.

The Series Component's Phase Angles {Parallel-Series RLC (Combination)}
With the impedances of both sets of components, the phase angle for each set can now be determined using the formulas $\cos\theta_{(RL)} = R_1/Z_{RL}$ and $\cos\theta_{(RC)} = R_2/Z_{RC}$, where

$$\theta_{(RL)} = \cos^{-1}\frac{R_1}{Z_{RL}} \quad \text{and} \quad \theta_{(RC)} = \cos^{-1}\frac{R_2}{Z_{RC}}$$

Using these formulas, the phase angles for both sets of components are

$$\theta_{(RL)} = \cos^{-1}\frac{17\Omega}{37.12\Omega} = \cos^{-1} .4579 = 62.75°$$

and

$$\theta_{(RC)} = \cos^{-1}\frac{9\Omega}{25.75\Omega} = \cos^{-1} .3495 = 69.54°$$

Because both sets of series components have different phase angles, the impedances Z_{RL} and Z_{RC} cannot be calculated as is by means of the parallel resistance method to determine the circuit's impedance. Later on, however, by the use of phasors, we'll be able to calculate the circuit's impedance by means of the parallel resistance method and determine the directions of the phase angles—that is, whether they are (+) or (–), resulting in a leading or lagging effect.

The Series Component's Currents {Parallel-Series RLC (Combination)}
With each series component's impedance and the circuit's applied voltage across each of the parallel-series components, the currents I_{RL} and I_{RC} can be determined, which will lead to finding the circuit's total current (I_T). Therefore,

$$I_{RL} = \frac{575V}{37.12\Omega} = 15.49A \quad \text{and} \quad I_{RC} = \frac{575V}{25.75\Omega} = 22.33A$$

The Series Component's Individual Currents {Parallel-Series RLC (Combination)}
All of the above information can now be used to continue the process of determining the circuit's total current. Referring now to the currents I_{RL} and I_{RC},which flow through the series component's RL and RC respectively, it should be clear that both currents consist of the in-phase currents I_{R1} and I_{R2} and the out-of-phase reactance currents I_L and $I_{C.}$ Using the phase angles of the impedances Z_{RL} and Z_{RC}, the two sets of current components can be calculated, which will, as previously stated, lead to the finding of the circuit's total current. Now refer to figure 40.

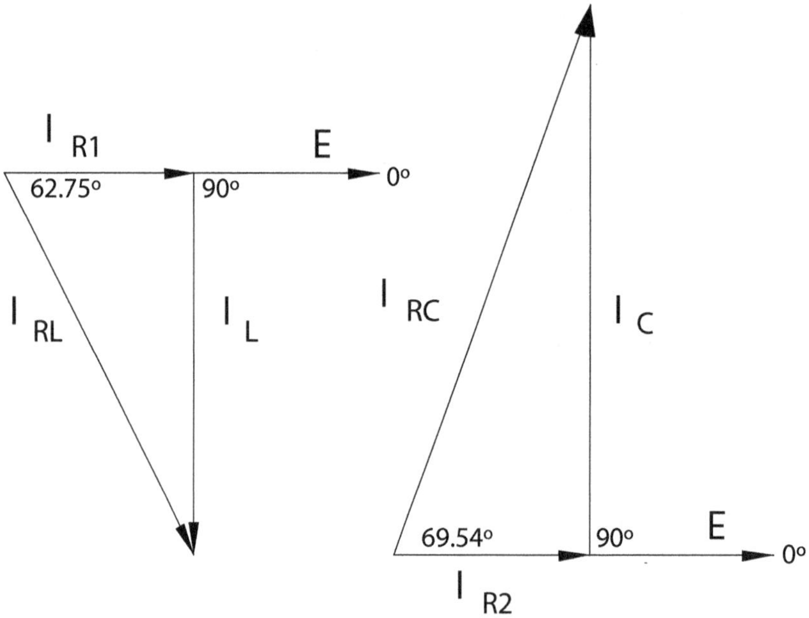

Fig. 40. RL and RC parallel-series components current triangles.

Figure 40 is an illustration of how the currents I_{RL} and I_{RC} and their in-phase and out-of-phase reactance currents appear graphically with respect to the impedances Z_{RL} and Z_{RC} phase angles and the circuit's applied voltage (E). Because the impedances Z_{RL} and Z_{RC} were derived from like components—that is, the components R_1 and L and R_2 and C, respectively—the use and application of both impedances are applicable. Based on the composition of each set of series components, the current I_{RL} lags the applied voltage E by 62.75°, while the current I_L lags the applied voltage E by 90°. In similar fashion, the current I_{RC} leads the applied voltage E by 69.54°, while the current I_C leads the applied voltage E by 90°. Having this illustration also makes it easier to see how the in-phase and out-of-phase currents can be determined using the trigonometric functions cosine and sine.

Use the cosine function to determine the in-phase currents I_{R1} and I_{R2}, where

$$I_{R1} = \cos 62.75° \times 15.49A = .4579 \times 15.49A = 7.09A$$

and

$$I_{R2} = \cos 69.54° \times 22.33A = .3496 \times 22.33A = 7.81A$$

Use the sine function to determine the out-of-phase currents I_L and I_C, where

$$I_L = \sin 62.75° \times 15.49A = .889 \times 15.49A = 13.77A$$

and

$$I_C = \sin 69.54° \times 22.33A = .9369 \times 22.33A = 20.92A$$

Now that both in-phase and out-of-phase current quantities are determined, the like quantities can finally be added together to derive the circuit's total current. However, before doing so, let's take the illustrations provided in figure 40 and merge them to illustrate the calculations that are about to take place between the two sets of current quantities. Refer now to figure 41.

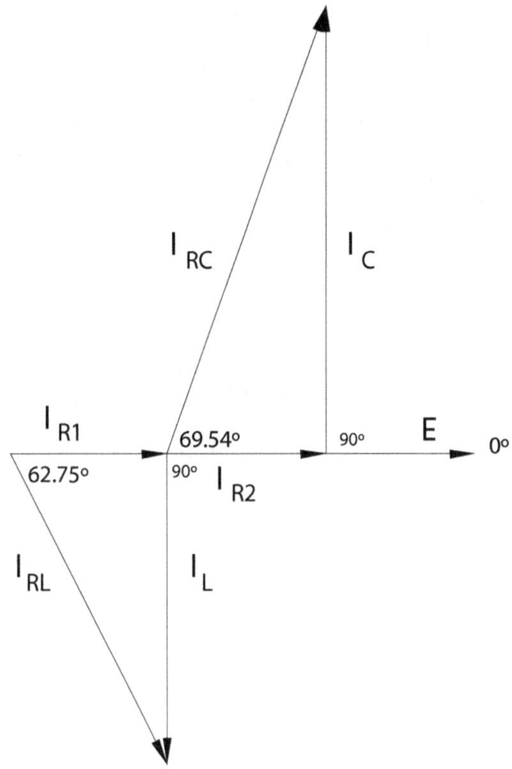

Fig. 41. Parallel-series vector currents.

Figure 41 provides a clear display of each current vector's displacements with respect to other current vectors. Now, with this illustration, let's calculate the additive sums of each set of like current components. As a result,

$$I_{(R1\&R2)} = 7.09A + 7.81A = 14.9A$$

$$I_{(CL)} = 20.92A + (-13.77A) = 7.15A$$

Notice how the current component I_L (13.77A) is shown as being a negative (-) quantity? Well, it is, and it's verified in figures 40 and 41 because the component I_L lies below the applied voltage vector E. When the application of phasors is applied later on, the type sign preceding a reactance component will be based on whether the component is being analyzed with respect to a reference current or voltage.

The Series Component's Voltage Drops {Parallel-Series RLC (Combination)}

With each set of series components and individual currents calculated, the voltage drops (E_{R1}, E_L, E_{R2}, E_C) across each component can be determined either collectively or individually. Although both resistance and reactance components are in parallel with the applied voltage (E), they are in series with each other. Therefore, the applied voltage across both sets of series resistance–reactance components will amount to the sum of the voltage drops across each set.

As a result, if calculated collectively, the series component's currents (I_{RL} and I_{RC}) can be used and the applied voltage verified with the formula $E = \sqrt{(E_R)^2 + (E_{L\ or\ C})^2}$ or the individual currents (I_{R1}, I_{R2}, I_L, and I_C) can be used and the applied voltage verified by adding the series components voltage drops together. Let's apply both means.

Voltage drops calculated collectively:

$$E_{R1} = I_{RL} \times R_1 = 15.49A \times 17\Omega = 263.33V$$

and

$$E_L = I_{RL} \times X_L = 15.49A \times 33\Omega = 511.17V$$

$$E = \sqrt{(E_{R1})^2 + (E_L)^2} = \sqrt{(263.33)^2 + (511.17)^2} = \sqrt{69{,}342.69 + 261{,}294.77} = \sqrt{330{,}637.46} = 575.01V$$

$$E_{R2} = I_{RC} \times R_2 = 22.33A \times 9\Omega = 200.97V$$

and

$$E_C = I_{RC} \times X_C = 22.33A \times 24.13\Omega = 538.82V$$

$$E = \sqrt{(E_{R2})^2 + (E_C)^2} = \sqrt{(200.97)^2 + (538.82)^2} = \sqrt{40{,}388.94 + 290{,}326.99} = \sqrt{330{,}715.93} = 575.08V$$

Voltage drops calculated individually:

$$E_{R1} = I_{R1} \times R_1 = 7.09A \times 17\Omega = 120.53V$$

and

$$E_L = I_L \times X_L = 13.77A \times 33\Omega = 454.41V$$

$$E_{R2} = I_{R2} \times R_2 = 7.81A \times 9\Omega = 70.3V$$

and

$$E_C = I_C \times X_C = 20.92A \times 24.13\Omega = 504.8V$$

To verify that the sums of the voltage drops (E_{R1} and E_L and E_{R2} and E_C) across each set of components are equal to the applied voltage, take note of the following:

$$120.53V + 454.41V = 574.94V \quad \text{and} \quad 70.3V + 504.8V = 575.1V$$

As expected, in each case, the calculated results are either approximate or the same as the applied voltage.

The Series Component's Apparent Powers (VA) {Parallel-Series RLC (Combination)}
Using that which was given as the circuit's voltage (E) and the calculated series component's currents (I_{RL} and I_{RC}) and impedances (Z_{RL} and Z_{RC}), the apparent power for each set of series components can be determined by use of the formulas $P_{VA(RL\ or\ RC)} = E \times I_{(RL\ or\ RC)}$, $I_{(RL\ or\ RC)}^2 \times Z_{(RL\ or\ RC)}$, and $E^2/Z_{(RL\ or\ RC)}$.

$P_{VA(RL)} = 575V \times 15.49A$
$= 8906.75VA$

$P_{VA(RL)} = (15.49A)^2 \times 37.12\Omega$
$= 239.94A \times 37.12\Omega$
$= 8906.57VA$

$P_{VA(RL)} = (575V)^2 / 37.12\Omega$
$= 330{,}625V/37.12\Omega$
$= 8906.92VA$

$P_{VA(RC)} = 575V \times 22.33A$
$= 12{,}839.75VA$

$P_{VA(RC)} = (22.33A)^2 \times 25.75\Omega$
$= 498.63A \times 25.75\Omega$
$= 12{,}839.72VA$

$P_{VA(RC)} = (575V)^2/25.75\Omega$
$= 330{,}625V/25.75\Omega$
$= 12{,}839.81VA$

The Series Component's True Powers (W) {Parallel-Series RLC (Combination)}
Based on the apparent power and phase angle for each set of series components, each set's true power can now be determined by use of the formulas, $P_{W(RL)} = P_{VA(RL)} \times \cos\theta_{(RL)}$ and $P_{W(RC)} = P_{VA(RC)} \times \cos\theta_{(RC)}$, where

$P_{W(RL)} = P_{VA(RL)} \times \cos\theta_{(RL)}$
$= 8906.75VA \times \cos 62.75°$
$= 8906.75VA \times .4579$
$= 4078.4W$

$P_{W(RC)} = P_{VA(RC)} \times \cos\theta_{(RC)}$
$= 12{,}839.75VA \times \cos 69.54°$
$= 12{,}839.75VA \times .3496$
$= 4488.78W$

Because they are like quantities, if the results of both sets are added, they will amount to the circuit's true power:

$$4078.4W + 4488.78W = 8567.18W$$

Later on, you will find that these results are approximately the same as that calculated for the circuit's true power.

The Series Component's Reactance Powers (VARs) {Parallel-Series RLC (Combination)}
Based on the apparent power and phase angle for each set of series components, each set's reactance power can now be determined by use of the formulas $P_{VARs(RL)} = P_{VA(RL)} \times \sin\theta_{(RL)}$ and $P_{VARs(RC)} = P_{VA(RC)} \times \sin\theta_{(RC)}$, where

$P_{VARs(RL)} = P_{VA(RL)} \times \sin\theta_{(RL)}$
$= 8906.75VA \times \sin 62.75°$
$= 8906.75VA \times .889$
$= 7918.1VARs$

$P_{VARs(RC)} = P_{VA(RC)} \times \sin\theta_{(RC)}$
$= 12{,}839.75VA \times \sin 69.54°$
$= 12{,}839.75VA \times .9369$
$= 12{,}029.56VARs$

Because both results are the products of the out-of-phase currents (I_L and I_C), and the currents are 180° out of phase with each other, the circuit's reactance power can be determined by subtracting the smaller reactance component from the larger reactance component of the two sets. Therefore, using this approach, the circuit's reactance power is

$$1{,}2029.56VARs - 7918.1VARs = 4111.46VARs$$

Just as the results of the true power, later on you will also find that these results are approximately the same as those calculated for the circuit's reactance power. However, to verify the results of both quantities, let's use the Pythagorean Theorem to determine whether they yield the same or approximate results as the circuit's apparent power as later calculated.

$$\begin{aligned} P_{VA} &= \sqrt{(8567.18)^2 + (4111.46)^2} \\ &= \sqrt{73{,}396{,}573.15 + 16{,}904{,}103.33} \\ &= \sqrt{90{,}300{,}676.48} \\ &= 9502.67\text{VA} \end{aligned}$$

The Series Component's Power Factors (PF)) {Parallel-Series RLC (Combination)}

To calculate the power factor for each set of series components, the formulas $PF_{(RL\ or\ RC)} = P_{W(RL\ or\ RC)}/P_{VA(RL\ or\ RC)}$ and $\cos\theta_{(RL\ or\ RC)} = PF_{(RL\ or\ RC)}$ are used. Therefore,

$$PF_{(RL)} = \frac{4078.4\text{W}}{8906.75\text{VA}} = .4579\ (45.79\%) \quad \text{and} \quad PF_{(RL)} = \cos 62.75° = .4579\ (45.79\%)$$

$$PF_{(RC)} = \frac{4488.78\text{W}}{12{,}839.75\text{VA}} = .3496\ (34.96\%) \quad \text{and} \quad PF_{(RC)} = \cos 69.54° = .3496\ (34.96\%)$$

Again, based on the resulting phase angles and the composition of each set of series components, the power factor about the RL components is 45.79% lagging, while the power factor about the RC components is 34.96% leading. Just as other situations, when the application of phasors is applied both leading and lagging effects will be more apparent.

The Circuit's Total Current {Parallel-Series RLC (Combination)}

Now that the results of the additive sums of each set of like currents are known, the illustration in figure 41 can be converted into a relative current triangle to finally derive the circuit's total current. Based on the results of the like quantities the circuit's total current can now be determined. As a result,

$$I_T = \sqrt{I_{(R1\&R2)}^{\ 2} + I_{(CL)}^{\ 2}}$$

where

$$\begin{aligned} I_T &= \sqrt{(14.9)^2 + (7.15)^2} \\ &= \sqrt{222.01 + 51.12} \\ &= \sqrt{273.13} \\ &= 16.53\text{A} \end{aligned}$$

Now observe figure 42.

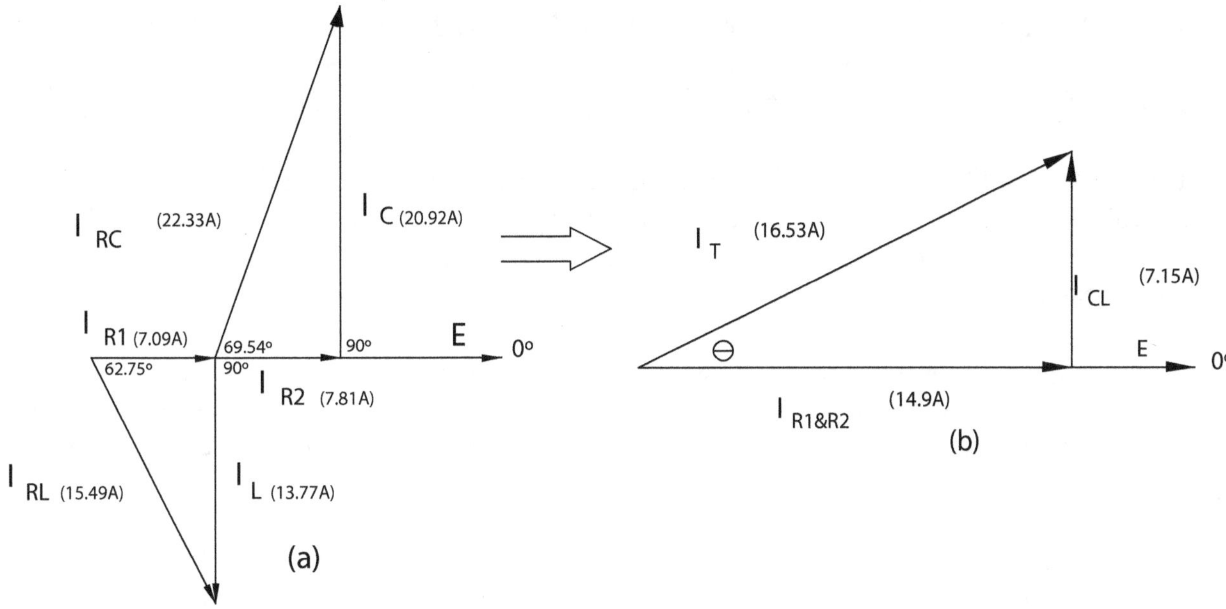

Fig. 42. Parallel-series circuit current triangle.

Figure 42 is the illustrated conclusion of figure 41 and the calculated results of the circuit's total current.

The Circuit's Impedance {Parallel-Series RLC (Combination)}

Now with the availability of the circuit's total current, the circuit's impedance can now be calculated using the formula $Z_T = E/I_T$. Therefore,

$$\begin{aligned} Z_T &= 575\text{V}/16.53\text{A} \\ &= 34.79\Omega \end{aligned}$$

The Circuit's Phase Angle {Parallel-Series RLC (Combination)

The circuit's phase angle can be determined using the current vector's relationships illustrated in figure 42(b), where the trigonometric functions

$$\cos\theta = \frac{I_{R1\&R2}}{I_T} \quad \text{and} \quad \sin\theta = \frac{I_{CL}}{I_T} \quad \text{and} \quad \tan\theta = \frac{I_{CL}}{I_{R1\&R2}}$$

when converted, result in

$$\theta = \cos^{-1}\frac{I_{R1\&R2}}{I_T} \quad \text{and} \quad \theta = \sin^{-1}\frac{I_{CL}}{I_T} \quad \text{and} \quad \theta = \tan^{-1}\frac{I_{CL}}{I_{R1\&R2}}$$

Therefore, applying either formula, the phase angle is

$$\theta = \cos^{-1} 14.9A/16.53A \quad \text{and} \quad \theta = \sin^{-1} 7.15A/16.53A \quad \text{and} \quad \theta = \tan^{-1} 7.15A/14.9A$$
$$= \cos^{-1} .9014 \qquad = \sin^{-1} .4325 \qquad = \tan^{-1} .4799$$
$$= 25.66° \qquad = 25.63° \qquad = 25.64°$$

Referring to figure 42(b), based on the resulting phase angle, the circuit's total current I_T leads the applied voltage E by 25.65° (θ), while the I_{CL} leads the applied voltage E by 90°. Although, the phase angle of the circuit's impedance has been determined later on by use of phasors we will define the direction of the phase angle.

The Circuit's Apparent Power (VA) {Parallel-Series RLC (Combination)}

Using that which was given as the circuit's voltage (E) and the calculated circuit's total current (I_T) and impedance (Z_T), the circuit's apparent power can be determined by use of the formulas $P_{VA} = E \times I_T$, $P_{VA} = I_T^2 \times Z$, and $P_{VA} = E^2/Z$.

$$P_{VA} = 575V \times 16.53A = 9504.75VA$$

$$P_{VA} = (16.53A)^2 \times 34.79\Omega = 273.24A \times 34.79\Omega = 9506.02VA$$

$$P_{VA} = (575V)^2/34.79\Omega = 330{,}625V/34.79\Omega = 9503.45VA$$

As you can see, the results are approximately the same compared to earlier calculations (9502.67VA).

The Circuit's True Power (W) {Parallel-Series RLC (Combination)}

Based on the circuit's apparent power and phase angle, the circuit's true power can now be determined by use of the formula $P_{W(T)} = P_{VA} \times \cos\theta$, where

$$P_{W(T)} = 9503.45VA \times \cos 25.63° = 9503.45VA \times .9016 = 8568.31W$$

compared to earlier calculations (8567.18W).

The Circuit's Reactance Power (VARs) {Parallel-Series RLC (Combination)}

Based on the circuit's apparent power and phase angle, the circuit's reactance power can also be determined by use of the formula $P_{VARs(T)} = P_{VA} \times \sin\theta$, where

$$P_{VARs(T)} = 9503.45VA \times \sin 25.63° = 9503.45VA \times .4326 = 4111.19VARs$$

compared to earlier calculations (4111.46VARs).

The Circuit's Power Factor (PF) {Parallel-Series RLC (Combination)}
To calculate the circuit's power factor for this parallel-series RLC combination circuit, the formulas $PF = P_{W(T)/}P_{VA}$ and $\cos\theta = PF$ are used. Therefore,

$$PF = \frac{8568.31W}{9503.45VA} = .9016\ (90.16\%) \quad \text{and} \quad PF = \cos 25.63^\circ = .9016\ (90.16\%)$$

Based on the resulting phase angle, the power factor for this circuit is 90.15% leading.

Method 2. Applying phasors.

We will now analyze the circuit applying phasors. However, now we will deal only with the circuit's impedance, total current, and phase angle.

The Circuit's Impedance {Parallel-Series RLC (Combination)}
From the information gathered from both sets of series components in Method 1, let's verify the magnitudes of the impedances Z_{RL} and Z_{RC} as earlier calculated by using the circuit's applied voltage E and the series component's currents I_{RL} and I_{RC}. As a result,

$$Z_{RL} = \frac{E}{I_{RL}} = \frac{575V}{15.49A} = 37.12\Omega \quad \text{and} \quad Z_{RC} = \frac{E}{I_{RC}} = \frac{575V}{22.33A} = 25.75\Omega$$

In comparison to the earlier calculated impedances, both magnitudes are the same (37.12Ω and 25.75Ω).

Refer to figure 42 to verify the positions of the phase angles relative to E, I_{RL}, and I_{RC}. With the illustrated phase angles, the impedances of Z_{RL} and Z_{RC} can be derived in both polar and rectangular forms, where

Form	Z_{RL}		Z_{RC}
Polar =	$37.12\underline{/-62.75^\circ}$	and	$25.75\underline{/69.54^\circ}$
Rectangular =	$17 - j33$		$9 + j24.13$

With the use of phasors, the circuit's impedance can now be determined simply by applying the product of the sum method. Observe:

$$Z_T = \frac{37.12\underline{/-62.75^\circ} \times 25.75\underline{/69.54^\circ}}{17 - j33 + 9 + j24.13}$$

$$\begin{array}{r} 17 - j33 \\ \underline{9 + j24.13} \\ 26 - j8.87 = 27.47\underline{/-18.84^\circ} \end{array}$$

$$= \frac{37.12 \times 25.75 \,/\underline{-62.75° + 69.54°}}{27.47/\underline{-18.84°}}$$

$$= \frac{955.84/6.79°}{27.47/\underline{-18.84°}}$$

$$= 34.796/\underline{6.79° - (-18.84°)}$$

$$= 34.796/\underline{25.63°} \quad \text{or} \quad 31.37 + j15.05$$

As you can see, the results are approximately the same as derived in Method 1 regarding the quantities of the magnitude and phase angle. Now you see how when using phasors, the direction of the resulting phase angle is always made known instead of having to ponder whether it's leading (+) or lagging (–) as realized in Method 1.

The Circuit's Total Current {Parallel-Series RLC (Combination)}
From what was gathered in Method 1 in reference to each component's in- and out-of-phase currents, the total current can be determined once the components of each phasor are written in rectangular form. Observe:

$$\begin{aligned} I_{R1} &= 7.09 + j0 \\ I_{R2} &= 7.81 + j0 \\ I_L &= 0 - j13.77A \\ I_C &= 0 + j20.92A \\ \hline I_T &= 14.9 + j7.15 \end{aligned}$$

Because the series components are in parallel, the sum of the total current is additive just like any other parallel circuit.

In polar form, the total current is

$$I_T = 16.52/\underline{25.65°}$$

The Circuit's Phase Angle {Parallel-Series RLC (Combination)}
Without saying, the circuit's impedance and total current in both situations concluded with the same (approximate) phase angle and they should have based on the total circuit. However, because the phase angle is positive, it signifies a leading effect, which results in the circuit having a leading power factor as expressed in Method 1. A leading power factor in this situation implies that the circuit's total current I_T leads the circuit's applied voltage E, which is consistent with figure 42(b). Figure 42(b) reflects a leading phase angle theta for the circuit where theta is 25.65°.

Series-Parallel RLC

Just as the objective in analyzing a parallel-series resistance circuit is to resolve all connected parallel components, to eventually transform the circuit into a series circuit, the objective is the same for analyzing a series-parallel RLC combination circuit. In the preceding parallel-series circuit, the impedances of the two resistance-reactance series components were eventually reduced to one impedance, that being the circuit's total impedance. Well, the same approach is used in analyzing a series-parallel RLC circuit. Once the multiple sets of parallel impedances are narrowed down into one total impedance, it is combined with the circuit's series components and analyzed as a series circuit.

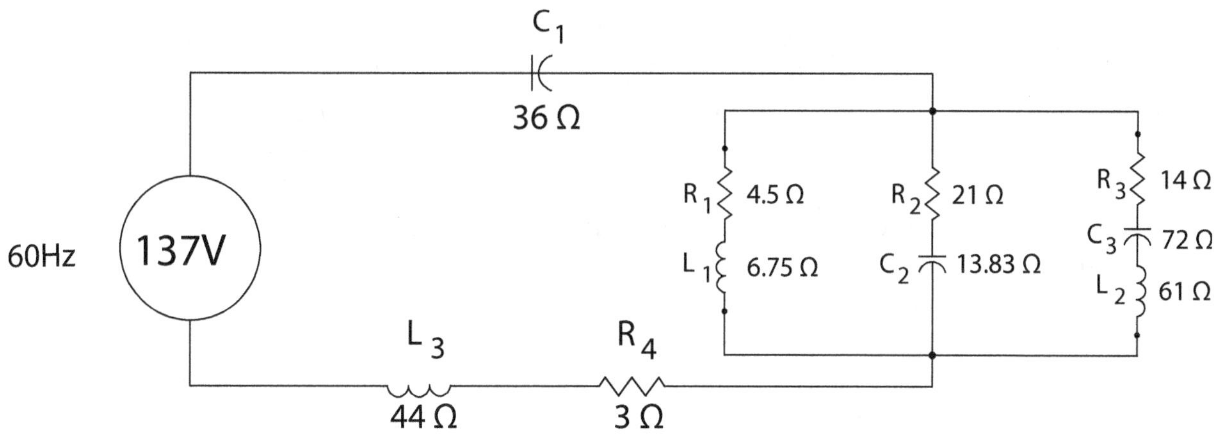

Fig. 43. Series-parallel RLC combination circuit.

Observe figure 43. Because the ohmic values of the reactance components are given, the impedance of each set of parallel-series components can be immediately determined. However, before proceeding, take an in-depth look at figure 43 again to acknowledge that once the total impedance of the multiple sets of parallel-series components is determined, the circuit will no longer be a combination circuit and can be treated as a series-RLC circuit instead. In addition, the parallel-series components of this circuit should also be a lot easier to analyze because they can be solved in the same manner as the previous parallel-series RLC combination circuit. Unlike the previous circuit, this circuit will be analyzed using the combination of both methods. Now let's get started.

Method 1. Applying the usual approach.

The Parallel-Series Component's Impedance {Series-Parallel RLC (Combination)}

Again, since these components are in series, the formula $\mathbf{Z} = \sqrt{(R)^2 + (X_{L \text{ or } C})^2}$ will be used. Beginning with the RL components and concluding with the RLC components, (the magnitudes of) the impedances are

$$\mathbf{Z}_{\mathbf{RL}} = \sqrt{(4.5)^2 + (6.75)^2} = \sqrt{20.25 + 45.56} = \sqrt{65.81} = 8.11\Omega$$

$$\mathbf{Z}_{\mathbf{RC}} = \sqrt{(21)^2 + (13.83)^2} = \sqrt{441 + 191.27} = \sqrt{632.27} = 25.14\Omega$$

$$\mathbf{Z}_{\mathbf{RCL}} = \sqrt{(14)^2 + (72 - 61)^2} = \sqrt{196 + 121} = \sqrt{317} = 17.80\Omega$$

The Parallel-Series Component's Phase Angles {Series-Parallel RLC (Combination)}
Applying the impedances of each set of components, the phase angle for each impedance can be determined using the formulas

$$\theta_{(RL)} = \cos^{-1}\frac{R_1}{Z_{RL}} \quad \text{and} \quad \theta_{(RC)} = \cos^{-1}\frac{R_2}{Z_{RC}} \quad \text{and} \quad \theta_{(RCL)} = \cos^{-1}\frac{R_3}{Z_{RCL}}$$

Using these formulas, the phase angle for each impedance is

$$\theta_{(RL)} = \cos^{-1}\frac{4.5\Omega}{8.11\Omega} = \cos^{-1} .5549 = 56.30°$$

and

$$\theta_{(RC)} = \cos^{-1}\frac{21\Omega}{25.14\Omega} = \cos^{-1} .8353 = 33.35°$$

and

$$\theta_{(RCL)} = \cos^{-1}\frac{14\Omega}{17.80\Omega} = \cos^{-1} .7865 = 38.14°$$

The Parallel-Series Resistance-Reactance Component's Voltage Drop {Series-Parallel RLC (Combination)}—Part I*
Because the three sets of parallel-series resistance-reactance components are in parallel with three individual series components—C_1, R_4, and L_3—the voltage drops across the three sets will not be the same as the circuit's applied voltage. Because this is the case, the only way to determine the actual voltage drops across the three sets is to have knowledge of some other known quantities, such as the current flow or the power consumption for any of the three sets. Being that they are in parallel and the voltage drops across each set of components would be the same, this portion of the circuit could be solved just as the previous parallel-series circuit. However, the fact remains, there are no other known quantities to work with, which results in this circuit being left at a standstill. So what do we do?

With the availability of phasors, the total impedance of the three sets of parallel-series resistance-reactance components can be determined, resulting in this circuit being converted into a series RLC circuit and then analyzed accordingly. So let's determined the total impedance of the three sets by use of phasors and then use the results to determine other unknown quantities to include the voltage drops across the parallel-series resistance-reactance components.

* Part I introduces the process that must take place before the results of this topic occurs. Refer to Part II for the conclusion.

The Parallel-Series Resistance-Reactance Component's Total Impedance {Series-Parallel RLC (Combination)}
With the product of the sum method, the total impedance of the parallel-series components can be determined. The only thing is, in using this method only two of the parallel-series components can be analyzed at one time. Refer to figure 44. Figure 44 is an illustration of the parallel-series resistance-reactance components and will be used to illustrate and derive the total impedance of the three parallel sets.

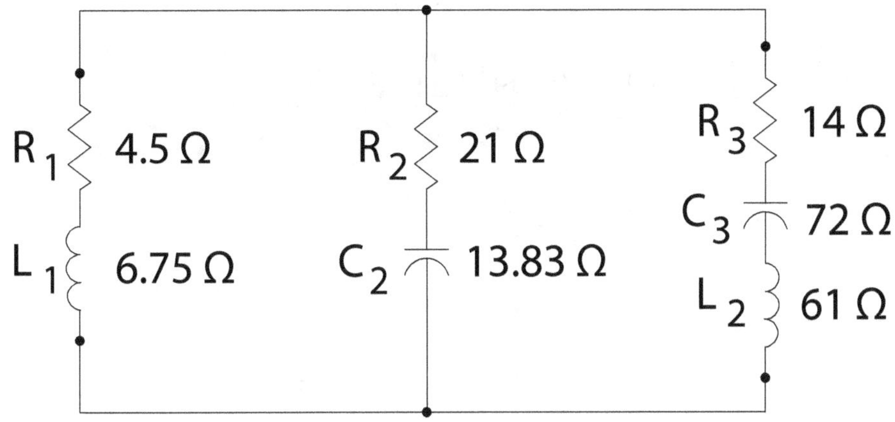

Fig. 44. Parallel-series resistance-reactance components.

Being that the magnitudes and phase angles of all three impedances have already been determined, we can use these quantities to convert them from polar to rectangular forms to allow the application of the implied (product of the sum) method. One thing to remember is, when current flows through L_1, it will lag the unknown voltage across the three parallel sets, which will cause the phase angle about the impedance Z_{RL} to lag also.

Now to get started, the impedances of the RL, RC and RLC components in polar forms are

$$Z_{RL} = 8.11 \underline{/-56.30°} \qquad Z_{RC} = 25.14 \underline{/33.35°} \qquad Z_{RCL} = 17.80 \underline{/38.14°}$$

To obtain the rectangular forms of each set of components the resistance-reactance quantities will be taken directly from the circuit to derive the equivalent phasors. Just remember the magnitudes and phase angles are rounded off quantities. Therefore,

$$\mathbf{Z}_{\mathbf{RL}} = 8.11 \underline{/-56.30°} = 4.5 - j6.75$$

$$\mathbf{Z}_{\mathbf{RC}} = 25.14 \underline{/33.35°} = 21 + j13.82$$

$$\begin{aligned}\mathbf{Z}_{\mathbf{RCL}} = 17.80 \underline{/38.14°} &= 14 + j72 - j61 \\ &= 14 + j11\end{aligned}$$

Working from right to left, we'll begin with the impedances $\mathbf{Z}_{\mathbf{RCL}}$ and $\mathbf{Z}_{\mathbf{RC}}$ to initiate finding the set's total impedance. Observe,

$$\mathbf{Z}_{\mathbf{RCL/RC}} = \frac{17.80 /38.14° \text{ x } 25.14/33.35°}{14 + j11 + 21 + j13.82}$$

$$\begin{array}{l} 14 + j11 \\ \underline{21 + j13.82} \\ 35 + j24.82 = 42.91 \underline{/35.34°} \end{array}$$

$$= \frac{17.80\angle 38.14° \times 25.14\angle 33.35°}{42.91\angle 35.34°}$$

$$= \frac{447.49\angle 38.14° + 33.35°}{42.91\angle 35.34°}$$

$$= \frac{447.49\angle 71.49°}{42.91\angle 35.34°}$$

$$= 10.43\angle 71.49° - 35.34°$$

$$= 10.43\angle 36.15° \quad \text{or} \quad 8.42 + j6.15$$

Now that we know the impedance between the two impedances, $\mathbf{Z_{RCL}}$ and $\mathbf{Z_{RC}}$, refer to figure 44(a) to review the results.

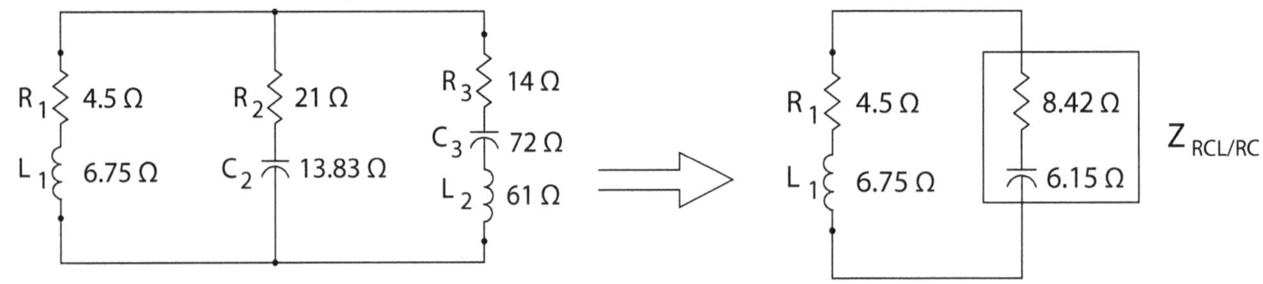

Fig. 44(a).

With the impedance $Z_{RCL/RC}$, we can apply it along with the impedance Z_{RL} to find the set's total impedance $Z_{RCL/RC/RL}$. Therefore,

$$Z_{RCL/RC/RL} = \frac{10.43\angle 36.15° \times 8.11\angle -56.30°}{8.42 + j6.15 + 4.5 - j6.75}$$

$$\begin{array}{l} 8.42 + j6.15 \\ \underline{4.5 \;\; - j6.75} \\ 12.92 - j.60 = 12.93\angle -2.66° \end{array}$$

$$= \frac{10.43\angle 36.15° \times 8.11\angle -56.30°}{12.93\angle -2.66°}$$

$$= \frac{84.59\angle 36.15° + (-56.30°)}{12.93\angle -2.66°}$$

$$= 6.54\angle -20.15° - (-2.66°)$$

$$= 6.54\angle -17.49° \quad \text{or} \quad 6.24 - j1.97$$

Refer now to Figure 44(b) to review the transitions of the impedances which led to determining the total impedance $Z_{RCL/RC/RL}$. Take notice that the derived total impedance has a negative (–) phase angle, which results in a lagging effect. As a result, current (I) will lag the voltage (E) across the total impedance ($Z_{RCL/RC/RL}$), meaning the application will take on the features of a RL circuit.

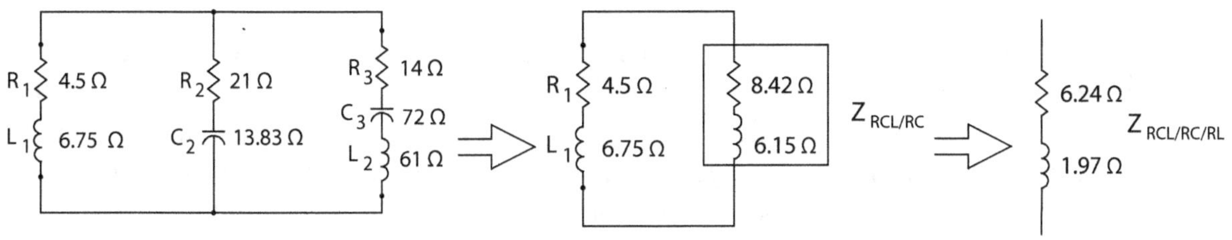

Fig. 44(b). Transitions of the impedances.

The Circuit's Total Impedance {Series-Parallel RLC (Combination)}

Now that we know the total impedance of the three parallel sets, the total can be substituted into the existing circuit to become an RLC series circuit. Refer to figure 45. Because the enclosed resistance and inductor are the equivalent results of the total impedance $Z_{RCL/RC/RL}$, they are referred to as R_{ZT} and L_{ZT}.

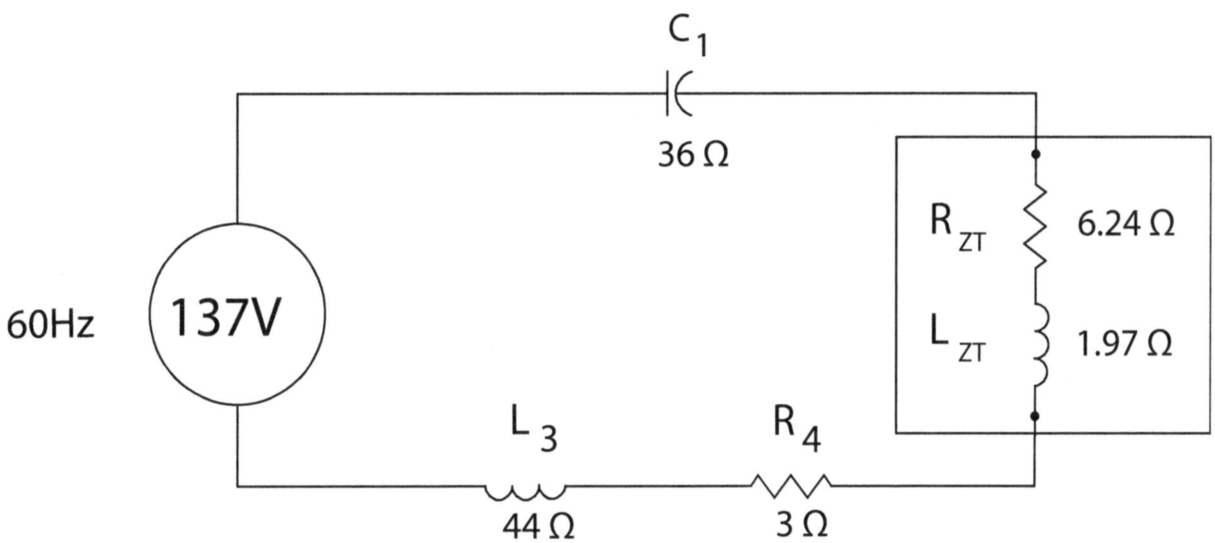

Fig. 45. Total impedance becomes an RLC series circuit.

Having the components in series will allow the circuit's total impedance to be determined simply by adding the components together, resulting in

$$
\begin{array}{llcl}
Z_T = & 0 & - & j36.0 \\
 & 6.24 & + & j0 \\
 & 0 & + & j\,1.97 \\
 & 3.0 & + & j0 \\
 & \underline{0} & \underline{+} & \underline{j44.0} \\
 & 9.24 & + & j9.97 = 13.59\underline{/47.18^\circ}
\end{array}
$$

Because the original series-parallel circuit has been converted into an equivalent RLC series circuit, there will only be one source of current I and multiple voltage drops E_{C1}, $E_{R(ZT)}$, $E_{L(ZT)}$, E_{R4} and E_{L3} across each component. As a result, the voltage drops $E_{L(ZT)}$ and E_{L3} across the inductors $L_{(ZT)}$ and L_3 will lead the current I by 90˚, while the voltage drop E_{C1} across the capacitor C_1 will lag the current I by 90°. This explains why the capacitance phasor component (36) is preceded with a negative (–) sign and the inductance phasor component (1.97) and (44) are preceded with positive (+) signs in the circuit's total impedance calculation, Z_T.

If inductance and capacitance reactances were used instead of phasors the same results would be obtained, for this was the approach used previously for the parallel-series RLC circuit in Figure 39 for calculating impedance. To verify the total impedance using the previous approach the out of phase (phasors) components are converted into reactance components and solved according. Just remember the circuit is still evaluated as an RLC series circuit. As a result, the phasor -j36 is the same as a 36Ω capacitance reactance and the phasors j1.97 and j44 are the same as 1.97Ω and 44Ω inductance reactances. As far as the resistances the total is still 9.24Ω (3Ω + 6.24Ω). Therefore, using the previous approach,

$$
\begin{aligned}
Z_T &= \sqrt{(9.24)^2 + (44 - 36 + 1.97)^2} \\
&= \sqrt{(9.24)^2 + (9.97)^2} \\
&= \sqrt{85.38 + 99.40} \\
&= \sqrt{184.78} \\
&= 13.59\Omega
\end{aligned}
$$

while the phase angle results in,

$$
\begin{aligned}
\theta &= \cos^{-1} \frac{9.24\Omega}{13.59\Omega} \\
&= \cos^{-1} .6799 \\
&= 47.16^\circ
\end{aligned}
$$

Although not the same, both derived phase angles are approximate. With the circuit's total impedance in place and verified, we can now move forward in finding other unknown quantities.

The Circuit's Current and Phase Angle {Series-Parallel RLC (Combination)}
Based on the circuit's total impedance and applied voltage the circuit's total current is,

$$I = \frac{137V}{13.59\Omega} = 10.08A$$

Because this circuit is being evaluated as an equivalent RLC series circuit and resulted to being more inductive than capacitive based on the total inductance reactances (45.97Ω) being larger than the capacitance reactance (36Ω), the circuit's current ($I_{10.08A}$) will lag the applied voltage (E_{T137V}), which, in terms, will yield a lagging phase angle.

The Circuit's Power Factor {Series-Parallel RLC (Combination)}
Based on the circuit's derived phase angle per calculated total current the power factor for this circuit is,

$$\text{Power Factor (PF)} = \cos 47.18° = .6797$$

The circuit's power factor is .6797 or 67.97% lagging.

The Individual Voltage Drops per Equivalent Series RLC circuit {Series-Parallel RLC (Combination)}
With the circuit's total current the individual voltage drops across each component illustrated in figure 45 can be determined. As a result,

$$E_{C1} = 10.08A \times 36\Omega = 362.88V$$
$$E_{R(ZT)} = 10.08A \times 6.24\Omega = 62.90V$$
$$E_{L(ZT)} = 10.08A \times 1.97\Omega = 19.86V$$
$$E_{R4} = 10.08A \times 3\Omega = 30.24V$$
$$E_{L3} = 10.08A \times 44\Omega = 443.52V$$

Although the voltage drops were calculated based on the equivalent RLC series circuit, the circuit's applied voltage cannot be determined simply by totaling the individual voltage drops due to the phase differences of the in-phase and out-of-phase components. In order to determine the circuit's applied voltage the formula, $E = \sqrt{(E_{RT})^2 + (E_{[L-C]T})^2}$ must again be used, where,

$$E_{RT} = 62.90V + 30.24V = 93.14V \quad \text{and} \quad E_{[L-C]T} = 443.52V + 19.86V - 362.88V = 100.5V$$

which results in

$$E = \sqrt{(93.14)^2 + (100.5)^2}$$
$$= \sqrt{8675.1 + 10{,}100.25}$$
$$= \sqrt{18{,}775.35}$$
$$= 137.02V$$

As you can see, the results are approximately the same as the given applied voltage (137V).

The Parallel-Series Resistance-Reactance Component's Voltage Drop {Series-Parallel RLC (Combination)}—Part II

Now that we know the total impedance of the parallel-series components and the circuit's total current, the voltage drop across the parallel-series components can be determined. Once the voltage drop is determined, the current flow through each set of components can then be determined also. With the calculated phasors $Z_{RCL/RC/RL} = 6.54\Omega$ and $I = 10.08A$, the product of the desired voltage drop is

$$\begin{aligned} E_{RCL/RC/RL} &= Z_{RCL/RC/RL} \times I_T \\ &= 6.54\Omega \times 10.08A \\ &= 65.92V \end{aligned}$$

The magnitude of this result can also be verified by using the individual voltage drops $E_{R(ZT)}$ and $E_{L(ZT)}$ which are products of the parallel-series resistance-reactance component's total impedance where

$$\begin{aligned} E_{R(ZT)/}E_{L(ZT)} &= \sqrt{(62.90)^2 + (19.86)^2} \\ &= \sqrt{3956.41 + 394.42} \\ &= \sqrt{4350.83} \\ &= 65.96V \end{aligned}$$

For clarity, $E_{RCL/RC/RL}$ and $E_{R(ZT)/}E_{L(ZT)}$ are the same component-wise and are approximately the same in magnitudes (65.92V and 65.96V).

The Parallel-Series Component's Currents {Series-Parallel RLC (Combination)}

With the voltage drop across the components in place and the impedances of each set of components being known, the parallel-series current quantities are

$$I_{RL} = \frac{65.96V}{8.11\Omega} = 8.13A \text{ (at } -56.30°)$$

$$I_{RC} = \frac{65.96V}{25.14\Omega} = 2.62A \text{ (at } 33.35°)$$

$$I_{RCL} = \frac{65.96V}{17.80\Omega} = 3.71A \text{ (at } 38.14°)$$

The Parallel-Series Component's Individual Currents {Series-Parallel RLC (Combination)}

With each current quantity, the individual currents or the in-phase resistance currents and the out-of-phase reactance currents can be determined.

$$\begin{aligned} I_{R1} &= 8.13A \times \cos -56.30° \\ &= 8.13A \times .5548 \\ &= 4.51A \end{aligned}$$

$$\begin{aligned} I_{R2} &= 2.62A \times \cos 33.35° \\ &= 2.62A \times .8353 \\ &= 2.19A \end{aligned}$$

$$\begin{aligned} I_{R3} &= 3.71A \times \cos 38.14° \\ &= 3.71A \times .7865 \\ &= 2.92A \end{aligned}$$

$I_{L1} = 8.13A \times \sin -56.30°$
$= 8.13A \times -.8320$
$= -6.76A$

$I_{C2} = 2.62A \times \sin 33.35°$
$= 2.62A \times .5498$
$= 1.44A$

$I_{C3L2} = 4.56 \times \sin 38.14°$
$= 3.71 \times .6176$
$= 2.29A$

Now let's take these calculated current quantities to validate the circuit's current (I), which will divide among the three sets of parallel-series components per figure 45. To do this, we'll begin by totaling the in-phase resistance currents and the out-of-phase reactance currents to derive and validate the desired results.

In-phase resistance current total ($I_{R(T)}$) = 4.51A + 2.19A + 2.92A = 9.62A.

Out-of-phase reactance current total ($I_{L/C(T)}$) = –6.76A + 1.44A + 2.29A = –3.03A.

$$I = \sqrt{(9.62)^2 + (-3.03)^2}$$
$$= \sqrt{92.544 + 9.181}$$
$$= \sqrt{101.73}$$
$$= 10.09A$$

Based on the derived results, the circuit's current is validated and shows to be approximately the same as that earlier quantity earlier derived (10.08A).

The Circuit's Apparent Power (VA), True Power (W) and Reactance Power (VARs) {Series-Parallel RLC (Combination)}

Using that which was given as the circuit's voltage (E), the calculated circuit's total current (I_T), and phase angle (θ), the circuit's apparent power, true power, and reactance power are determined.

$P_{VA} = 137V \times 10.08A$
$= 1380.96VA$

$P_{W(T)} = 1380.96VA \times \cos 47.18°$
$= 1380.96VA \times .6797$
$= 938.64W$

$P_{VARs(T)} = 1380.96VA \times \sin 47.18°$
$= 1380.96VA \times .7335$
$= 1012.93VARs$

This concludes the study of alternating current circuits. Overall, twelve types of alternating current circuits have been covered, ranging from a single-component series-RL circuit to a series-parallel RLC combination circuit.

SINGLE-PHASE (1ϕ) POWER–THREE-PHASE (3ϕ) POWER UNDERSTANDING THE DIFFERENCE

Single-phase. Three-phase. Without a doubt, the difference only appears to be two phases. You would think that something as simple as this would not be difficult to understand. Actually, what it all comes down to is one single conductor. Considering an exclusive line-to-line single-phase and three-phase system, clearly the difference is that the single-phase system has two line conductors, while the three-phase system has three line conductors. Now there's no need to become alarmed, because all of this will be covered in this section. But speaking of difficulty, for many, the most difficult part of understanding three-phase power and how it differs from single-phase power is applying the square root of three, the only difference. Because three-phase power is usually the most difficult of the two types of power sources to understand and apply, more focus will be placed on three-phase systems.

Single-Phase (1ϕ) Power

The output voltage of a (single-phase) rotating alternating current *(ac)* generator* is the resulting effects of an interaction that occurs between a produced alternating magnetic field and the generator's stator windings. Voltage, if you remember is the electrical force, which causes the movement of electrons, which produces the flow of current. Once the output voltage and the flow of current are produced the availability of single phase electrical power exist.

Regardless of whether it's a single-phase *ac* generator or a single-phase transformer, only one set of windings is used. If it's a single-phase transformer, the secondary windings are used to supply electrical voltage and power to the end user, which is referred to as single-phase service. In a single-phase service, two line conductors and a neutral (or grounded) conductor are provided to produce a 240/120V, 3-wire, single-phase service via transformer. Typically, because transformers are not manufactured to produce a secondary single-phase, 3-wire, 208/120V secondary voltage independently, when such voltage needs are required, it must originate from a three-phase, 4-wire, 208/120V source. Figure 46 is an illustration of an *ac* sine wave depicting a single-phase output voltage.

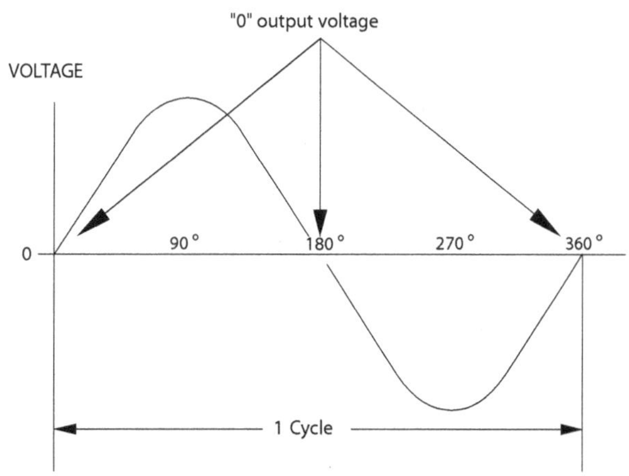

Fig. 46. An *ac* sine wave depicting a single-phase output voltage.

Take notice of the three times the output voltage produces zero volts during the completion of one full cycle. Although not as efficient as three-phase voltage (power), there will always be a need for single-phase voltage and single-phase systems, as not all electrical installations justify the need for three-phase power.

*An alternating current *(ac)* generator is also called an alternator.

Three-Phase (3ϕ) Power

In contrast, three-phase power is produced by the same principles as single-phase power, except a three-phase generator is used, which consists of three separate (phase) windings displaced 120° apart. In the same manner, an alternating magnetic field (amf), as shown in figure 47, causes an interacting effect in the windings (A,B,C), which produces by induction three output voltages that vary in magnitude and angular direction with respect to time throughout the rotation of the generator. The end results are the availability of three-phase power. Figure 48 is an illustration depicting the sine waves of each output voltage according to phase, along with an isolated depiction of the output voltage at a given time and angle.

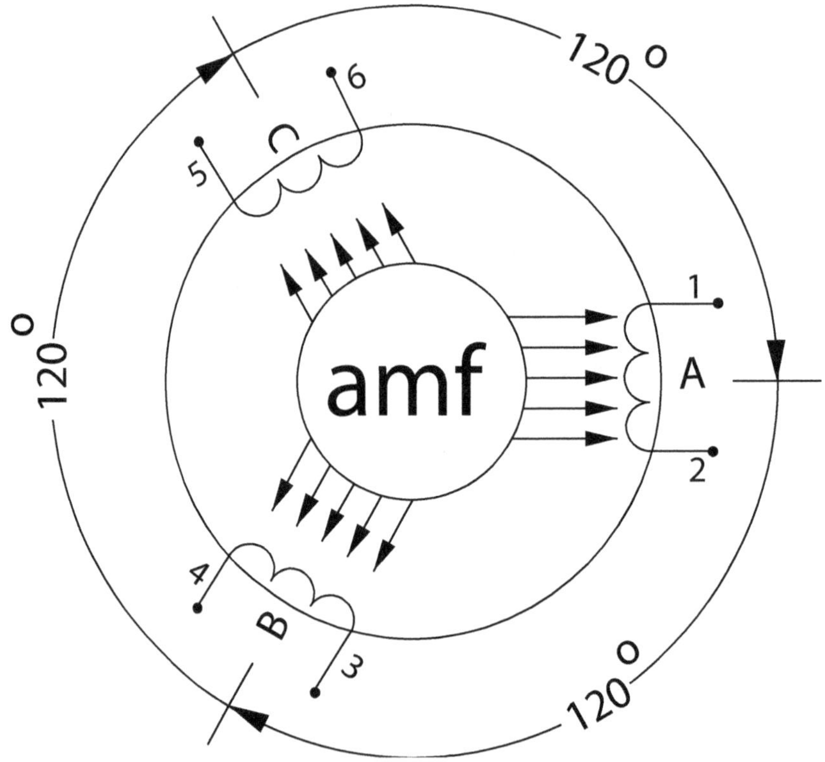

Fig. 47. Alternating magnetic field.

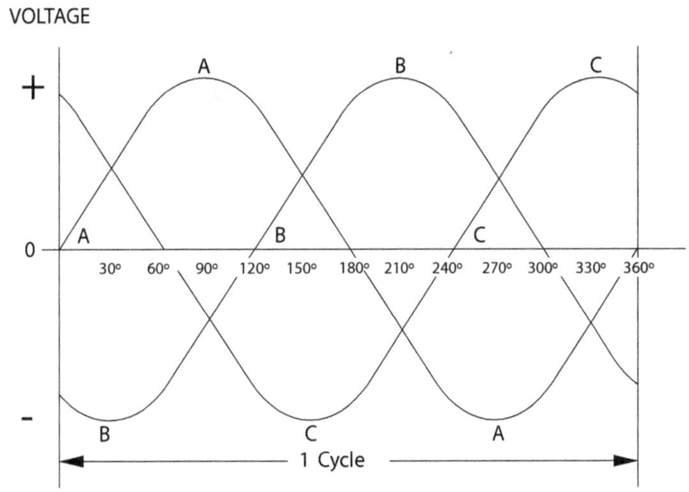

Fig. 48. Sine waves of output voltage according to phases.

To better understand the origin and production of three-phase power, let's begin with the three individual windings in figure 47. Observe how windings A, B, and C each have dual endpoints labeled 1-2, 3-4, and 5-6 . If endpoints 1, 3, and 5 were connected together, leaving endpoints 2, 4, and 6 free, such connections would be recognized as a "Y" connection (spelled w-y-e). Endpoints 2, 4, and 6 are where the three-phase output voltages (OV) would be supplied from the generator when wye-connected. On the other hand, if endpoints 2 and 3, 4 and 5, and 6 and 1 were connected in series, and three separate conductors were connected in parallel between each set of endpoint connections, this type connection would be recognized as a delta connection. The free ends of the parallel-connected conductors are where the three-phase output voltage is supplied from the generator when delta-connected. Observe figure 49. Where high voltage is needed, a wye connection is used, and where high current is needed, a delta connection is used.

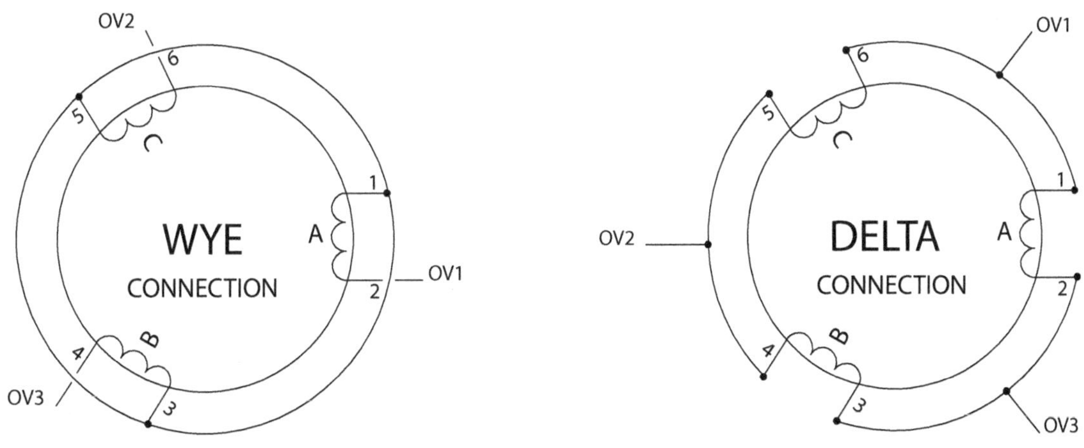

Fig. 49. Wye connection or delta connection.

When three-phase wye-connected loads are supplied from a three-phase generator, it is normally illustrated as shown in figure 50.

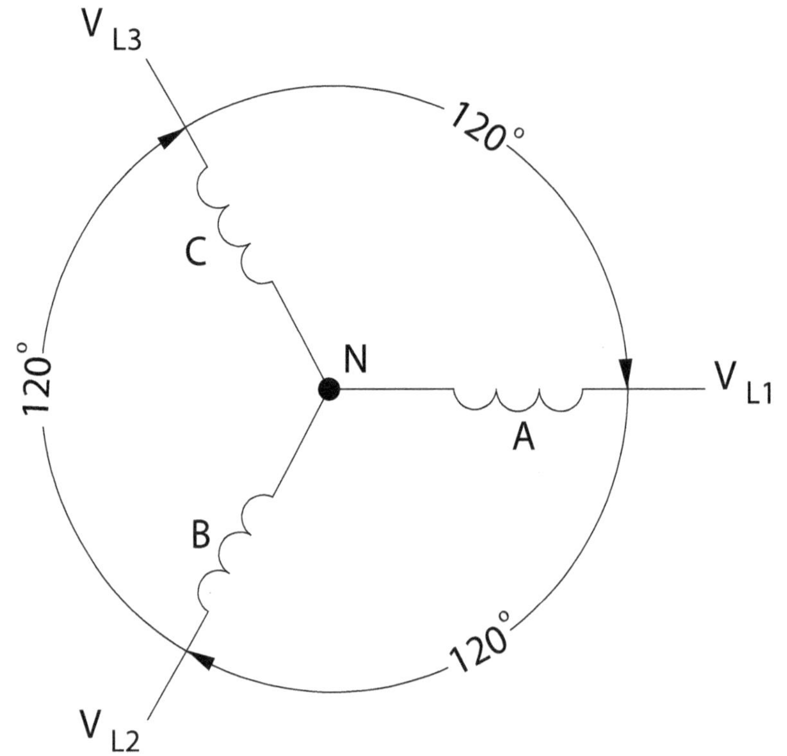

Fig. 50. Three-phase wye-connected loads supplied by a three-phase generator.

Notice how the physical displacements of the windings at 120° apart cause the overall configuration to resemble the letter "Y," making the name more obvious now. With a wye connection, there are three line voltages and three neutral voltages. The line voltages as shown in figure 50 are V_{L1}, V_{L2}, and V_{L3}, whereas the neutral voltages are V_{L1N}, V_{L2N}, and V_{L3N}. The voltages between V_{L1} and V_{L2}, V_{L2} and V_{L3}, and V_{L1} and V_{L3} are called line-to-line voltages. The voltages between V_{L1}, V_{L2}, V_{L3}, and N are called line-to-neutral voltages, or just neutral voltages. As you can see, the line and neutral voltages both have a mutual relationship, but the relationship most often not seen is the one that causes them to differ by the square root of three ($\sqrt{3}$).

Deriving the Square of Three ($\sqrt{3}$)

Perhaps from the first time you were introduced or made aware of three-phase systems, you may have wondered where the $\sqrt{3}$ came from or why was it required. Now that we have a much better understanding of vectors and right triangles, let's use both to see how the $\sqrt{3}$ came into play for three-phase applications. Because each line and neutral voltage of a wye connection has magnitude and direction, both are considered vectors. Now as far as the right triangle is concerned, we'll have to create one but that won't be too difficult. Because the line voltages of a wye connection are all the same and differ in phases by 120°, any part of the configuration shown in figure 50 can be used. Being the case, let's use phases A and C and their related line voltages $\mathbf{V}_{L1}$ and $\mathbf{V}_{L3}$. Figure 51(a) is the resulting replica. Figure 51(b) is a transformation of Figure 51(a).

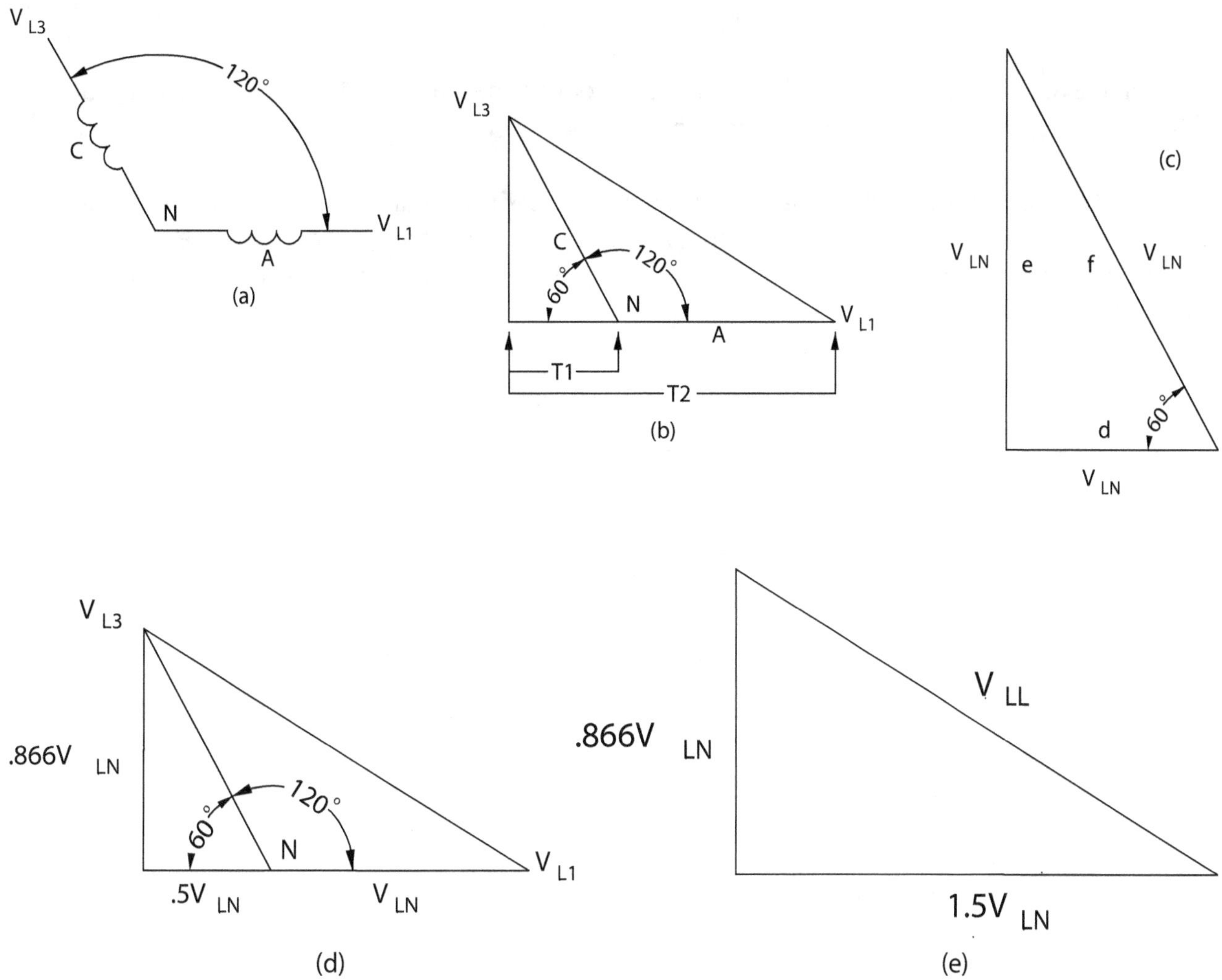

Fig. 51(a– e).

Notice how the neutral voltage vector V_{L1N} was extended a half of a distance to the left, and a line extended perpendicular from V_{L3}, to form a right triangle. Now two right triangles (T1 and T2) are formed, but we'll concentrate on the smaller triangle (T1) to define its extended sides. Figure 51(c) is a replica of the smaller triangle. Because the triangle was derived from the neutral voltage vectors V_{L1N} and V_{L3N}, we'll refer to it as the neutral triangle. Because it is a neutral triangle, its outer sides are commonly identified as V_{LN}, while its inner sides will be referred to as d, e, and f. Because sides d and e are respectively adjacent and opposite to the 60° angle, it can be said that

$$d = V_{LN}(f) \times \cos 60° \quad \text{and} \quad e = V_{LN}(f) \times \sin 60°$$
$$= V_{LN}.5 \qquad\qquad\qquad\qquad = V_{LN}.866$$

Now that sides d and e have been identified, it is safe to say that figure 51(d) is an extension of figure 51(b). Being that the original side V_{L1N} is also a related neutral voltage, it too can be labeled V_{LN}. As a result, this original side, when added to side d, results in

$$V_{LN} + .5V_{LN} = 1.5\ V_{LN}$$

while side e remains as is. The conclusion of it all produces figure 51(e). Because V_{L1} and V_{L3} are line voltages, the resultant vector between the two can be identified as V_{LL}.

Figure 51(e) illustrates the relationship between the line-to-line voltages and the line-to-neutral voltages of a wye connection, but most importantly it paves the way for "deriving the $\sqrt{3}$." Just like any other right triangle, and in the same manner (applying the Pythagorean Theorem), the line-to line voltage (V_{LL}) can be determined as

$$\begin{aligned} V_{LL} &= \sqrt{(1.5\ V_{LN})^2 + (.866\ V_{LN})^2} \\ &= \sqrt{2.25(V_{LN})^2 + .75(V_{LN})^2} \\ &= \sqrt{2.25 + .75\ (V_{LN})^2} \\ &= \sqrt{3\ (V_{LN})^2} \\ &= \sqrt{3} \text{ x } \sqrt{(V_{LN})^2} \\ &= \sqrt{3} \text{ x } V_{LN}^{2 \text{ x } 1/2} * \\ &= \sqrt{3} \text{ x } V_{LN} \end{aligned}$$

*Because the square root of any number is the same as raising a number to the half power, this procedure was used to isolate V_{LN}.

And there you have it, the relationship between the line-to-line voltage and the line-to-neutral voltage of a wye connection, where

$$V_{LL} = \sqrt{3} \text{ x } V_{LN}$$

and the realization that the ratio between the line-to-line voltage and the line-to-neutral voltage is $\sqrt{3}$ ($V_{LL}/V_{LN} = \sqrt{3}$). In any case, the line-to-neutral voltage will always be dependent upon the line-to-line voltage, where

$$V_{LN} = \frac{V_{LL}}{\sqrt{3}}$$

Now the only question remaining is this: "What about current as it relates to a wye connection?" Using figure 50 as reference, when current flows from the line voltages ($V_{L1\text{-}3}$) through each phase winding (A-C) of a wye connection, the current will flow in series, meaning the current that flows from the line voltage will be the same as that which flows through the phase windings. Therefore, it can be said that the relationship between the line current and the phase current of a wye connection is simply

$$I_L = I_P$$

An altered view of the wye connection in figure 50 would reveal the three parts of the Y-connected system being commonly connected at each endpoint, which would then resemble the Greek symbol delta or an equilateral triangle as shown in figure 52.

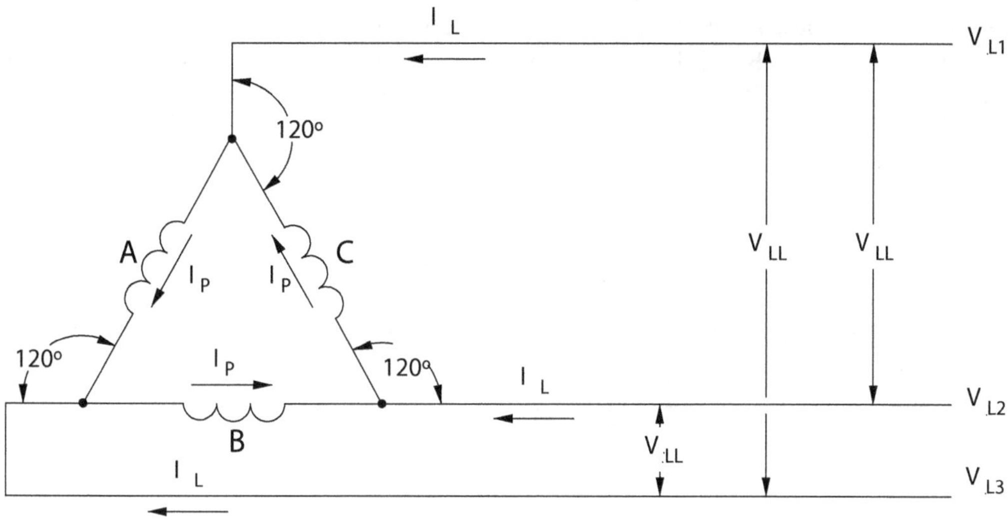

Fig. 52. Three-phase delta connection.

Take notice that the 120° displacements still exist but between the line and phase currents. Because of its resemblance, this type configuration is said to be delta-connected. In total opposition to that of a wye connection, the line-to-line voltage and the voltage across the phase windings of a delta connection are both the same:

$$V_{LL} = V_P$$

Because the line currents and the phase currents are 120° out of phase, the relationship between both currents can be treated in the same manner as the line-to-line voltage and the line-to-neutral voltage of a wye-connected system. As a result,

$$I_L = \sqrt{3} \times I_P$$

By using the exact procedures where the $\sqrt{3}$ was derived using V_{LL} and V_{LN}, the same results can be found if I_L and I_P were used instead. Similarly, the ratio between the line current and the phase current is also $\sqrt{3}$ ($I_L/I_P = \sqrt{3}$). Just as the neutral voltage is dependent upon the line voltage in a wye system, the phase current is dependent upon the line current in a delta system, where

$$I_P = \frac{I_L}{\sqrt{3}}$$

Portions of this topic will again be covered in the section entitled “Transformers.”

Based on that which was learned about single-phase and three-phase power, as well as about wye and delta-connected three-phase systems, that knowledge will be used to solve the following problems.

Practical Problems Pertaining to Single-Phase (1ϕ) Power

1. Determine the operating current of a 157kW load if energized by a 240V single-phase source where the power factor is 90%.

   Because the power factor is the ratio of the consumed power and the apparent power, the apparent power can be determined by dividing the power factor into the electrical power consumed by the load; therefore,

$$P_{VA(1\phi)} = \frac{157kW}{.90}$$
$$= 174.44kVA$$

   Now that the apparent power has been determined, the operating current can be calculated by first considering the formula $P_{VA} = E \times I$ and then converting the formula to derive $I = P_{VA}/E$. As a result, the generator's operating current is

$$I = \frac{174.44kVA}{240V}$$
$$= 726.83A$$

2. A single-phase alternator is used to supply a 350kW load. If the load pulls 735A of current at an 88% power factor, determine the alternator's apparent power, output voltage, and phase angle.

   Using the same concept as used in problem No. 1 to determine the alternator's apparent power,

$$P_{VA(1\phi)} = \frac{350kW}{.88}$$
$$= 397.73kVA\ (397{,}730VA)$$

   Now that the apparent power has been determined, the generator's output voltage can be calculated by again considering the formula $P_{VA} = E \times I$ and converting the formula to derive $E = P_{VA}/I$. As a result, the generator's output voltage is

$$E = \frac{397.73kVA}{735A}$$
$$= 541.13V$$

   Based on the formula $\cos\theta = PF$, the phase angle is determined by use of the converted formula, $\theta = \cos^{-1} PF$, where

$$\theta = \cos^{-1} .88$$
$$= 28.36°$$

3. A 480V single-phase generator delivers electrical power to a 135A load at a power factor rating of 64%. Determine the power consumed by the load, the apparent power provided by the generator, and the reactance power.

The power consumed by the load can be determined by the formula $P_W = E \times I \times PF/1000$, which results in

$$P_{W(1\phi)} = \frac{480V \times 135A \times .64}{1000}$$
$$= 41.472kW\ (41{,}472W)$$

Using the calculated power consumption, the apparent power amounts to

$$P_{kVA\,(1\phi)} = \frac{41.472kW}{.64}$$
$$= 64.8kVA\ (64{,}800VA)$$

or, as an alternate method,

$$P_{kVA\,(1\phi)} = 480V \times 135A\,/\,1000$$
$$= 64.8kVA$$

Use the formula $P_{VA} = \sqrt{P_W^{\,2} + P_{VARs}^{\,2}}$ and convert it into a usable formula to determine the reactance power, where $P_{VARs} = \sqrt{P_{VA}^{\,2} - P_W^{\,2}}$. As a result, the reactance power is

$$P_{VARs} = \sqrt{(64.8)^2 - (41.472)^2}$$
$$= \sqrt{4199.04 - 1719.93}$$
$$= \sqrt{2479.11}$$
$$= 49.79kVARs\ (49{,}790VARs)$$

4. If a single-phase motor pulls 76.37A at 208V while driving a 12.5kW load, determine the power factor of the motor and the phase angle.

   Using the formula, $PF = \cos\theta = kW / kVA$, the motor's power factor is

$$\cos\theta = \frac{12.5kW}{208V \times 76.37A}$$
$$= \frac{12.5kW}{15.88kVA}$$
$$= .7872$$

   With the calculated power factor, the phase angle is

$$\theta = \cos^{-1} .7872$$
$$= 38.08°$$

Practical Problems Pertaining to Three-Phase (3ϕ) Wye (Y) Connections

(The value 1.732 is used for all problems involving the $\sqrt{3}$.)

1. Determine the line-to-line voltage of a 3-phase wye-connected generator that has a line-to-neutral voltage rated for 1385V.

   Because the line-to-neutral voltage (V_{LN}) in a 3-phase wye-connected system is linked to the line-to-line voltage (V_{LL}) by the square root of 3 ($\sqrt{3}$),

$$\begin{aligned} V_{LL} &= \sqrt{3} \text{ x } 1385\text{V} \\ &= 2398.82\text{V} \end{aligned}$$

2. If the output voltage of a 3-phase wye-connected generator produces 382A per line at 4270V, determine the generator's phase current and the voltage across each phase of the generator.

   With any three-phase wye-connected system, the phase current(s) will always be the same as that of the line current(s) because both line and phase components are connected in series with each other; therefore, $I_L = I_P$, which results in the phase current being 382A also.

   Because the phase voltage will always be dependent upon the line voltage in any 3-phase wye-connected system regardless of whether the system is supplying the voltage (generator) or receiving the voltage (load), the phase voltage will always be the smaller of the two voltages and thus reduced in quantity by the $\sqrt{3}$. Therefore, $V_{LN} = V_{LL}/\sqrt{3}$, where

$$\begin{aligned} V_{LN} &= 4270\text{V}/\sqrt{3} \\ &= 2465.4\text{V} \end{aligned}$$

3. How much electrical power does a 460V three-phase wye-connected load consume if the load pulls 48A and operates at an 83% power factor?

   The electrical power consumed by a load that is supplied by an alternating current source is always measured in units of watts or kilowatts. But the most important thing to always understand is that this power is supplied by an apparent power, which is the product of the system's operating voltage and current. However, because a power factor is involved, the consumed power will only be a portion of the apparent power, that portion being 83% of the apparent power. As a result, the power consumed by the three-phase load is determined by the formula $P_W = \sqrt{3} \text{ x V x I x PF}$ where

$$\begin{aligned} P_{W(3\phi)} &= \sqrt{3} \text{ x } 460\text{V x } 48\text{A x } .83 \\ &= 31{,}741.32\text{W or } \frac{31{,}741.32\text{W}}{1000} = 31.74\text{kW (rounded down)} \end{aligned}$$

4. A three-phase motor is supplied by a wye-connected generator that produces 268V from line-to-neutral. If the motor pulls 137A at a 91% power factor, determine the generator's output voltage, the per-phase power, and the three-phase power consumed by the motor.

If the generator produces a 268V line-to-neutral voltage, then the generator's output voltage will result in being the line-to-neutral voltage increased by the $\sqrt{3}$. As a result, the generator's output voltage is

$$V_{LL}\ (V_{OUTPUT}) = \sqrt{3} \text{ x } 268V$$
$$= 464.18V$$

In a wye-connected system, the line current and the phase current are equal: $I_L = I_P$. Because of this the motor's per-phase power consumption is

$$P_{W(1\phi)} = 268V \text{ x } 137A \text{ x } .91$$
$$= 33{,}411.56W$$

The three-phase power consumed by the motor is

$$P_{W(3\phi)} = \sqrt{3} \text{ x } 464.18V \text{ x } 137A \text{ x } .91$$
$$= 100{,}229.7W \text{ (100.23kW rounded up)}$$

The results of the three-phase power can be verified by multiplying the motor's per-phase power by three. Check it out:

$$33{,}411.56W \text{ x } 3 = 100{,}234.68W$$

As you can see, the results are approximately the same.

5. A 13.8kV wye-connected load draws 163A while consuming 2.86MW (megawatts) of power. Determine the load's apparent power (in VA, kVA, and MVA), power factor, and the resulting phase angle.

The apparent power results in

$$P_{VA(3\phi)} = \sqrt{3} \text{ x } 13.8kV\ (13{,}800VA) \text{ x } 163A$$
$$= 3{,}895{,}960.8VA$$

or

$$= \frac{3{,}895{,}960.8VA}{1000} = 3895.96kVA$$

or

$$= \frac{3{,}895{,}960.8VA}{1{,}000{,}000} = 3.896MVA \text{ (rounded up)}$$

The power factor can be determined based on the calculated apparent power, which results in

$$\frac{2.86\text{MW}}{3.896\text{MVA}} = .7341$$

Based on the derived power factor, the resulting phase angle is determined by the formula $\theta = \cos^{-1}$ PF, where

$$\theta = \cos^{-1} .7341$$
$$= 42.77°$$

6. A three-phase wye-connected system operating at unity power factor is rated 72.5kV at 1500kVA. Determine the system's phase voltage, power factor, and power consumption.

Again, using the formula $V_{LN} = V_{LL}/\sqrt{3}$, the system's phase voltage is

$$V_{LN} = 72.5\text{kV}/\sqrt{3}$$
$$= 41.86\text{kV } (41{,}860\text{V})$$

Because the system is operating at a unity power factor, the power factor is simply one (1) or 100%. A unity power factor means that the system's power consumption is the same as the apparent power or, in other words, all of the apparent power is being consumed by the system. Take a look:

$$\text{Power factor} = \frac{1500\text{kW}}{1500\text{kVA}} = 1$$

7. Three individual resistive loads are rated for 13.67Ω each at 120V. If the loads are supplied from a wye-connected system, determine the current flow through each load, the voltage between each load, the power consumed by each load, and the overall three-phase power consumption.

The current flow through each load can be determined by dividing the quantity of the resistive load into the operating voltage:

$$I = 120\text{V} / 13.67\Omega$$
$$= 8.78\text{A}$$

Because the loads are being supplied by a wye-connected system, the voltage between each load would be the line-to-line voltage. Therefore, the voltage between each load is

$$V_{LL} = \sqrt{3} \times 120\text{V}$$
$$= 208\text{V (rounded up)}$$

As a result, the power consumed by each load, which can also be called the single-phase power, is

$$P_{W(1\phi)} = 120V \times 8.78A$$
$$= 1053.6W$$

And the overall three-phase power consumption amounts to either of the following:

$$P_{W(3\phi)} = 3 \times 1053.6W$$
$$= 3160.8W$$

or

$$P_{W(3\phi)} = \sqrt{3} \times 208V \times 8.78A$$
$$= 3163.05W$$

As you can see, using either method will provide approximate results. As a reminder, because the line and phase currents are equal in a wye-connected system, there's only one source of current. If the latter result were divided by three, it would also yield a single-phase power that is approximate to 1053.6W. Observe:

$$3163.05W / 3 = 1054.35W$$

8. The phase windings of a three-phase wye-connected alternator delivers 276.3V. If the alternator supplies three wye-connected loads, where the series impedance for each load is 15.66Ω resistive and 9.34Ω inductive reactance, determine the impedance of each load, the alternator's line voltage, the current being supplied by the alternator, the power consumed by each load, and the alternator's operating power factor and phase angle. Also determine the three-phase reactance power produced by the loads.

The impedance for each load is

$$Z = \sqrt{(15.66)^2 + (9.34)^2}$$
$$= \sqrt{245.24 + 87.24}$$
$$= \sqrt{332.5}$$
$$= 18.23\Omega$$

The alternator's line voltage is

$$V_{LL} = \sqrt{3} \times 276.3V$$
$$= 478.55V$$

Using the phase voltage and the impedance of each load, the current being supplied by the alternator can be determined as so because that same current will flow through each phase of a wye-connected source; therefore,

$$I_L = I_P = 276.3V/18.23\Omega$$
$$= 15.16A$$

Because the power consumed by each load is a derivative of the load resistance, the power consumption for each load is

$$P_{W(1\phi)} = (15.16A)^2 \times 15.66\Omega$$
$$= 3599.07W$$

The power factor (pf) could simply be determined by dividing the load's impedance into the load's resistance ($15.66\Omega / 18.23\Omega = .8590$), and the phase angle derived by means of the power factor ($\theta = \cos^{-1} .8590 = 30.80^\circ$). However, let's derive the alternator's apparent power and true power and use both results to verify the already derived power factor. The alternator's apparent power is

$$P_{VA(3\phi)} = \sqrt{3} \times 478.55V \times 15.16A$$
$$= 12{,}565.34VA$$

Using the results of the calculated power consumption per load, the alternator's true power is

$$P_{W(3\phi)} = 3 \times 3599.07W$$
$$= 10{,}797.21W$$

With the apparent power and the true power, the alternator's operating power factor can now be verified.

$$pf = \frac{10{,}797.21W}{12{,}565.34VA} = .8593$$

As you can see, the results are approximately the same.

Using the apparent power and previously determined phase angle, the three-phase reactance power produced by the loads is

$$P_{VARs(3\phi)} = 12{,}565.34VA \times \sin 30.80^\circ$$
$$= 12{,}565.34VA \times .5120$$
$$= 6433.45VARs$$

or

$$P_{VARs(3\phi)} = \sqrt{(12{,}565.34)^2 - (10{,}797.21)^2}$$
$$= \sqrt{41{,}308{,}025.5}$$
$$= 6427.13VARs$$

Practical Problems Pertaining to Three-Phase (3ϕ) Delta (Δ) Connections

(Practical problems 3–8 were taken from the practical problems pertaining to three-phase wye-connections to compare the similarities and differences of both delta and wye-connected systems. The value 1.732 is used for all problems involving the $\sqrt{3}$.)

1. Determine the phase current of a delta-connected 3-phase generator when the line current measures 81.7 amps.

   Because the phase current (I_P) in a 3-phase delta-connected system is a portion of the line current (I_L) and differs by the reciprocal of the square root of 3 ($1/\sqrt{3}$),

   $$I_P = 1/\sqrt{3} \times 81.7\text{A or } 81.7\text{A}/\sqrt{3}$$
   $$= 47.17\text{A}$$

   Similar to the phase voltage being dependent upon the line voltage in a 3-phase wye-connected system, the phase current in a delta-connected system will always be dependent upon the line current.

2. If the same generator in problem No. 1 later experiences a phase current of 55.33A, how much current will flow through the line?

   Because the phase current (I_P) in a 3-phase delta-connected system is linked to the line current (I_L) by the square root of 3 ($\sqrt{3}$),

   $$I_L = \sqrt{3} \times 55.33\text{A}$$
   $$= 95.83\text{A}$$

3. How much electrical power does a 460V three-phase delta-connected load consume if each phase current of the load pulls 48A and operates at an 83% power factor?

   To calculate the electrical power consumed by the load, the load's line current must first be determined because this is a delta-connected load. Therefore,

   $$I_L = \sqrt{3} \times 48\text{A}$$
   $$= 83.14\text{A}$$

   As a result, the power consumed by this three-phase delta-connected load is determined by the same formula as that used for a three-phase wye-connected load, $P_W = \sqrt{3} \times E \times I \times PF$, where

   $$P_{W(3\phi)} = \sqrt{3} \times 460\text{V} \times 83.14\text{A} \times .83$$
   $$= 54{,}978.62\text{W or } \frac{54{,}978.62\text{W}}{1000} = 54.979\text{kW (rounded up)}$$

4. A three-phase motor is supplied by a delta-connected generator that produces 268V across each phase. If the motor pulls 137A at a 91% power factor, determine the generator's output voltage, the per phase power, and the three-power consumed by the motor.

   Because the line-to-line voltage (V_{LL}) in any delta-connected system is equal to the phase voltage (V_P) the generator's output voltage would be the same as the 268V phase voltage.

   Because the motor pulls 137A, which is the line current to determine the per-phase power, the motor's phase current must be determined:

$$I_P = 137A/\sqrt{3}$$
$$= 79.1A$$

   Based on the motor's calculated phase current, the motor's per-phase power consumption is

$$P_{W(1\phi)} = 268V \times 79.1A \times .91$$
$$= 19{,}290.91W$$

   The three-phase power consumed by the motor is

$$P_{W(3\phi)} = \sqrt{3} \times 268V \times 137A \times .91$$
$$= 57{,}868.82W \text{ or } 57.8688kW \text{ (57.87kW rounded up)}$$

   The results of the three-phase power can be verified by multiplying the motor's per-phase power by three. Check it out:

$$19{,}290.91W \times 3 = 57{,}872.73W$$

   As you can see, the results are approximately the same.

5. A 13.8kV delta-connected load draws 163A while consuming 2.86MW (megawatts) of power. Determine the load's apparent power (in VA, kVA, and MVA), its power factor, and the resulting phase angle.

   The 163A identifies the delta-connected load's line current. Therefore, the apparent power, the power factor, and the resulting phase angle for this delta-connected load will produce the same results as when wye-connected.

   The apparent power

$$P_{VA(3\phi)} = \sqrt{3} \times 13.8kV \text{ (13,800VA)} \times 163A$$
$$= 3{,}895{,}960.8VA$$

or

$$= \frac{3{,}895{,}960.8\text{VA}}{1000} = 3895.96\text{kVA}$$

or

$$= \frac{3{,}895{,}960.8\text{VA}}{1{,}000{,}000} = 3.896\text{MVA}$$

The power factor is:

$$\frac{2.86\text{MW}}{3.896\text{MVA}} = .7341$$

The resulting phase angle is:

$$\theta = \cos^{-1} .7341$$
$$= 42.77°$$

6. A three-phase delta-connected system operating at unity power factor is rated 72.5kV at 1500kVA. Determine the system's power consumption, power factor, and the phase voltage.

   Because a delta-connected system is now being considered, the line voltage and phase voltage are the same; therefore, V_{LL} (72.5kV) = V_P(72.5kV). This is the only difference between the wye-connected system and the delta-connected system. The power consumption for this delta-connected system will also be the same as the wye-connected system because of the unity power factor.

7. Three individual resistive loads are rated for 13.67Ω, each at 240V. If the loads are supplied from a delta-connected system, determine the current flow through each load, the line current, the voltage across each load, the power consumed by each load, and the overall three-phase power consumption.

   The current flow through each load can be determined by dividing the quantity of the resistive load into the operating voltage:

$$I_P = 240\text{V} / 13.67\Omega$$
$$= 17.56\text{A}$$

   As a result, the line current amounts to

$$I_L = \sqrt{3} \times 17.56\text{A}$$
$$= 30.41\text{A}$$

   Because the loads are being supplied by a delta-connected system, the voltage across each load would be the same as the 240V line voltage.

The power consumed by each load, which again can also be called the single-phase power, is

$$P_{W(1\phi)} = 240V \times 17.56A = 4214.4W \quad (E \times I)$$

or

$$P_{W(1\phi)} = (17.56)^2 \times 13.67\Omega = 308.35 \times 13.67\Omega = 4215.14W \quad (I^2 \times R)$$

And the overall three-phase power consumption amounts to either of the following:

$$P_{W(3\phi)} = 3 \times 4214.4W = 12{,}643.2W$$

or

$$P_{W(3\phi)} = \sqrt{3} \times 240V \times 30.41A = 12{,}640.83W$$

As you can see, using either method will provide approximate results. As a reminder, because the line and phase voltages are equal in a delta-connected system, there's only one source of voltage. If the latter results were divided by three, it would also yield a single-phase power that is approximate to 4214.4W. Observe:

$$12{,}640.83W/3 = 4213.61W$$

8. The phase windings of a three-phase delta-connected alternator deliver 276.3V. If the alternator supplies three delta-connected loads, where the series impedance for each load is 15.66Ω resistive and 9.34Ω inductive reactance, determine the impedance of each load, the alternator's line voltage, the current being supplied by the alternator, the power consumed by each load, and the alternator's operating power factor and phase angle. Also, determine the three-phase reactance power produced by loads.

The impedance for each load is

$$Z = \sqrt{(15.66)^2 + (9.34)^2} = \sqrt{245.24 + 87.24} = \sqrt{332.5} = 18.23\Omega$$

The alternator's line voltage is the same as the phase voltage: 276.3V.

Using the phase voltage and the impedance of each load, the current being supplied by the alternator can be determined by first finding the phase current, which is

$$I_P = 276.3V / 18.23\Omega = 15.16A$$

Because the current being supplied by the alternator is actually the line current, that value is

$$I_L = \sqrt{3} \times 15.16A$$
$$= 26.26A$$

Because the power consumed by each load is a derivative of the load resistance, the power consumption for each load that is delta-connected is also

$$P_{W(1\phi)} = (15.16A)^2 \times 15.66\Omega$$
$$= 3599.07W$$

Because the resistance and impedance quantities for these delta-connected loads are the same as those for the wye-connected load, the power factor (pf) and phase angle are the same ($15.66\Omega / 18.23\Omega = .8590$) and ($\theta = \cos^{-1} .8590 = 30.80°$). However, let's derive the alternator's apparent power and true power for this delta-connection and use both results to again verify the already derived power factor. The alternator's apparent power is

$$P_{VA(3\phi)} = \sqrt{3} \times 276.3V \times 26.26A$$
$$= 12{,}566.77VA$$

Using the results of the calculated power consumption per load, the alternator's true power is

$$P_{W(3\phi)} = 3 \times 3599.07W$$
$$= 10{,}797.21W$$

Notice how the apparent power is almost exactly the same as that calculated for the wye-connection, and the three-phase power consumption is exactly the same. With the apparent power and the true power, the alternator's operating power factor can now be verified.

$$pf = \frac{10{,}797.21W}{12{,}566.77VA} = .8592$$

As you can see, the results are approximately the same for both connections. Therefore, by using the apparent power and previously determined phase angle, the three-phase reactance power produced by the delta-connected loads will also be approximately the same as the loads for the wye-connection.

$$P_{VARs(3\phi)} = 12{,}566.77VA \times \sin 30.80°$$
$$= 12{,}566.77VA \times .5120$$
$$= 6434.19VARs$$

9. A three-phase delta-connected motor consumes 65.47kW of electrical power. If the motor pulls 22.19A while operating at a 72% power factor, determine the motor's supply voltage, the phase voltage across the motor's windings, the current flow through the windings, the winding's impedance, and the resistance and inductance reactance about each winding.

Using the formula $kW = \sqrt{3} \times E \times I \times PF$ and converting it to $E = \dfrac{kW}{\sqrt{3} \times A \times PF}$,

the motor's supply voltage can be determined, which results in

$$E = \frac{65.47kW}{\sqrt{3} \times 22.19A \times .72}$$

$$= \frac{65{,}470W}{27.672}$$

$$= 2365.93V$$

Again, since this is a delta-connected load, the phase voltage across the motor's windings will be the same as the 2365.93V line voltage, while the current flow through the windings will be inversely proportional to the line current by the $\sqrt{3}$. As a result,

$$I_P = 22.19A/\sqrt{3}$$
$$= 12.81A$$

To determine the winding's impedance, the phase current is divided into the voltage source, where

$$Z = 2365.93V/12.81A$$
$$= 184.7\Omega$$

To determine the corresponding components of the impedance (i.e., the resistance and the inductance reactance about each winding), the formulas $R = Z \times \cos\theta$ (PF) and $X_L = Z \times \sin\theta$ will be used where θ is equal to 43.95°, which was derived from the formula $\theta = \cos^{-1} PF$. As it stands,

$$R = 184.7\Omega \times .72 \quad \text{and} \quad X_L = 184.7\Omega \times \sin 43.95°$$
$$= 132.98\Omega \qquad\qquad = 184.7\Omega \times .694$$
$$= 128.18\Omega$$

TRANSFORMERS

The everyday presence and use of transformers is as common as automobiles. With the use of alternating current (*ac*), a transformer changes one level of voltage to a different level of voltage. The most unique feature of a transformer is that it has no moving parts. Fundamentally, a transformer (figure 53) consists of two separate windings, which are wound (coiled) around a common iron core that's either solid or laminated. The number of times a winding is wound around the core is referred to as winding turns or turns.

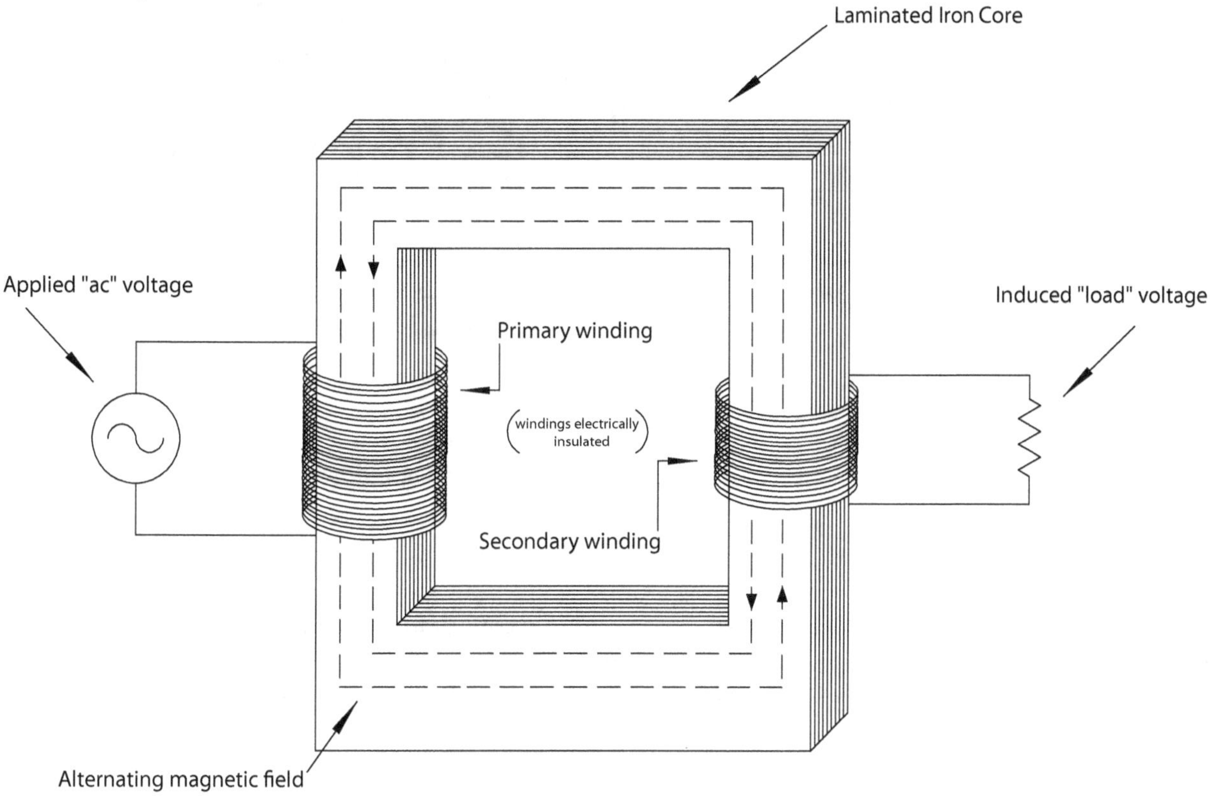

Fig. 53. The transformer.

The primary windings of a transformer usually receive power from an electrical source, while the secondary windings deliver power to a supplied load (refer to figure 54). When operating in such manner, a transformer can operate as either a step-up or a step-down transformer.

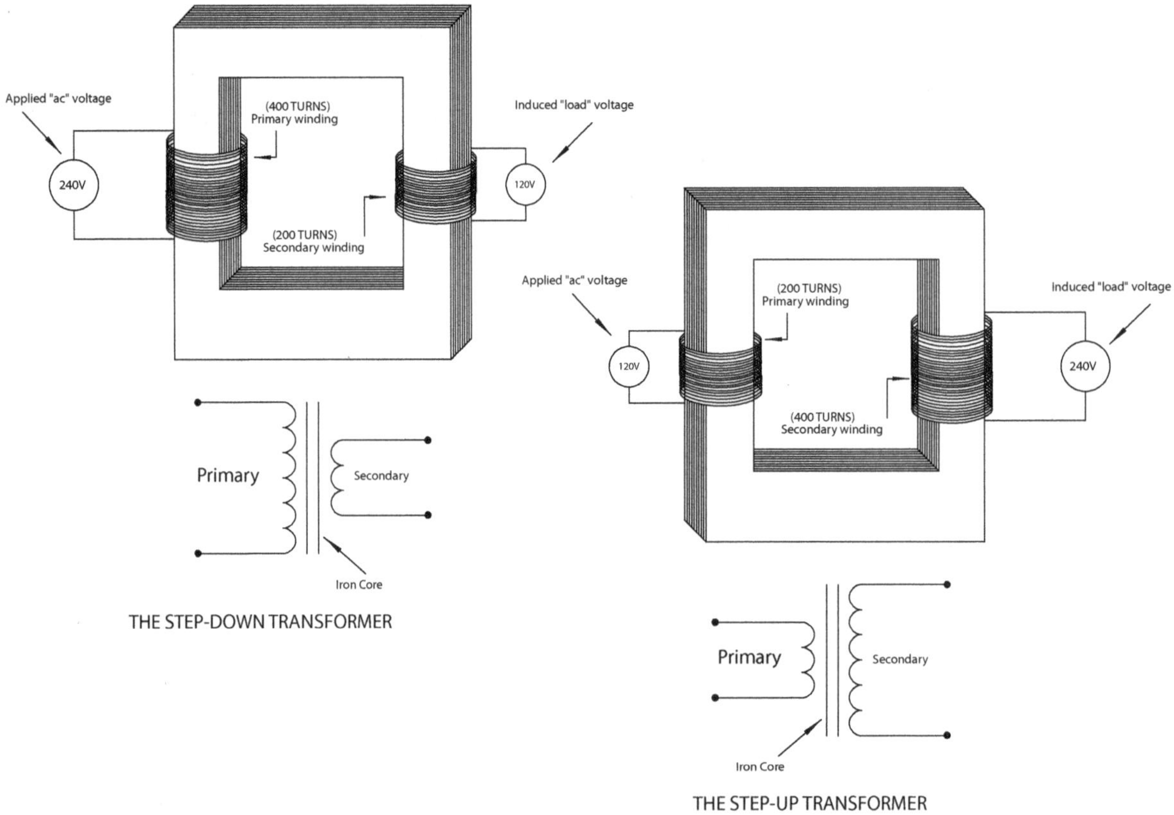

Fig. 54. The step-down/step-up transformer.

When the primary windings of a transformer are energized with an *ac* voltage source, an alternating magnetic field or flux is produced in the iron core. The alternating magnetic field causes a lower alternating voltage to be induced in the secondary windings of the transformer. This process identifies the principles of transformation.

In practical terms, if the *ac* applied source voltage is 480 volts, then the voltage about the primary windings will be 480 volts also. However, as for the voltage produced in the secondary windings, it depends upon the number of secondary turns in proportion to the number of primary turns. If both primary and secondary turns are the same, the voltage will be the same in both windings.

Again, referring to figure 54, if the primary windings of a transformer have more turns than the secondary windings, the primary voltage will be higher than the secondary voltage, and the transformer is referred to as a step-down transformer. On the other hand, if the secondary windings of a transformer have more turns than the primary windings, the secondary voltage will be higher than the primary voltage, and the transformer is referred to as a step-up transformer. The first application is generally how all transformers are used.

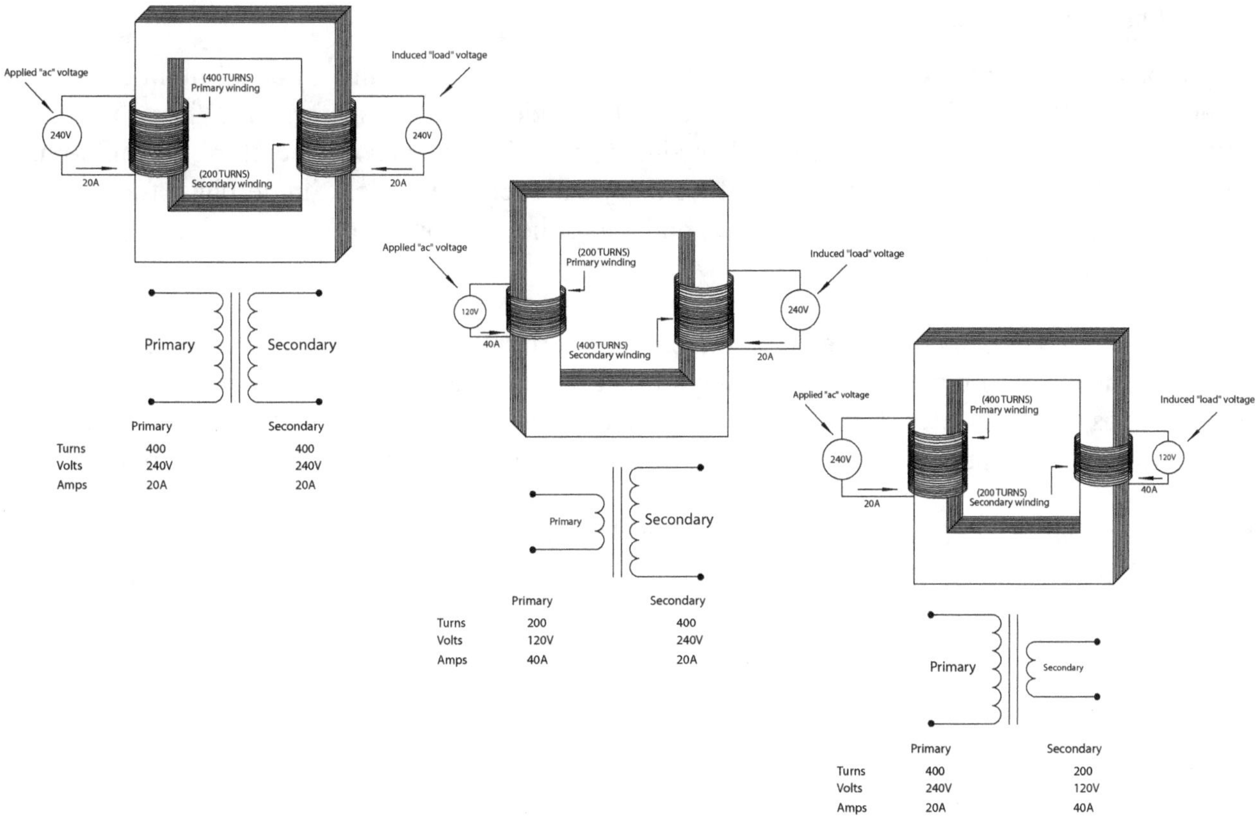

Fig. 55. Turns/voltage/current relationship.

Refer now to figure 55. In relationship to current, if a transformer's secondary has more turns than its primary, the secondary voltage will be higher than the primary voltage, but the amount of current flowing through the primary will be higher than the amount of current flowing through the secondary. On the other hand, if a transformer's primary has more turns than the secondary, the primary voltage will be higher than the secondary voltage but the amount of current flowing through the secondary will be higher than the amount of current flowing through the primary. In either case, the volt-amperes of the primary will always be equal to the volt-amperes of the secondary. Therefore, regardless of whether a load operates at a higher voltage and a lower current or vice-visa, as for as cost, the customer will still pay the same amount because the cost of power is based on the product of voltage and current opposed to just current alone. In everyday usage, voltage will remain fairly constant, while current changes with demand. For such applications, refer to "Transformer Formulas and Calculations."

Transformer Connections

Depending upon the windings or the combination of windings involved, a transformer can be connected in a variety of ways to obtain a desired voltage. In this section, we will look at connecting single-phase, two-phase, and three-phase transformers. The term phase is usually substituted with the Greek symbol ϕ, which identifies the Greek letter phi. Abbreviated, the terms single-phase, two-phase, and three-phase are respectively identified as 1ϕ, 2ϕ, and 3ϕ.

Single-Phase (1ϕ) Transformers

Single-phase transformers usually consist of two identical coils having an equal number of turns (evenly wound) on both the primary and secondary sides of the transformer. When joined, the coils form the transformer's primary and secondary windings or phase. Therefore, a single-phase transformer has one phase of windings on its primary and secondary sides. Single-phase transformers are used for both power and lighting needs and can either be connected in series or in parallel to produce two different voltages.

When the two inner ends of the coils are joined, they form a common connection between the coils while leaving the two outer ends free to be appropriately connected. The two outer ends of the coils are referred to as line conductors because they are connected to either the conductors extending from the source voltage to the primary side or the conductors extending from the secondary side to the load.

So often you hear the line conductors being referred to as Phase A and Phase B. Because there is only one set of phase windings on a single-phase transformer, using the terms Phase A and Phase B implies that a single-phase transformer has two phases as opposed to one. Therefore, to be specific, the line conductors associated with the windings of a single-phase transformer should be referred to as Line 1 (L1) and Line 2 (L2) as opposed to Phase A and Phase B.

In the study of series and parallel resistances, it is understood that series resistance is always additive, and parallel resistance is always less than the smallest resistance of a combination. When the two identical coils of a single-phase transformer are connected in series, the resistance between the two coils is totaled and therefore increased. In contrast, when the two identical coils of a single-phase transformer are connected in parallel, the total resistance between the two coils is reduced by half.

In comparison, the series/parallel resistance relationship and the relationship between the number of turns and voltage are actually the same. When the two coils of a single-phase transformer are connected in series, the voltage between the line conductors is equal to the source voltage (primary) or the induced voltage (secondary). If connected in parallel, the voltage between the line conductors is reduced in half. Voltage between the line conductors is referred to as line-to-line voltage: L1 to L2. When only two line conductors are involved, it is called a 2-wire system.

Consider the 480/240V single-phase transformer in figure 56. When the two coils on the primary and secondary sides of the transformer are connected in series, the voltage across the windings (line-to-line) on the primary side is 480 volts, while the voltage across the windings (line-to-line) on the secondary side is 240 volts.

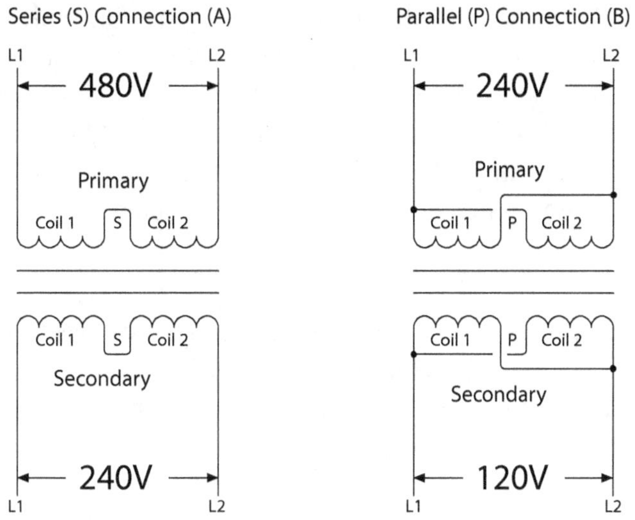

Fig. 56. Series/parallel connections.

Now when the two coils are connected in parallel, the voltage across the winding (line-to-line) on the primary side is reduced to 240 volts, and the voltage across the windings (line-to-line) on the secondary side is reduced to 120 volts.

Up until now, only line-to-line voltage has been discussed. In figure 57, when a conductor is placed between the two evenly wound coils on the transformer's secondary side, the connection is referred to as a center-tap. A center-tapped conductor is most often identified as a neutral conductor.

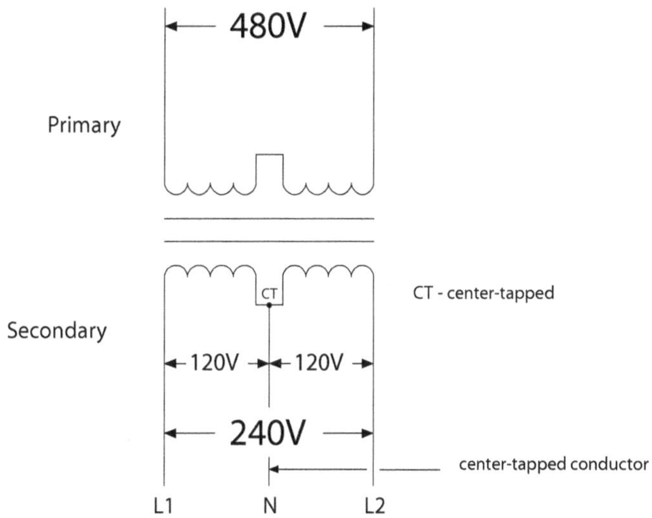

Fig. 57. The 3-wire, 240/120V, single-phase system.

From the neutral conductor to either of the line conductors (LI or L2), a line-to-neutral voltage is obtained, which is 120 volts, one-half of the 240-volts line-to-line voltage. With the inclusion of the neutral conductor, we have now gone from a 2-wire to a 3-wire system. In a 3-wire system, if the load currents flowing through L1 and L2 are unbalanced, the neutral conductor will always carry the unbalanced current between the two line conductors.

Unlike the secondary 240 volts and 120 volts obtained in figure 56, which are solely line-to-line voltages where only a single source of voltage is provided, the 3-wire system in figure 57 provides a line-to-voltage along with two line-to-neutral voltages. Of all single-phase transformer connections, the secondary 3-wire system is the most popular and the most frequently used for providing power and lighting in dwellings and small commercial buildings. With this type 3-wire system, 120V and 240V (2-wire) and 240/120V (3-wire) equipment and loads can be supplied.

Two-Phase (2ϕ) Transformers

Although not as popular or as frequently used as single-phase and three-phase power systems, two-phase systems are still used in certain areas of this country. Two-phase transformers are also manufactured in the same manner as whole unit (or pad-mounted) three-phase transformers. However, two individual single-phase transformers can also be used to produce the same results.

The distribution of two-phase power is made possible by use of two single-phase voltages, which are 90° out of phase with each other. See figure 58.

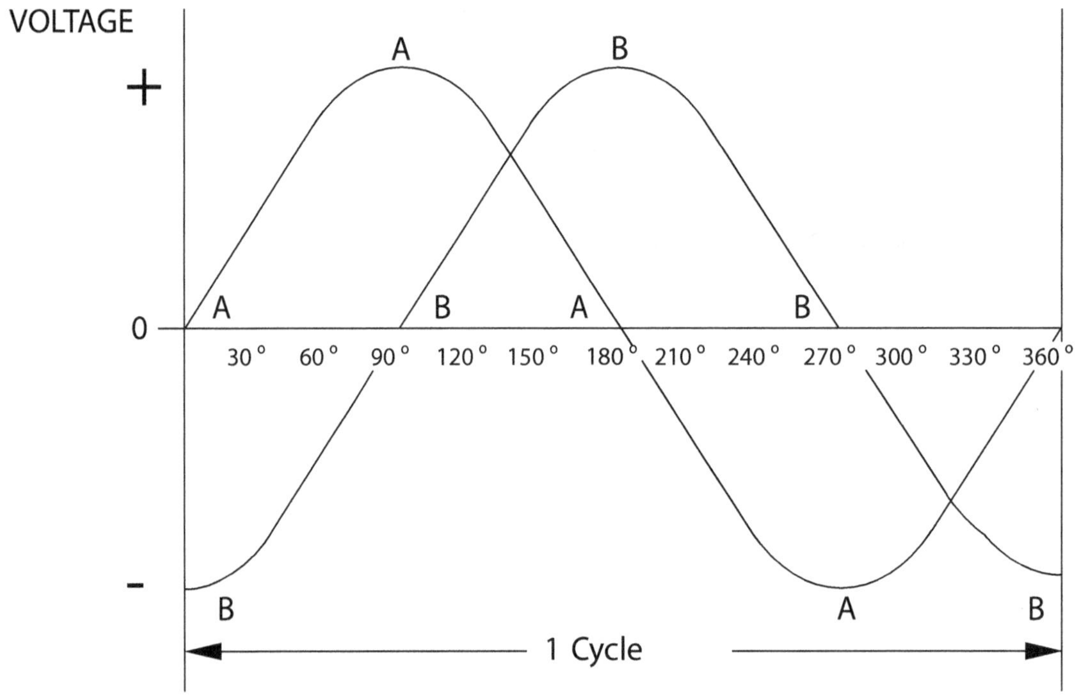

Fig. 58. Two-phase voltage sine waves.

With the use of these source voltages, a voltage is induced in the secondary windings of two individual single-phase transformers. The secondary windings can be either individual windings or dual windings, which are connected 90° out of phase with each other. Regardless of combination, both types will supply two-phase power. See figure 59.

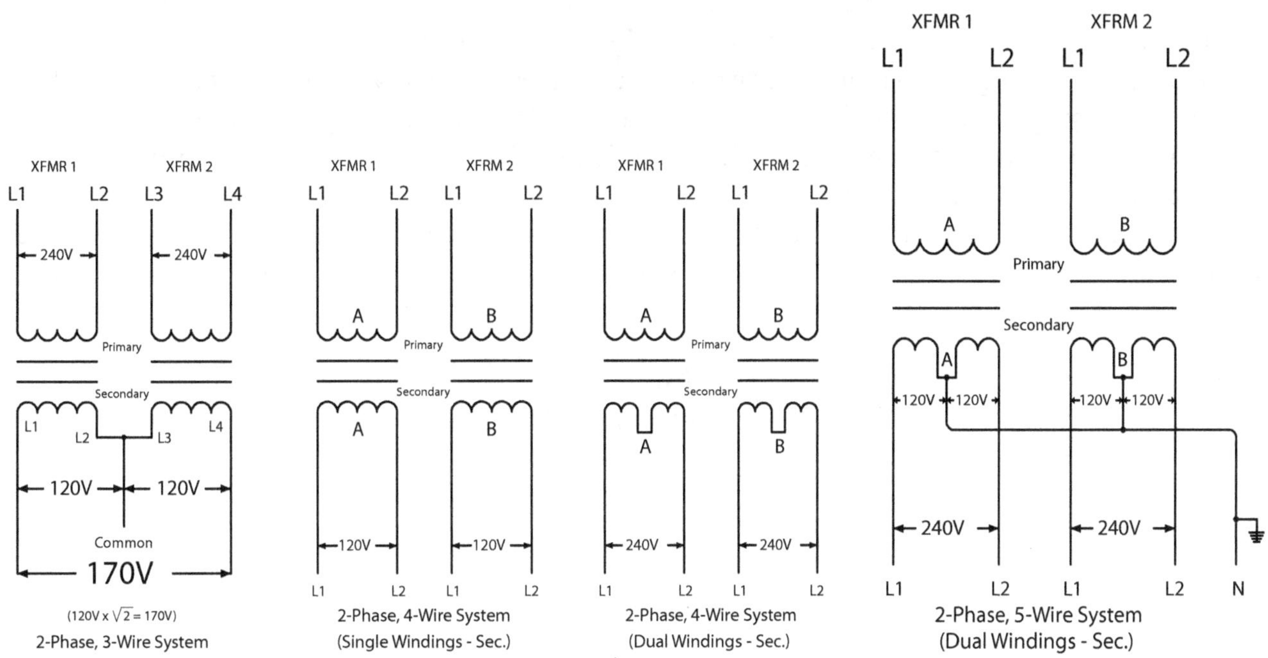

Fig. 59. Two-phase, 3-, 4-, and 5-wire systems.

Power distribution for two-phase systems can be either 4- or 5-wire systems with secondary source voltages at 120V, 240V, 480V, 575V, and—where two-phase motors (synchronous and induction) with high horsepower ratings are used—2300V.

To establish a grounded system, a fifth wire (conductor) is added to a 4-wire system, thus making it a 5-wire system. With the use of a grounded conductor, a phase-to-ground (neutral) voltage is acquired. The grounded conductor also provides means for the flow of unbalanced current between the two transformers and a ground-fault clearing path in the event of such occurrence.

Because two single-phase transformers are used, the total kVA rating of the transformer bank is the sum of each transformer's kVA rating; that is, two 25kVA single-phase transformers have a total rating of 50kVA (25kVA x 2).

For a 2-phase, 3-wire system where the transformer's secondary windings are interconnected, the secondary power rating can be determined as such:

$$P_{VA(2\phi)} = \sqrt{2} \text{ x E x I}$$

For a 2-phase, 4-wire system where the transformer's secondary windings are separate, the secondary power rating can be determined as such:

$$P_{VA(2\phi)} = 2 \text{ x E x I}$$

With a 2-phase, 3-wire system, the voltage between the outside conductors is equal to the winding voltage times the square root of 2 ($\sqrt{2}$). Because the current in the common conductor of a 2-phase, 3-wire system is greater than that of the current in the outside conductors, the National Electrical Code® (NEC®) makes reference of such system in the heading of Table 430.249 for determining the current in the common conductor. The full-load current values given in Table 430.249 is based on two-phase motors being supplied by a 4-wire system. Unlike the synonymous usage of the terms neutral or common, the referenced common in this situation is just that: a common connection between two transformer windings to produce a 2-phase, 3-wire system. This conductor is referred to as a common because it provides means for the use of three-pole devices such as circuit breakers and switches rather than carrying unbalanced current. To ease the understanding of this application, observe figure 60.

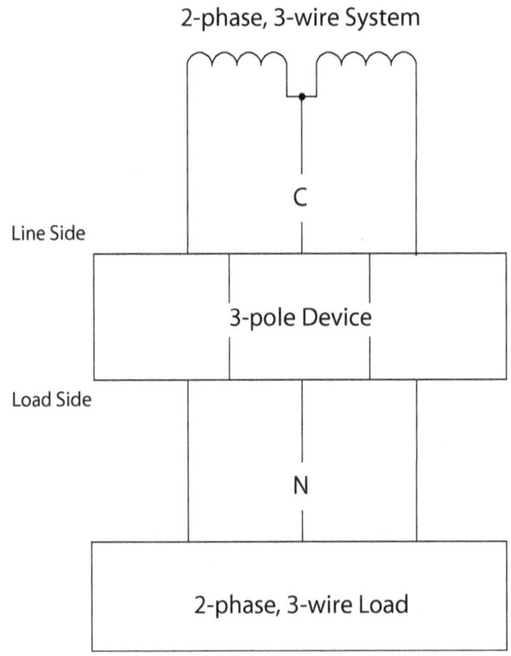

Fig. 60. Application of common-neutral conductors.

In making such observations, notice how the conductor that is referred to on the line side of the device as a *common* is now referred to as a *neutral* on the load side of device. This is because the neutral and the two outside conductors are now used to supply a 2-phase, 3-wire load instead of a device. Where such application exists, the neutral conductor will carry more current than the adjoining outside conductors. This is why the Exception to NEC 220.61(A) references the neutral conductor and any ungrounded conductor as it relates to the maximum unbalanced load such combination can produce. Although different in applications, the current experienced by both common and neutral conductors is still determined by similar factors—the square root of 2 ($\sqrt{2}$ or 1.41) or an approximation of the square root of 2 (1.40 or 140%).

Three-Phase (3ϕ) Transformers

Similar to single-phase transformers, three-phase transformers are used throughout the world in various capacities to provide means for three-phase transformation and power.

To obtain three-phase power, the option exists for using three individual single-phase transformers to form a three-phase transformer bank or using one whole unit three-phase transformer. A whole unit (or pad-mounted) transformer is constructed by winding three individual phase windings onto a single iron core that can be either solid or laminated.

With a three-phase transformer, four types of connections are available. They are

1. The delta-to-delta connection.
2. The delta-to-wye connection.
3. The wye-to-delta connection.
4. The wye-to-wye connection.

The dual sided connections of a three-phase transformer, whether delta or wye, identify the connection of the primary side of the transformer first followed by the secondary side. For simplicity, the Greek delta symbol (Δ) is used to identify a delta connection, and the capital letter (Y) is used to identify a wye connection. A wye connection is also referred to as a star connection. A delta-connected transformer takes on the image of an equilateral triangle. Again, a wye-connected transformer takes on the image of the capital letter Y, where three adjoining phase windings are commonly connected at one end, while the three remaining ends are positioned 120° apart. Of the four possible connections, the delta-to-wye connection is the one most commonly used for commercial and industrial applications. Depending upon the type of loads being supplied, the secondary wye connection will experience harmonic currents. Wye-to-wye connections are seldom used because of harmonic and load balancing difficulties. The delta-to-delta connection is most commonly used for industrial applications. For high voltage transmission, wye-to-delta connections are used.

Delta (Δ) Connections (3W)

A three-phase delta-connected transformer can be connected in either a closed or an open configuration. As covered earlier, with a three-phase delta-connected transformer, only one voltage exists because the line-to-line voltage and the voltage across each phase are the same, which again results in

$$V_{LL} = V_P$$

In a closed delta system, three transformer windings are used and connected in series. Although similar in connections, an open delta system only uses two transformers. The use of two transformers can be done intentionally when the need for three-phase power is moderate or due to the loss of one transformer. In any case, where an open delta system exists, its output capacity is approximately 87% ($\sqrt{3}/2$ or .866) of that of a closed delta system or approximately 58% ($1/\sqrt{3}$ or .577) of a delta-connected three-phase transformer bank. For example, if an open delta system consisted of two 75kVA single-phase transformers, the total output capacity would be 130.5kVA (75kVA x 2 x .87) opposed to 150kVA; or, where three single-phase transformers were used and one transformer was lost, the total output capacity would still be 130.5kVA (75kVA x 3 x .58). For most delta-connected systems, at least one transformer is always sized larger than the other(s). The transformers shown in figure 61 display the wiring configurations for a closed and an open delta three-phase connected system.

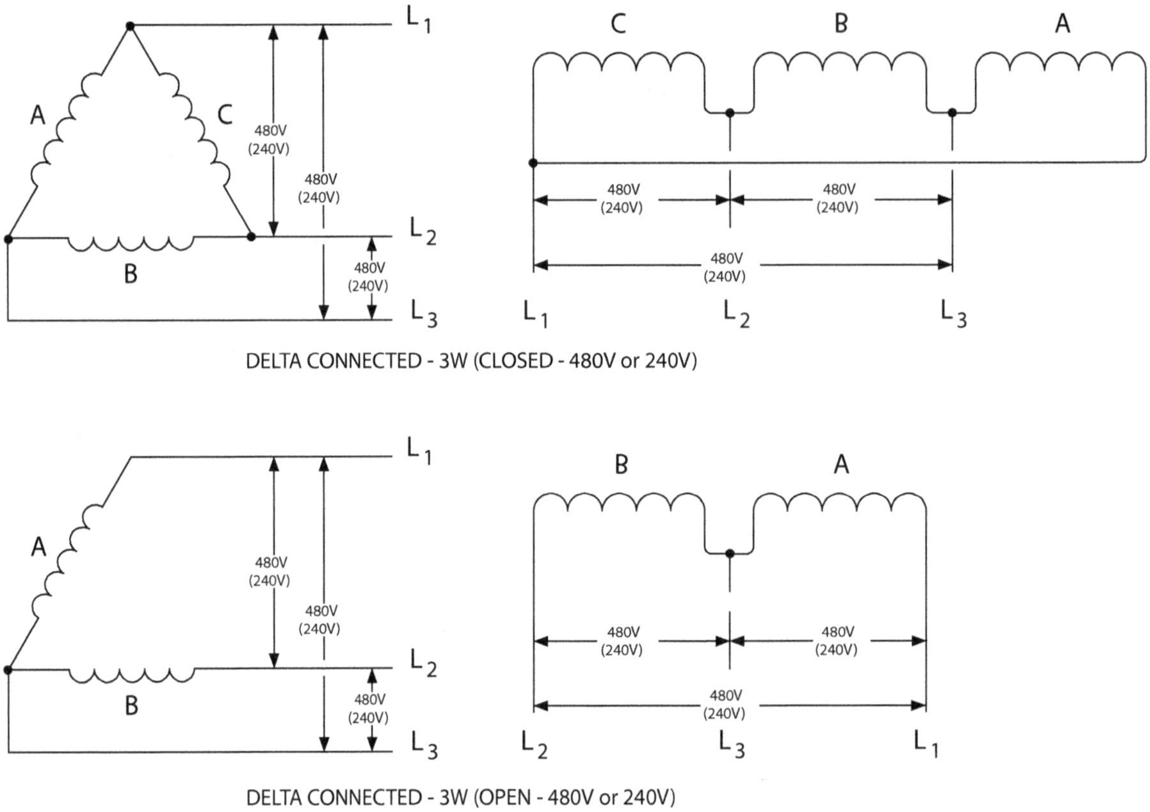

Fig. 61. Closed and open delta systems.

Because there are only three line conductors involved with each combination of transformers, both configurations are categorized as 3-wire delta-connected or, simply abbreviated, 3W-Δ.

Sometimes one of the line conductors is intentionally grounded at one corner of the delta-connected system. When this happens, the system is referred to as a corner-grounded system. See figure 62.

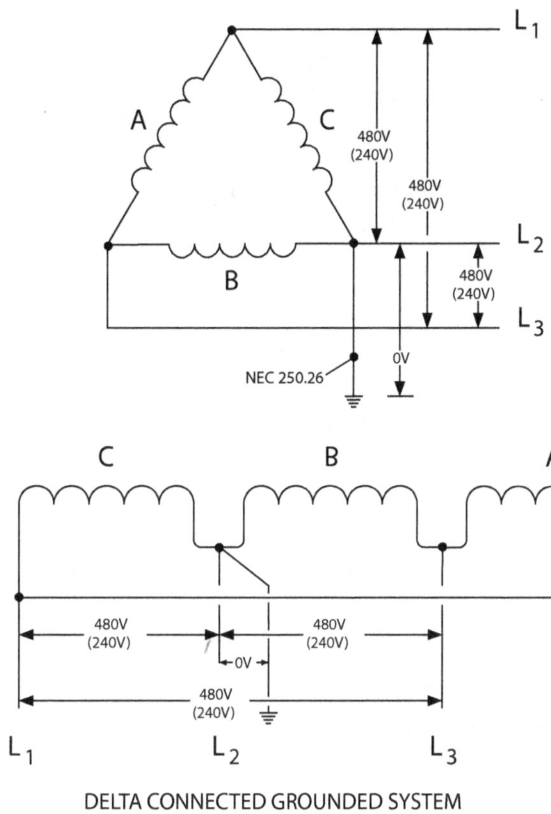

Fig. 62. Corner-grounded system.

Although grounded, the conductor is not a neutral and therefore does not serve the same purpose. The purpose of a "corner-ground" is to provide a ground-fault return path for the other line conductors (in such case L_1 and L_3 of figure 62) in the event they accidentally make contact with a grounded surface of a building or other structure. Grounding the conductor does not alter the voltage between the line conductors, it only stabilizes the potential to ground at "0" volts with respect to L_2 while the other line conductors' potential to ground remains the same as the source voltage. Unlike the two remaining line conductors, the grounded conductors must never be connected in series with a fuse or circuit breaker at the service equipment. The installation of these conductors must be done in accordance with NEC 250.26.

Regardless of application, 3-wire, delta-connected systems can only supply single and three-phase line-to-line loads or equipment.

Referring back to figure 52, because the line conductors are connected between each set of phase windings when the system is energized, the current flow through any line conductor (line current [I_L]) is the vector sum of adjacent phase winding currents [I_P]. This again is due to the 120° phase differences that occur between any two line currents. As a result, the line and the phase currents are not the same, unlike the line and phase voltage of a delta-connected system. In a closed delta system, the division of current occurs at all three line-to-phase connections and only once in an open delta system. At any rate, when the line currents are balanced, the current flowing through the phase windings (phase current) is 58% ($1/\sqrt{3}$ or .577) of the line currents, which again results in the same relationships as covered earlier, where

$$I_L = I_P \times \sqrt{3} \quad \text{or} \quad I_P = \frac{I_L}{\sqrt{3}}$$

Unbalanced line currents can also be determined when phase currents are known. This is performed by using the phase current values that are adjacent to a corresponding line current. Referring to figure 52 once again, based on the following relationship, unbalanced line currents can be determined, where

$$I_{L1} = \sqrt{(I_{P1})^2 + (I_{P2})^2 + I_{P1} \times I_{P2}} \quad \text{(where } I_{P1} \text{ and } I_{P2} \text{ are adjacent phase currents)}$$

$$I_{L2} = \sqrt{(I_{P2})^2 + (I_{P3})^2 + I_{P2} \times I_{P3}} \quad \text{(where } I_{P2} \text{ and } I_{P3} \text{ are adjacent phase currents)}$$

and

$$I_{L3} = \sqrt{(I_{P3})^2 + (I_{P1})^2 + I_{P3} \times I_{P1}} \quad \text{(where } I_{P3} \text{ and } I_{P1} \text{ are adjacent phase currents)}$$

Delta (Δ) Connections (4W)

As mentioned, a 3-wire, delta-connected system can only supply single and three-phase line-to-line equipment and loads. To supply single-phase, 120V or 240/120V neutral loads or equipment, a fourth conductor must be connected. By center tapping a conductor to one of the single-phase transformers of a 3-wire closed or open delta-connected system, the same results are obtained as with a dedicated 3-wire single phase (transformer) system. But now, the three line conductors, along with an added neutral conductor, are used with this three-phase delta-connected system to obtain a 120V or a 240/120V neutral voltage similar to that of a 3W-single phase system. As a result, the system is now converted from a 3-wire (3W) to a 4-wire (4W) delta-connected system. Refer to figure 63.

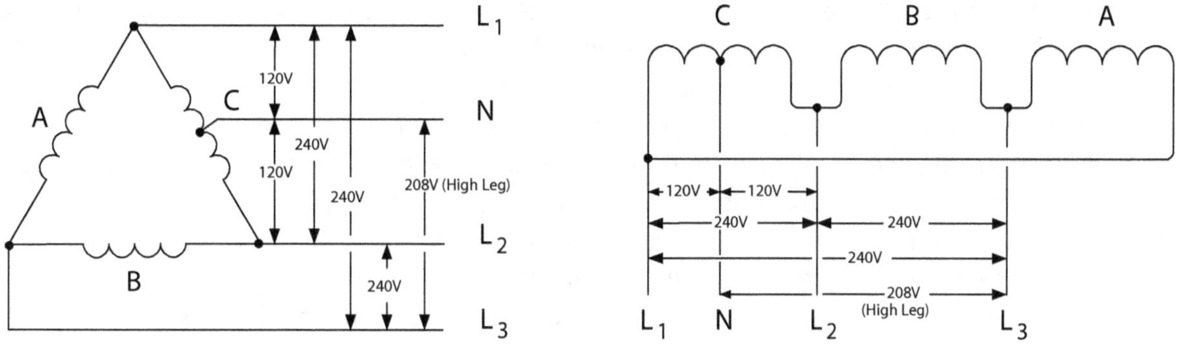

DELTA CONNECTED - 4W (CLOSED) 240/120V

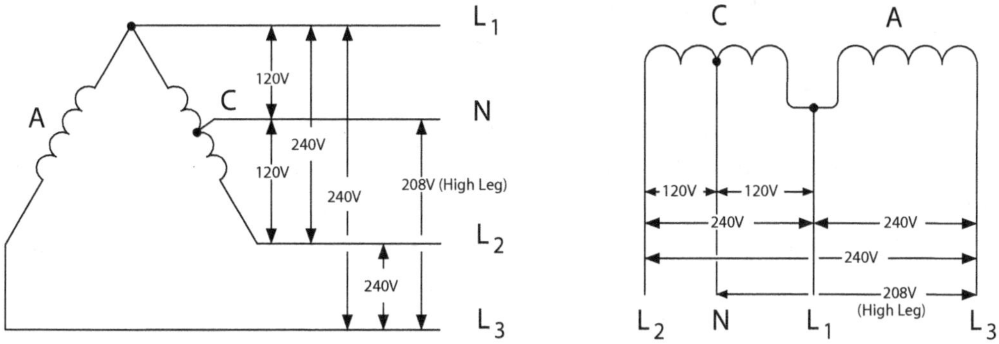

DELTA CONNECTED - 4W (OPEN) 240/120V

Fig. 63. 4W delta-connected systems.

Figure 63 displays a 4-wire (4W), three-phase (3ϕ), 240/120V closed and open delta (Δ) system. As you can see, there are multiple source voltages available: 240V-3ϕ, 240V-1ϕ, 240/120V-1ϕ multiwire, and line-to-neutral voltages that are either 120V or 208V. In all, a 4-wire, three-phase delta-connected system has three line-to-line voltages and two usable 120V line-to-neutral voltages. Of all voltages, the 208V line-to-neutral voltage is the most respected. For this reason, let's take an in-depth look at this voltage before moving on.

Again, because this voltage is the most respected, it has a specific name and is referred to as the "high leg," meaning higher neutral voltage or higher voltage from line to neutral. The "high leg" is also called "red leg," "red leg delta," "stinger," "wild leg," or some other term, according to territory. Referring to figure 63, notice that the two usable 120V line-to-neutral voltages are located between L_1 and L_2 and the center-tapped neutral conductor (N) on C phase. Because the 208V line-to-neutral voltage is between L_3 and N, it is derived by going from L_3 to L_2, which is 240 volts, and then from L_2 to N, which is 120 volts. By adding the two respective voltages, the resulting voltage is 360 volts. Because the overall system is three-phase, the resulting voltage is divided by the square root of 3. Hence, this high leg line-to-neutral voltage is so derived:

$$V_{L\text{-}N} = \frac{240V + 120V}{\sqrt{3}} = 208V$$

Given that this high-leg 208V product is a neutral voltage, it is not usable solely because there are no loads or equipment that operate at 208V line-to-neutral.

Often enough, intended 120V circuits are either accidentally or for lack of knowledge connected to the high leg, causing excessive or irreparable damage to the equipment being supplied. It is also assumed that a 208V single-phase motor can operate off the high leg, but such equipment must be operated from a 208V line-to-line source opposed to a 208V line-to-neutral. A 208V line-to-line source provides more electrical force than a 208V neutral; thus, a 208V neutral source will require a load to consume more current than rated for. Because of the potential hazards involved with the high leg, the NEC [110.15, 230.56] requires the affiliated line conductor to be durably and permanently marked by an outer finish that is orange in color or by other effective means. In the next section, you will see that 208 volts are only usable when used as a line-to-line voltage.

Wye or Star (Y) Connections (4W)

Unlike a delta-connected transformer, the phase windings of a wye- or star-connected transformer are connected in parallel. Because each individual phase winding has two end connections, one connection from each winding is commonly joined together, while the three remaining end connections are left to be connected to the three line conductors needed to supply a three-phase transformer. To complete the connection, a grounded conductor (also referred to as a neutral conductor) is connected to the winding connections that are commonly joined, resulting in a 4-wire (4W), three-phase (3ϕ), wye- or star (Y)-connected system—simply abbreviated 4W-3ϕ-Y-connected. See figure 64.

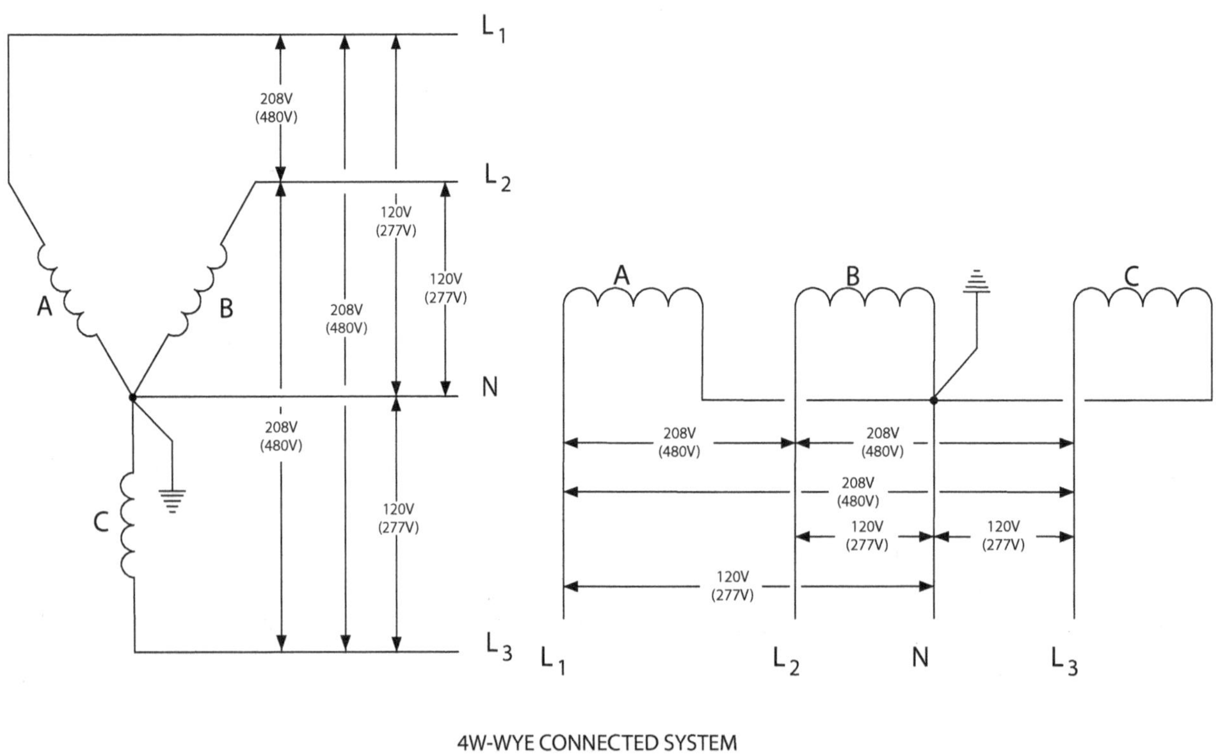

Fig. 64. Wye system.

In contrast with a delta-connected system, all 4-wire, wye-connected systems will produce two supply voltages. Between any two-phase windings of the three-phase transformer, a line-to-line (L-L) voltage exists; and from any phase winding to the grounded conductor, a line-to-neutral (L-N) voltage exists. In all, a wye-connected system will consist of three line-to-line voltages and three line-to-neutral voltages, which are dependent upon the line-to-line source voltage. The two different voltages come as a result of the 120° phase difference between each line-to-line voltage, resulting in the same relationships as covered earlier, where

$$V_{LL} = V_{LN} \times \sqrt{3} \quad \text{or} \quad V_{LN} = \frac{V_{LL}}{\sqrt{3}}$$

Contrary to what one might have thought, it has now been made clear that the line-to-neutral voltage is not simply one-half of the line-to-line voltage, and neither is the line-to-line voltage twice that of the line-to-neutral voltage. Similar to the line and phase currents associated with a delta-connected transformer, the line-to-line voltage is the vector sum of the voltages across any two adjacent phase windings.

A 4-wire, wye-connected transformer is used when 208/120V and 480/277V 3-phase supply voltages are desired. Depending upon the type loads being served, wye-connected systems are extremely susceptible to harmonic currents.

Because the line conductors and phase windings of a wye-connected system are connected in series, the current flow through both is the same, where again

$$I_L = I_P$$

To calculate the unbalanced neutral current for a three-phase, 4W-wye connected system consisting only of linear loads, the following formula is applied:

$$I_{Neutral} = \sqrt{[(L_1^2 + L_2^2 + L_3^2)] - [(L_1 \times L_2) + (L_2 \times L_3) + (L_1 \times L_3)]}$$

Autotransformers

A transformer consisting only of a single winding is called an autotransformer.

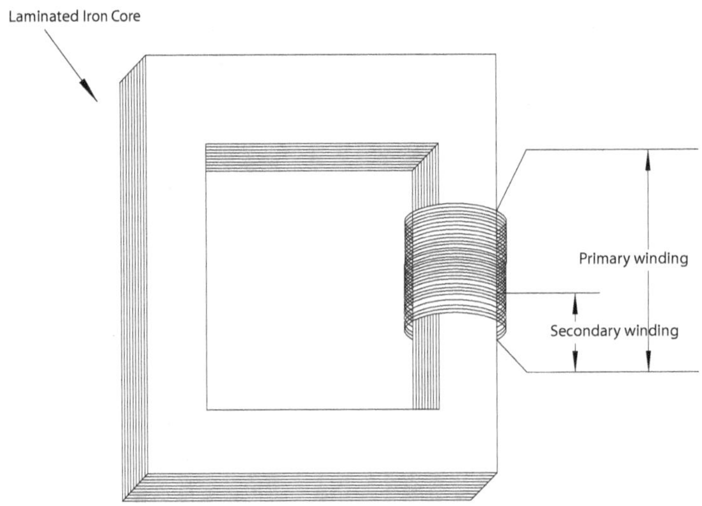

Fig. 65. The autotransformer.

As with regular type transformers, where the primary and secondary windings are nonconnected, there are no separate primary and secondary windings for an autotransformer, therefore, classifying such a transformer as interconnected.

Part of an autotransformer's single winding is used for the primary, while the other part is used as the secondary.

An autotransformer can be used as a step-up or step-down transformer and operates on the same principle as any other transformer because the applied primary voltage causes an induced voltage in the secondary. Refer to figure 66.

When used as a step-down autotransformer, the primary windings will have more turns of wire than the secondary, thus causing the autotransformer's primary voltage to be higher than the secondary voltage.

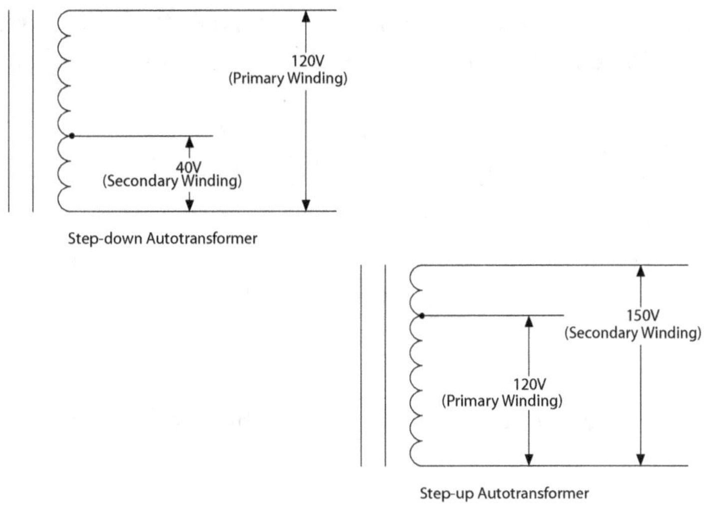

Fig. 66. The step-down and step-up autotransformer.

The lower secondary voltage is tapped at a desired voltage and can be either increased by using more turns for the secondary winding or decreased by using fewer turns for the secondary winding.

When used as a step-up autotransformer, the order of windings is reversed, where only a portion of the single winding is used for the primary and the two outside leads of the entire winding are used to increase the secondary voltage.

Transformer Formulas and Calculations

A. Transformer's Primary (P) and Secondary (S) Voltage (V) and Current (I)

Note: A transformer's primary power rating (kVA) equals its secondary power rating (kVA).

Original Formula: $V_P \times I_P = V_S \times I_S$

V_P = Primary Voltage I_P = Primary Current
V_S = Secondary Voltage I_S = Secondary Current

Converted Formulas

$$\mathbf{V_P} = \frac{V_S \times I_S}{I_P} \qquad \mathbf{I_P} = \frac{V_S \times I_S}{V_P}$$

$$\mathbf{I_S} = \frac{V_P \times I_P}{V_S} \qquad \mathbf{V_S} = \frac{V_P \times I_P}{I_S}$$

1. The secondary voltage of a transformer is 208 volts, and it has a full-load current of 87 amps. If the transformer's primary voltage is 465 volts, what is the primary's full-load current?

   Use the formula $I_P = \frac{V_S \times I_S}{V_P}$, where I_P = transformer's primary full-load current, to determine primary full-load current:

$$I_P = \frac{208V \times 87A}{465V} = 38.92A$$

   Notice the relationship between voltage and current: the higher the voltage, the lower the current and vice versa.

2. The secondary current of a transformer is 35.8 amps. If the transformer's primary full-load current is 9.3 amps at 550 volts, determine the transformer's secondary voltage.

   Use the formula $V_S = \frac{V_P \times I_P}{I_S}$, where V_S = transformer's secondary voltage

   to determine secondary voltage:

$$V_S = \frac{550V \times 9.3A}{35.8A} = 142.88V$$

Alternative Method (when formulas are not available)
Mathematically, the relationship between the primary and secondary ratings of a transformer is simply a constant value. In questions number 1 and 2, the given voltage and current ratings of both transformers yielded an unknowingly constant value when like ratings were divided into another.

For example, when the transformer's primary voltage rating is divided by the transformer's secondary voltage rating, a constant value was unknowingly derived and applied. Take a look:

$$\frac{208V}{465V} = .45 \text{ (value rounded off)}$$

Now if the voltage ratings were reversed another constant value could also be derived,

$$\frac{465V}{208V} = 2.24 \text{ (value rounded off)}$$

These constant values can be used to derive answers that are approximate to those gathered in both questions, therefore providing an alternate method to evaluate the results of your calculations. Just remember that when values are rounded off, the degree of accuracy is compromised, and when used in calculations, the results of an answer can only be approximated in comparison to another.

The constant value .45 indicates that the secondary voltage is 45% of the primary voltage, while the primary current is .45 times less than the secondary current. The constant value 2.24 indicates that the primary voltage (465V) is 2.24 times greater than the secondary voltage (208V), while the secondary current is 2.24 times greater than the primary current.

In terms of ratios, the constant value .45 represents a .45-to-1 ratio, meaning that for every .45 volts of secondary voltage, there is 1 volt of primary voltage. The constant value 2.24 represents a 2.24-to-1 ratio, meaning that for every 2.24 volts of primary voltage, there is 1 volt of secondary voltage.

Now let's use the constant values to manipulate the alternate method.

If the secondary voltage is .45 times less than the primary voltage, then the primary current must be .45 times less than the secondary current; therefore,

$$I_P = 87A \times .45 = 39.15A$$

which is a hair higher than 38.92A as calculated in question No. 1 but close enough to compare and check your original answer.

Although given, the secondary current (I_S) can be approximated just for practice by using the results of I_P based on the constant value 2.24. Observe:

$$I_S = 39.15A \times 2.24 = 87.7A$$

which is close to 87A, the given value.

Now in question No. 2, the secondary voltage was determined using a given formula, so now let's derive a constant value using the two known like units (current) to approximate the secondary voltage (V_S).

$$\frac{35.8A}{9.3A} = 3.85 \text{ or } \frac{9.3A}{35.8A} = .26 \text{ (both values rounded off)}$$

Based on the results of the constant values, the secondary current is 3.85 times greater than the primary current, and the primary current is .26 times less than the secondary current; therefore, the primary voltage is 3.85 times greater than the secondary voltage, and the secondary voltage is .26 times less than the primary voltage. Take a look:

$$I_S = 9.3A \text{ x } 3.85 = 35.81A$$

$$I_P = 35.8A \text{ x } .26 = 9.31A$$

$$V_S = 550V \text{ x } .26 = 143V$$

$$V_P = 143 \text{ x } 3.85 = 550.55V$$

As you can see, the values (rounded up) derived by using the alternate method are approximately the same as the original values.

In actuality, the alternate method is what the formulas actually represents. This alternate method can always be used where similar units are involved.

B. Transformer's Voltage (V) and Turns (T)

Note: A transformer's primary voltage (V_P) x secondary turns (T_S) equals secondary voltage (V_S) x primary turns (T_P).

$$\text{Original Formula: } V_P \text{ x } T_S = V_S \text{ x } T_P$$

Converted Formulas

$$V_P = \frac{V_S \text{ x } T_P}{T_S} \qquad T_S = \frac{V_S \text{ x } T_P}{V_P}$$

$$T_P = \frac{V_P \text{ x } T_S}{V_S} \qquad V_S = \frac{V_P \text{ x } T_S}{T_P}$$

3. A single-phase transformer's primary windings have 3000 turns while rated for 480 volts. The transformer's secondary voltage is rated for 24 volts. Determine the turns ratio of the transformer.

Use the formula $T_S = \frac{T_P \text{ x } V_S}{V_P}$,where T_S = number of turns on secondary winding

to determine the number of secondary turns:

$$T_S = \frac{3000 \text{ turns x } 24V}{480V} = 150 \text{ turns}$$

The turns ratio is determined by dividing the number of secondary turns into the number of primary turns: 3000 turns/150 turns = 20. The constant value (20) means that for every one turn on the secondary winding, there are twenty turns on the primary winding. Therefore, the transformer's turns ratio is 20-to-1 in this case. Nevertheless, in all cases, a transformer's turn and voltage ratio will always be the same. The more turns, the higher the voltage (the smaller the current); the fewer turns, the lower the voltage (the higher the current).

The calculated result is just as that referenced in the concluding comments of question No. 1.

4. A three-phase 500kVA transformer is rated for 13.8kV-480/277V. If the secondary windings of the transformer have 3000 turns, determine the number of turns in the primary windings.

To determine the number of primary turns,

$$T_P = \frac{13{,}800V \text{ x } 3000 \text{ turns}}{480V} = 86{,}250 \text{ turns}$$

The turns ratio of the transformer can be determined based on the primary-to-secondary voltage, which is

$$\frac{13{,}800V}{480V} = 28.75$$

The value 28.75 represents the turns ratio of the transformer, meaning for every turn of the transformer's secondary windings, there are 28.75 turns on the transformer's primary windings. Therefore,

$$28.75 \text{ x } 3000 \text{ turns (secondary)} = 86{,}250 \text{ (primary turns)}$$

C. Transformer's Amperes (A) and Turns (T)

Note: A transformer's primary amperes (A_P) x primary turns (T_P) equals its secondary amperes (A_S) x secondary turns (T_S).

$$\text{Original Formula: } A_P \text{ x } T_P = A_S \text{ x } T_S$$

Converted Formulas

$$A_P = \frac{A_S \text{ x } T_S}{T_P} \qquad T_P = \frac{A_S \text{ x } T_S}{A_P}$$

$$A_S = \frac{A_P \text{ x } T_P}{T_S} \qquad T_S = \frac{A_P \text{ x } T_P}{A_S}$$

5. The secondary windings of a transformer have 1237 turns and produce 132 amps. If the primary windings of the transformer produce 17 amps, how many turns do the primary windings have?

Use the formula $T_P = \frac{A_S \text{ x } T_S}{A_P}$, where T_P = number of turns on primary winding to determine the number of primary turns:

$$T_P = \frac{132A \text{ x } 1237 \text{ turns}}{17A} = 9605 \text{ turns}$$

D. Transformer's Apparent Power

6. The primary of a single-phase transformer is rated for 7200 volts and has a full-load current rating of 7 amps. What is the apparent power of the transformer?

   To determine the apparent power of the single-phase transformer, use the following formula:

$$P_{VA(1\phi)} = E \times I$$

$$= 7200V \times 7A$$

$$= 50{,}400VA$$

   The transformer's apparent power is 50,400VA or 50.4kVA.

7. The common conductor of a two-phase, 3-wire system carries 37.56A. If the system's line to common voltage is rated for 340 volts, determine the minimum apparent power of the transformers needed to supply this system.

   Because the common conductor of this system carries 37.56A, the line conductors will carry the current of the common conductor divided by the square root of 2 ($\sqrt{2}$ or 1.41). As a result,

$$I_L = 37.56A / 1.41$$
$$= 26.64A$$

   Because the line voltage in this situation would equal the phase voltage (or line to common voltage as stated) times $\sqrt{2}$, the given voltage can be used as is because the $\sqrt{2}$ is already factored into the formula needed to derive an answer; therefore, using the formula $P_{VA} = \sqrt{2} \times E \times I$,

$$P_{VA(2\phi)} = \sqrt{2} \times 340V \times 26.64A$$
$$= 12{,}771VA \text{ (rounded off) or } 12.771kVA$$

   The needed transformer's apparent power must be no less than 12.771kVA combined.

8. If the line current of a two-phase, 4-wire system is rated for 17.8A at 240V, determine the minimum apparent power of the transformers needed to supply this system.

   Using the formula $P_{VA} = 2 \times E \times I$, the minimum apparent power amounts to

$$P_{VA} = 2 \times 240V \times 17.8A$$
$$= 8544VA \text{ or } 8.544kVA$$

9. The secondary of a three-phase transformer is rated for 480 volts and has a full-load current rating of 54 amps. What is the apparent power of the transformer?

   To determine the apparent power of the three-phase transformer, use the following formula:

$$\begin{aligned} P_{VA(3\phi)} &= \sqrt{3} \text{ x E x I}, \text{ where } \sqrt{3} = 1.732 \\ &= 1.732 \text{ x } 480\text{V x } 54\text{A} \\ &= 44{,}893.44\text{VA} \end{aligned}$$

   The transformer's apparent power is 44,893.44VA or 44.893kVA.

10. A 208 volt, three-phase motor has a total load of 12,650 watts with an 83% (.83) power factor. If supplied by a three-phase transformer, determine the minimum kVA capacity (apparent power) needed to run the motor. Also, determine the wattage consumed in each phase of the transformer.

$$\begin{aligned} P_{VA(3\phi)} &= 12{,}650 \text{ watts} / .83 \\ &= 15{,}240.96\text{kVA (minimum)} \end{aligned}$$

$$\begin{aligned} P_{W(1\phi)} &= 12{,}650 \text{ watts} / 3 \\ &= 4216.7 \text{ watts (consumed per phase)} \end{aligned}$$

E. Transformer's Primary and Secondary Full-Load Currents

11. What are the primary and secondary full-load current ratings for a 3ϕ, 112.5kVA transformer rated for 4250-208/120V?

   Use the formula $I = \frac{\text{kVA x } 1000}{\text{V x } \sqrt{3}}$, where I will determine the transformer's primary or secondary full-load current.

   To calculate the transformer's primary (I_P) and secondary (I_S) full load currents, only the line-to-line voltages are used—4250V (primary) and 208V (secondary)—120V is a line-to-neutral voltage.

$$I_P = \frac{112.5\text{VA x } 1000}{4250\text{V x } 1.732} = 15.28\text{A}$$

$$I_S = \frac{112.5\text{VA x } 1000}{208\text{V x } 1.732} = 312.28\text{A}$$

   JUST A REMINDER: It should be understood that "k" represents a thousand (1000) and should be multiplied by the value preceding it, in this case 112.5. Therefore, 112.5kVA is actually 112.5(x 1000)VA.

12. A single-phase transformer has an overall load of 50kVA. The transformer's secondary voltage is 240/120V. Determine the transformer's neutral current if the 240 volt loads are 36.57kVA.

To find the neutral current, the line-to-line 240-volt loads must be subtracted from the transformer's total load:

$$50\text{kVA} - 36.57\text{kVA} = 13.43\text{kVA (transformer's neutral [120V] load)}$$

To find the neutral current:

$$\frac{13.43\text{kVA}}{240\text{V}} = 55.96\text{A (neutral current)}$$

The kVA value is divided by 240 volts instead of 120 volts because the neutral load will be distributed between the line-to-line voltages (L_1 to N, L_2 to N). Dividing the value by 120 volts implies that the entire neutral load will be totally connected across either L_1 or L_2.

F. Wye- and Delta-Connected Transformers

13. Determine the line currents of a balanced three-phase delta (Δ)-connected transformer if the phase currents are 67.3 amps.

The line current can be derived by using the formula

$$I_L = I_P \text{ x } \sqrt{3}, \text{ where } I_P = 67.3 \text{ amps}$$
$$= 67.3 \text{ amps x } \sqrt{3} = 116.56 \text{ amps}$$

Each line current is calculated at 116.57 amps. However, just remember that when working with three-phase delta-connected systems, the phase current is always dependent upon the line current or, to put it another way, the line current is always independent of the phase current.

14. For a 4W-3ϕ 480/277V system, the current flowing through L_1 = 43A, through L_2 = 57A, and through L_3 = 34A. Determine the system's neutral current if all loads are linear.

$$I_{NEUTRAL} = \sqrt{[(43)^2 + (57)^2 + (34)^2)] - [(43 \text{ x } 57) + (57 \text{ x } 34) + (43 \text{ x } 34)]}$$
$$= \sqrt{[1849 + 3249 + 1156] - [2451 + 1938 + 1462]}$$
$$= \sqrt{6254 - 5851}$$
$$= \sqrt{403}$$
$$= 20\text{A (rounded off)}$$

15. Assume the phase currents of a three-phase delta-connected transformer measure as follows: A phase = 13.67A, B phase = 22.16A, and C phase = 18.9A. Determine the transformer's line currents.

With respect to phases A and B, the line current is

$$I_L = \sqrt{(13.67)^2 + (22.16)^2 + (13.67 \text{ x } 22.16)}$$
$$= \sqrt{677.93 + 302.93}$$
$$= 31.32\text{A}$$

With respect to phases B and C, the line current is

$$I_L = \sqrt{(22.16)^2 + (18.9)^2 + (22.16 \text{ x } 18.9)}$$
$$= \sqrt{848.28 + 418.82}$$
$$= 35.6\text{A}$$

With respect to phases C and A, the line current is

$$I_L = \sqrt{(18.9)^2 + (13.67)^2 + (18.9 \text{ x } 13.67)}$$
$$= \sqrt{544.08 + 258.36}$$
$$= 28.33\text{A}$$

16. Determine the neutral voltage of a three-phase wye (Y)-connected transformer if the line voltage is 500 volts.

The neutral voltage can be derived by using the formula

$$V_{LN} = V_{LL} / \sqrt{3}, \text{ where } V_{LL} = 500 \text{ volts}$$
$$= 500 \text{ volts} / \sqrt{3} = 288.68 \text{ volts}$$

The neutral voltage is 288.68 volts. Here again, just remember that when working with three-phase wye-connected systems, the neutral voltage is always dependent upon the line voltage. To put it another way, the line voltage is always independent of the neutral voltage.

17. A delta (Δ)-wye (Y)–connected transformer bank consisting of three single-phase transformers is used to step the transformer's voltage down from 13.3kV to 208/120V. What is the transformer's voltage ratio?

$$(13.3\text{kV}) \frac{13{,}300\text{V}}{208\text{V}} = 63.94\ (64)$$

The transformer's voltage ratio is approximately 64 to 1.

18. The voltage-turns ratio of a transformer is 8:1. If the transformer's primary has 15 ampere-turns, determine the transformer's secondary current.

The transformer's 8:1 voltage-turns ratio simply means turns-wise. For every 8 turns on the transformer's primary windings, there is 1 turn on the secondary winding. Having more turns on a transformer's primary windings equates to a higher winding resistance compared to that of the secondary windings. More current will flow through the secondary windings of a transformer as a result. Therefore, the transformer's secondary current should be eight times that of the primary current, and the number of secondary windings should be eight times less than those of the primary windings. Therefore,

15A x 8 = 120A (transformer's secondary current)

19. A 12-to-1 step-down transformer has a primary voltage rating of 37.5kV and a secondary voltage rating of 4.1kV. If the transformer's taps are changed to increase the number of primary turns by 3.2%, determine the adjusted rating of the transformer's primary and secondary voltages.

    Because a transformer's voltage and turns ratios are proportional, if the number of primary turns are increased by 3.2% (.032), so are the voltage ratings of both primary and secondary sides. As a result, the transformer's primary (V_P) and secondary (V_S) voltages are

    $$V_P = (37.5\text{kV})\ 37{,}500\text{V} \times 1.032 = 38{,}700\text{V}\ (38.7\text{kV})$$

    $$V_S = (4.1\text{kV})\ 4100\text{V} \times 1.032 = 4231.2\text{V}\ (4.23\text{kV})$$

20. Three 25kVA pole-mounted single-phase transformers are used to produce a 75kVA three-phase cluster. The transformers are delta (Δ)–delta (Δ) connected and, subsequently, one of the transformers is lost to lightning damage. Determine the kVA capacity of the two remaining transformers.

    In a situation such as this, the damaged transformer is removed from service and the two remaining transformers are connected to operate as an open delta system. When this happens, the transformer's once total (75 kVA) capacity is reduced by 57.7% . As a result, the kVA capacity between the two transformers is 43.28kVA.

    75kVA x .577 = 43.28kVA

Load Factors

Power factor implies a phase angle and, without a phase angle, there is neither a power factor nor a reactance. When an electrical load consumes all the electrical power that is made available by an electrical system, it is said that the electrical load and system are operating in unity. This explains the term unity power factor. When a unity power factor exists, there is nothing out of phase and therefore no out-of-phase reactance quantities to consider. To understand this, consider the following analogy. If a glass is filled with drinking water, and the glass of drinking water is totally consumed by an individual, it can be said that the water was transferred from the glass to the individual, leaving no water behind. What has occurred is that which was made available has all now been consumed. In this analogy, the filled glass represented the available or apparent power, and the transfer (or consumption) of water represented the true power of an electrical load. This is how an electrical system or load operates when both power elements are in unity.

Now let's consider that same glass of water if only 80% of the water was consumed by the individual and the remainder was thrown out. When this occurs, there are now other things that have to be considered. The first and foremost is that 100% of that filled glass of water was made available and 20% of that was wasted because all of the water was not consumed. Typically, this is how an electrical system operates. A certain amount of electrical power is made available, in this case perhaps by an utility provider, and that electrical power that is made available is consumed in portions by the customer. In particular some situations to recover their losses, a utility provider will access a penalty upon the consumer if a certain percentage of that which was made available is not consumed by the customer.

Now this is where the term power factor really comes into play because it can be used to determine whether there will be a penalty accessed or not. Depending upon the situation, when the operating load(s) of an electrical system consume less than 80% of the electrical power made available by the utility provider, a power factor penalty is assessed upon the customer. The wasted or lost power experienced by the utility provider is what we know as reactance power. Reactance power (VARs) is that third element of the power triangle that's completely opposite of true power (W), and it's another takeoff of the apparent power (kVA). By now, you should be quite familiar with the term power factor as to how it is defined and calculated. Therefore, in our final approach we will take an in depth look at its application and how its used to analyze and improve the performance of an electrical system. To get started, consider the electrical system in figure 67.

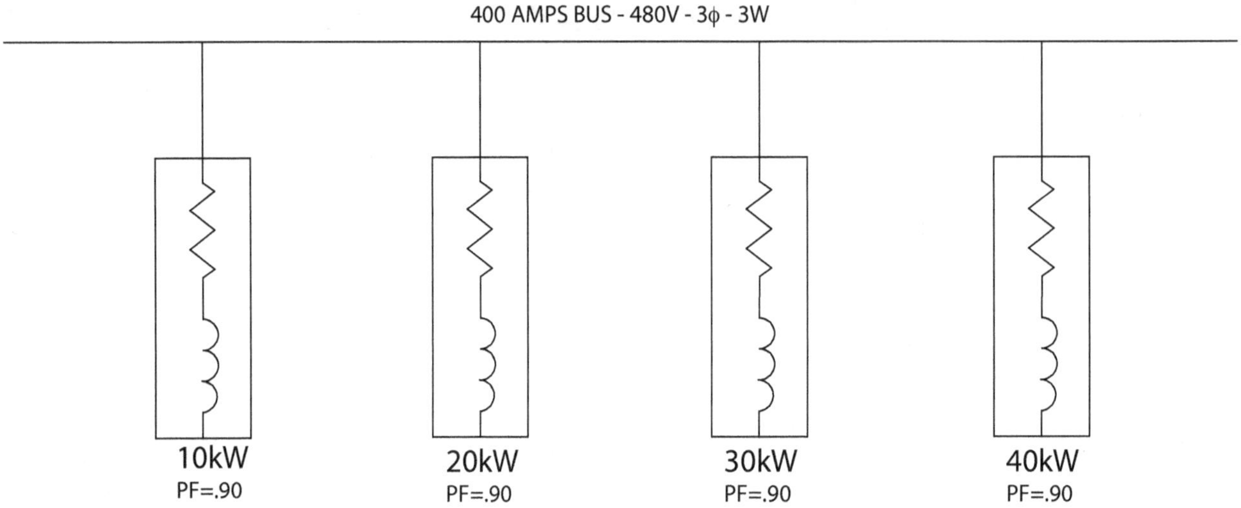

Fig. 67.

The loads are all 480V three-phase induction loads that are connected to a 400 amps bus. Because each load has the same power factor, determining the system's total current can be initiated by first totaling each load's true power. As you know, the true power is the electrical power that is consumed by an electrical load. This is why the true power is also referred to as the working power. With the total of each load's true power, the system's total current can easily be determined by using the formula $kW = \sqrt{3} \times E \times I \times PF$ and converting it to $I = \dfrac{kW}{\sqrt{3} \times E \times PF}$.

With this formula, the system's total current results in

$$I = \frac{100kW}{\sqrt{3} \times 480V \times .90}$$

$$= \frac{100kW}{748.22}$$

$$= 133.65A$$

Now let's consider the induction loads in figure 68.

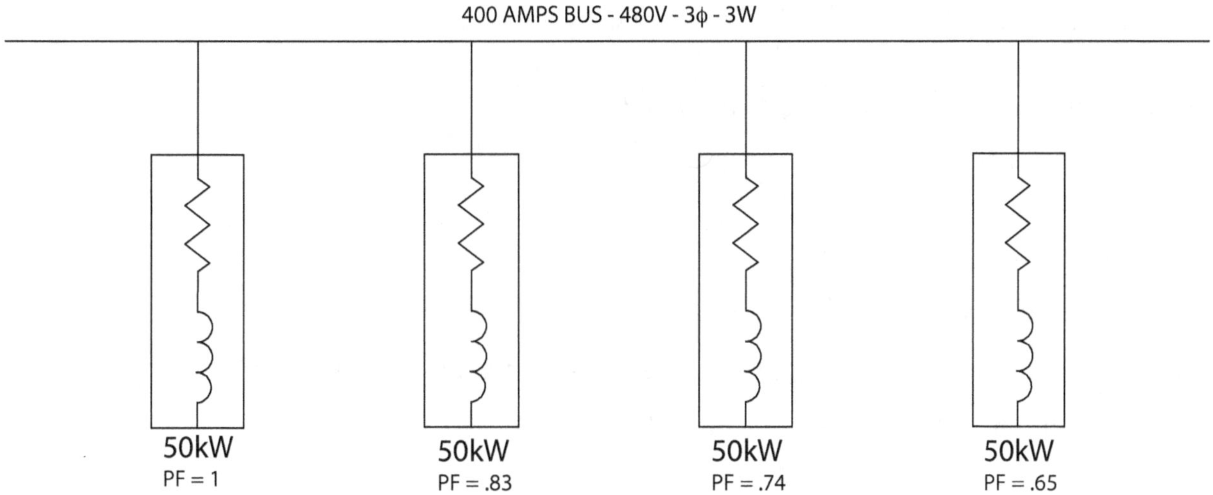

Fig. 68.

Although the load quantities are all the same, this system's total current cannot be determined in a fashion similar to the system in figure 67 because of the unequal power factors. Having unequal power factors will cause each load to experience a different phase angle, apparent power, reactance power, reactance current, and total current. As a result, these individual quantities must first be determined per load and then resolved collectively before the system's total current can be calculated. So now let's begin by evaluating each load.

Load No. 1

Because Load No. 1 has a unity power factor, the apparent power is the same as the true power. When a unity power factor exists, there is no reactance power, and the resulting phase angle is zero.

The phase angle $(\theta) = \cos^{-1} 1 = 0°$

The apparent power (kVA) $= \dfrac{50\text{kW}}{1} = 50\text{kVA}$

The reactance power (kVARs) = 50kVA x sin 0° = 50kVA x 0 = 0VARs

The resistance current $(I_R) = \dfrac{50\text{kW}}{\sqrt{3} \times 480\text{V}} = 60.14\text{A}$

The reactance current $(I_L) = 0$

The total current $(I_T) = \sqrt{(60.14)^2 + 0} = 60.14\text{A}$

Load No. 2

The phase angle $(\theta) = \cos^{-1} .83 = 33.9°$

The apparent power (kVA) $= \dfrac{50\text{kW}}{.83} = 60.24\text{kVA}$

The reactance power (kVARs) = 60.24kVA x sin 33.9° = 60.24kVA x .558 = 33.61kVARs

The resistance current $(I_R) = \dfrac{50\text{kW}}{\sqrt{3} \times 480\text{V}} = 60.14\text{A}$

The reactance current (I_L) = $\frac{33.61\text{kVARs}}{\sqrt{3} \times 480\text{V}}$ = 40.43A

The total current (I_T) = $\sqrt{(60.14)^2 + (40.43)^2}$ = 72.47A

Load No. 3

The phase angle (θ) = $\cos^{-1} .74 = 42.27°$

The apparent power (kVA) = $\frac{50\text{kW}}{.74}$ = 67.57kVA

The reactance power (kVARs) = 67.57kVA x sin 42.27° = 67.57kVA x .673 = 45.47kVARs

The resistance current (I_R) = $\frac{50\text{kW}}{\sqrt{3} \times 480\text{V}}$ = 60.14A

The reactance current (I_L) = $\frac{45.47\text{kVARs}}{\sqrt{3} \times 480\text{V}}$ = 54.69A

Total Current (I_T) = $\sqrt{(60.14)^2 + (54.69)^2}$ = 81.29A

Load No. 4

The phase angle (θ) = $\cos^{-1} .65 = 49.46°$

The apparent power (kVA) = $\frac{50\text{kW}}{.65}$ = 76.92kVA

The reactance power (kVARs) = 76.92kVA x sin 49.46° = 76.92kVA x .76 = 58.46kVARs

The resistance current (I_R) = $\frac{50\text{kW}}{\sqrt{3} \times 480\text{V}}$ = 60.14A

The reactance current (I_L) = $\frac{58.46\text{kVARs}}{\sqrt{3} \times 480\text{V}}$ = 70.32A

The total current (I_T) = $\sqrt{(60.14)^2 + (70.32)^2}$ = 92.53A

System's Total

The apparent power (kVA) = 50kVA + 60.24kVA + 67.57kVA + 76.92kVA = 254.73kVA

The true power = 50kW x 4 = 200kW

The power factor = $\frac{200\text{kW}}{254.73\text{kVA}}$ = .7851

The phase angle (θ) = $\cos^{-1} .7851 = 38.27°$

The reactance power (kVARs) = 254.73kVA x sin 38.27°
= 254.73kVA x .6194 = 157.78kVARs

The resistance current (I_R) = $\frac{200\text{kW}}{\sqrt{3} \times 480\text{V}}$ = 240.57A

The reactance current (I_L) = $\frac{157.78\text{kVARs}}{\sqrt{3} \times 480\text{V}}$ = 189.79A

The total current (I_T) = $\sqrt{240.57)^2 + (189.79)^2}$ = 306.42A

For obvious reasons, did you notice that the resistance current was the same for each load? Also, did you notice that as the power factor decreased from Load No. 2 to Load No. 4, the reactance current increased per load in the opposite manner?

As you can see, the apparent power for each constant 50kW load is totally dependent upon each load's power factor. Because the apparent power is the resulting power element ($kVA = \sqrt{kW^2 + kVARs^2}$) of a constant true power and a varying reactance power, the apparent power will change as the reactance power changes. Again, take notice of how the apparent power is a dependent element. Now consider the total current of each load. In each situation, the total currents were dependent upon the reactance currents because each resistance current was the products of a constant load or true power. In any case, the performance or efficiency of an electrical system or load and the demands of a utility provider weigh heavily upon the power factor. As stated earlier, if a unity power factor exists, reactance is not a factor, and where a power factor does exist, reactance is not only a factor but the cause. The bottom line is that when a utility provider has to supply more demand to an electrical system due to a low power factor, larger conductors, transformers, and associated equipment are required, which leads to higher construction cost. For these reasons alone, when an undesirable power factor exists, the assessment of a power factor penalty is or can be imposed.

Because an induction load will produce a lagging current, the resulting reactance power will also be lagging. When an induction load has a low power factor, the contributing lagging reactance power can be offset with a leading reactance power to improve the power factor. If a leading reactance power is equal to that of the lagging reactance power, a unity power factor will exist; regardless, any quantity of leading reactance power will increase a low power factor. Because capacitors and synchronous motors produce a leading current and reactance power, they are used to improve the power factor of an electrical system or load. So now let's continue this topic by considering how capacitors and synchronous motors are used for power factor corrections without considering harmonic currents and other phenomena that could affect the operation of the capacitors.

Capacitors and Synchronous Motors Used for Power Factor Correction

By now it should be clearly understood that the power factor is the ratio between the true power and the apparent power: PF = kW/kVA. Another way of displaying this expression is

$$PF = \frac{kW}{\sqrt{(kW)^2 + (kVARs)^2}} = \frac{\text{True Power}}{\sqrt{(\text{True Power})^2 + (\text{Reactance Power})^2}}$$

Having the formula written in this manner allows a better view of how the power factor can be changed if the reactance power is likewise changed. Because the true power is directly related to the load and once it is set in place the load remains constant, the only means of correcting the power factor is by way of the reactance power.

Consider a single-phase 77kW induction load that operates with a lagging power factor at 70%. To support such load and power factor, the apparent power has to be 110kVA. Now, if it was desired to increase the load's power factor to 80%, the existing apparent power would be reduced by 87.5% (.875). This is derived by dividing the desired power factor into the load's existing power factor,

.70/.80. If reduced by 87.5%, the apparent power needed to support the load would now amount to 96.25kVA (110kVA x .875). Even after being increased to 80%, the power factor will still encounter a lagging effect, which will cause the reactance current and the reactance power about the load to also encounter the same lagging effect.

Because the apparent power and the power factor are indirectly proportional, this explains why one decreases as the other increases. To see it as it really is, if there is a change in the apparent power, the change occurred resulting from a change in the reactance power, because the true power remains constant based on the fixed induction load. To see this more clearly, let's compare the resulting quantities associated with both power factors.

Power Factor at 70%

The phase angle (θ) = $\cos^{-1} .70 = 45.57°$
The true power = 77kW
The apparent power = 110kVA
The reactance power
110kVA x sin 45.57° = 78.55kVARs

Power Factor at 80%

The phase angle (θ) = $\cos^{-1} .80 = 36.87°$
The true power = 77kW
The apparent power = 96.25kVA
The reactance power
96.25kVA x sin 36.87° = 57.75kVARs

Based on the individual results of both quantities, figure 69 provides an illustration of each power triangle per power factor.

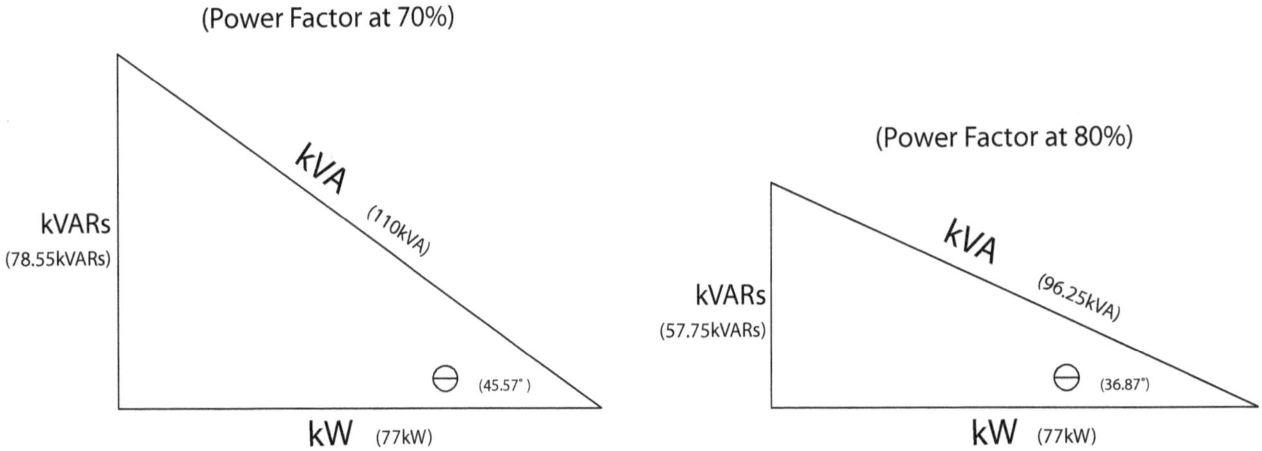

Fig. 69.

Because the lagging reactance power can be offset with a leading reactance power, this is where the use of capacitors enters the picture. To increase the power factor, the difference between the two reactance powers as calculated is used to determine the amount of leading reactance power needed to make such correction by use of a capacitor or a combination of capacitors. Therefore, the leading reactance power needed to correct this power factor amounts to

$$78.55\text{kVARs} - 57.75\text{kVARs} = 20.8\text{kVARs}$$

Another way of deriving these results is to apply the load's true power (kW), reactance powers (kVARs), and power factor by means of the trigonometric function tan θ = kVARs/kW, which when converted results in the formula kVARs = kW x tan θ. Refer to figure 70.

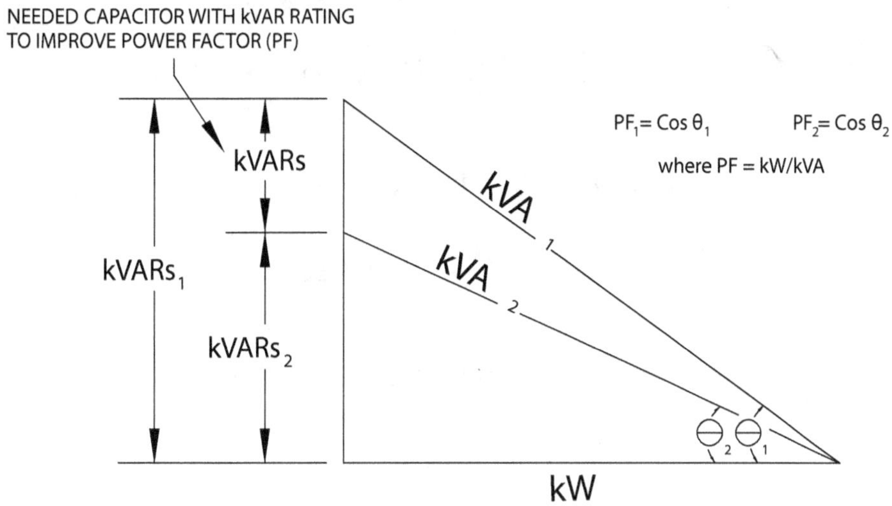

Fig. 70.

Having a formula in place, the given quantities in figure 70 can be expressed as

$$kVARs_1 = kW \times \tan\theta_1 \quad \text{and} \quad kVARs_2 = kW \times \tan\theta_2$$

With both formulas, the needed leading reactance power can again be determined by calculating the difference between the formulas. As a result,

$$\begin{aligned} kVARs_{\text{(Needed)}} &= kVARs_{1\text{ (Exising)}} - kVARs_{2\text{ (Desired)}} \\ &= kW \times \tan\theta_{1\text{ (Existing)}} - kW \times \tan\theta_{2\text{ (Desired)}} \end{aligned}$$

Because the latter formulas involve the same true power (kW), an abbreviated formula can be derived, which proves to be more useful:

$$kVARs_{\text{(Needed)}} = kW \times [\tan\theta_{1\text{ (Existing)}} - \tan\theta_{2\text{ (Desired)}}]$$

With the use of this formula,

$$\begin{aligned} kVARs_{\text{(Needed)}} &= 77kW \times (\tan 45.57° - \tan 36.87°) \\ &= 77kW \times (1.02 - .75) \\ &= 77kW \times .27 \\ &= 20.79kVARs \end{aligned}$$

Capacitors that are used for power factor corrections are rated according to their reactance power capacity in units of kVARs per operating voltage and frequency. Therefore, to determine the capacitance of the needed capacitor, additional calculations are required.

Based on the calculated reactance power needed to correct the power factor of the 77kW load from 70% to 80%, let's assume that the load will be supplied from a 237V source at 60Hz. With this information, the capacitance of the needed capacitor can be determined. At 237V, the needed reactance power will produce the following amount of leading reactance current (I_C). Observe:

$$I_C = \frac{20.79\text{kVARs}}{237\text{V}} = 87.72\text{A}$$

With the reactance current, the capacitance reactance is

$$X_C = \frac{237\text{V}}{87.72\text{A}} = 2.70\Omega$$

Using the formula $X_C = 1/2\pi fC$, it can be converted to $C = 1/2\pi f X_C$ to determine the capacitance of the needed capacitor, which results in

$$C = \frac{1}{2\pi f X_C} = \frac{1}{2 \times 3.14 \times 60\text{Hz} \times 2.70\Omega} = .00098294 \text{ or } 982.94\mu\text{F}$$

So in terms of capacitance, a 982.94μF capacitor is required to produce 20.79kVARs of reactance power.

Because the 237V source will possibly vary during the operation of the load, the capacitor's kVARs will vary likewise. When the voltage rating of a capacitor is exceeded, the capacitor's rated kVARs will increase based on the square of the ratio between the source voltage and the capacitor's rated manufactured voltage. To determine the actual kVARs output per voltage increase, the following formula is applied:

$$\text{kVARs}_{\text{(Output)}} = \text{kVARs}_{\text{(rating)}} \times (\text{source voltage/voltage rating})^2$$

Using the formula, let's assume the capacitor is rated for 240V, and the source voltage during certain intervals of the 77kW load increases from 237V to 250V (based on using the calculated kVARs as the capacitor's kVARs rating).

$$\begin{aligned}\text{kVARs}_{\text{(Output)}} &= 20.79\text{kVARs} \times (250\text{V}/240\text{V})^2 \\ &= 20.79\text{kVARs} \times (1.042)^2 \\ &= 20.79\text{kVARs} \times 1.086 \\ &= 22.58\text{kVARs}\end{aligned}$$

When operating at a 250V source, the capacitor's kVARs will be increased to 22.58kVARs.

Considering all that has been discussed, let's use the gathered data to consider two possible situations that could occur in figure 68, which will involve increasing the power factor to 95% (.95) and the placements of such needed capacitors. In the first situation, the overall electrical system will be considered, and in the second situation, each load will be considered individually.

Situation 1. The Overall Electrical System [Capacitor(s) at Source]

Power Factor at 78.51% (.7851)

Power Factor at 95% (.95)

.7851/.95 = .826

From previously gathered data:
The phase angle (θ) = $\cos^{-1}$ .7851 = 38.27°
The true power = 200kW
The apparent power = 254.73kVA

The reactance power = 157.78kVARs

The phase angle (θ) = $\cos^{-1}$ .95 = 18.19°.
The true power = 200kW
The apparent power
200kW / .95 = 210.53kVA or
(254.73kVA x .826 = 210.41kVA)
The reactance power
210.53kVA x sin 18.19° = 65.72kVARs

$$\begin{aligned} kVARs_{(needed)} &= 200kW \times (\tan 38.27° - \tan 18.19°) \\ &= 200kW \times (.789 - .329) \\ &= 200kW \times .460 \\ &= 92kVARs \end{aligned}$$

The result is the total capacitance needed to supply the overall 480V system. However, because this is a three-phase source, the total capacitance can be obtained from a single capacitance source or by three individual capacitance sources. If three individual capacitance sources are used, they must be rated for at least 30.67kVARs, a rating derived by dividing the total capacitance by three. If the individual line loads of this three-phase 200kW load are not balanced, individual capacitors of the same kVARs ratings may prove to be insufficient. For best results, it is always good to consult a manufacturer for deciding the type combination(s) that will best serve an installation. Now refer to figure 71 to reference the installation of the capacitors selected to supply the overall system.

This type application is normally used when installing a capacitor at certain type loads will produce adverse conditions.

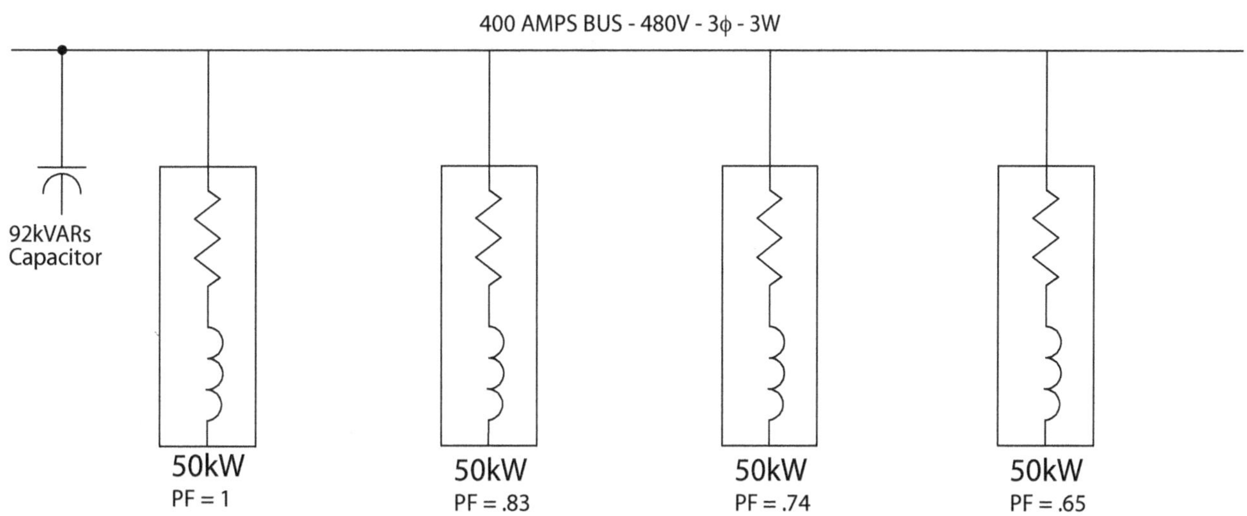

Fig. 71. Capacitor installed at source.

Situation 2. Individual Loads [Capacitor(s) at Loads]

Load No. 1
[Unity Power Factor exists—capacitor not required.]

Load No. 2
From previously gathered data:

Power Factor at 83% (.83)

The phase angle (θ) = $\cos^{-1}$ .83 = 33.9°
The true power = 50kW
The apparent power = 60.24kVA
The reactance power = 33.61kVARs

Power Factor at 95% (.95)

The phase angle (θ) = $\cos^{-1}$ .95 = 18.19°
The true power = 50kW
The apparent power
50kW / .95 = 52.63kW or
(60.24kVA x .874 = 52.65kVA)
The reactance power
52.63kVA x sin 18.19° = 16.43kVARs

.83/.95 = .874

$$\begin{aligned} \text{kVARs}_{\text{(needed)}} &= 50\text{kW} \times (\tan 33.9° - \tan 18.19°) \\ &= 50\text{kW} \times (.672 - .329) \\ &= 50\text{kW} \times .343 \\ &= 17.15\text{kVARs} \end{aligned}$$

Needed Capacitance
Single = 17.15kVARs
Individual = 5.717kVARs (3)

Load No. 3 [Capacitor(s) at Load]
From previously gathered data:

Power Factor at 74% (.74)

The phase angle (θ) = $\cos^{-1}$ .74 = 42.27°
The true power = 50kW
The apparent power = 67.57kVA
The reactance power = 45.47kVARs

Power Factor at 95% (.95)

The phase angle (θ) = $\cos^{-1}$ .95 = 18.19°
The true power = 50kW
The apparent power
50kW / .95 = 52.63kW or
(67.57kVA x .779 = 52.64kVA)
The reactance power
52.63kVA x sin 18.19° = 16.43kVARs

.74/.95 = .779

$$
\begin{aligned}
kVARs_{(needed)} &= 50kW \times (\tan 42.27° - \tan 18.19°) \\
&= 50kW \times (.909 - .329) \\
&= 50kW \times .58 \\
&= 29kVARs
\end{aligned}
$$

Needed Capacitance
Single = 29kVARs
Individual = 9.67kVARs (3)

Load No. 4 [Capacitor(s) at Load]
(from previously gathered data)
Power Factor at 65% (.65)

Power Factor at 95% (.95)

.65/.95 = .684

The phase angle (θ) = $\cos^{-1}$ .65 = 49.46°
The true power = 50kW
The apparent power = 76.92kVA
The reactance power = 58.46kVARs

The phase angle (θ) = $\cos^{-1}$ .95 = 18.19°
The true power = 50kW
The apparent power
50kW / .95 = 52.63kW or
(76.92kVA x .684 = 52.61kVA)
The reactance power
52.63kVA x sin 18.19° = 16.43kVARs

$$
\begin{aligned}
kVARs_{(needed)} &= 50kW \times (\tan 49.46° - \tan 18.19°) \\
&= 50kW \times (1.169 - .329) \\
&= 50kW \times .84 \\
&= 42kVARs
\end{aligned}
$$

Needed Capacitance
Single = 42kVARs
Individual = 14kVARs (3)

Now refer to figure 72 to reference the installation of the capacitors selected to supply each individual load.

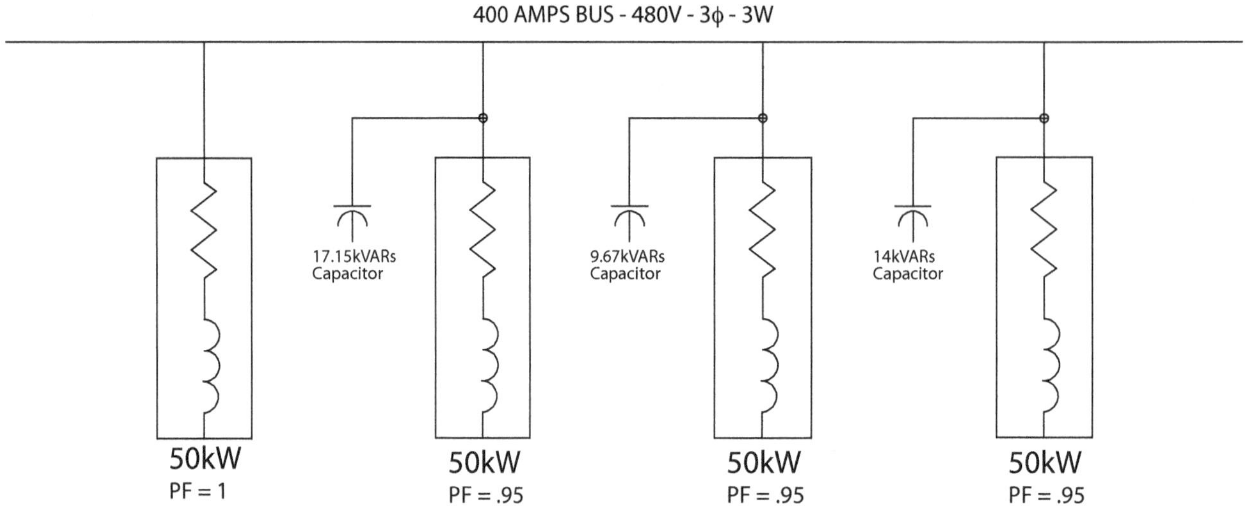

Fig. 72. Capacitors installed at loads.

When a synchronous motor is connected in parallel with an electrical system, the motor's field current can be adjusted to produce a reactance power, where the resulting reactance current leads the operating voltage by a certain phase angle. Similar to a capacitor, this reactance power causes a lagging reactance current to be offset, thereby increasing an electrical system's overall power factor. When used to correct the power factor, a synchronous motor is sometimes referred to as a synchronous or rotating capacitor (condenser).

When a capacitor was used to correct the power factor for the overall electrical system in figure 71, approximately 92kVARs of reactance power was required to increase the overall power factor from 76.84% to 95%. With this source of reactance power, a leading reactance current at approximately 111A (92kVARs/√3 x 480V) would offset the lagging reactance current about the electrical system. Therefore, to correct the power factor of the overall electrical system using a 480V 3-phase synchronous motor, the motor's field current will need to be adjusted to pull approximately 111A.

Practical Problems Pertaining to Power Factor (PF) and Power Factor Corrections

1. A three-phase watt-hour meter reflects an electrical load consuming 121.68kW of power. A corresponding ammeter reads 157A per phase. If the source voltage is 467 volts, determine the load's power factor and the phase angle.

$$PF = \frac{kW}{kVA} = \frac{121.68kW}{467V \times 157A \times 1.732} = .958$$

Having calculated the power factor of the load, the phase angle can be determined.

$$PF = \cos\theta \text{ (cosine theta)}$$

where cos θ is the ratio between the actual power consumed in kW and the apparent power measured in kVA. Theta (θ) represents the desired phase angle.

$$\cos\theta = .958$$

$$\theta = \cos^{-1}.958$$

$$= 16.66°$$

The phase angle is equal to 16.66°.

2. Refer to figure 73. Based on the given information, determine the circuit's true power, reactance power, apparent power, power factor, and operating current.

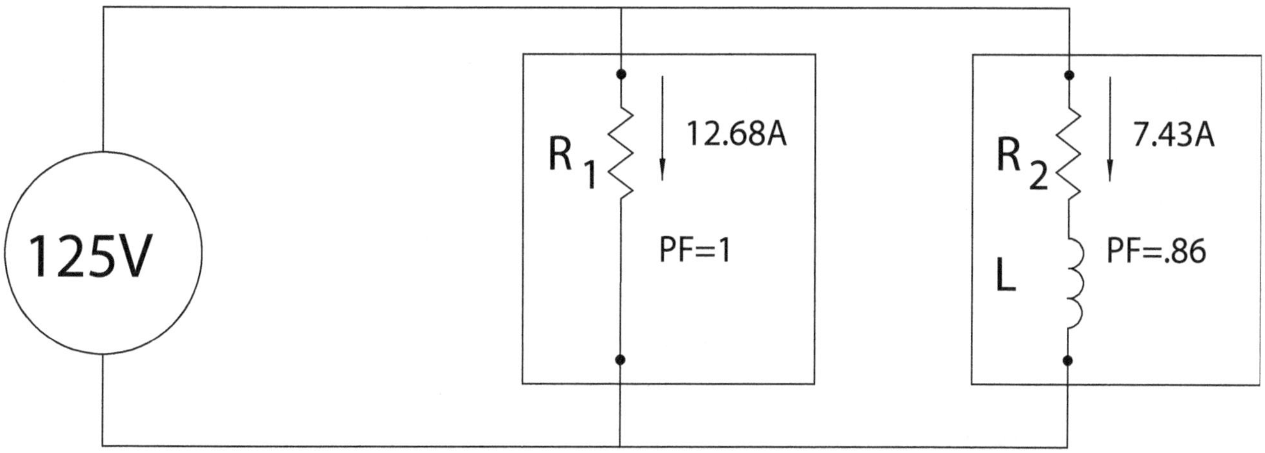

Fig. 73. Resistive/induction loads.

The Circuit's True Power (W)
To determine the circuit's true power, the power consumed by both loads must first be determined. Therefore, the power consumed by the resistive load is

$$P_{W(R1)} = 125V \times 12.68A = 1585W\ (1.585kW)$$

while the power consumed by the induction (RL) load is

$$P_{W(R2)} = 125V \times 7.43A \times .86 = 798.73W$$

To completely understand how all required power quantities are to be determined, vectors diagrams will be used to illustrate the calculated quantities. Refer to figure 74.

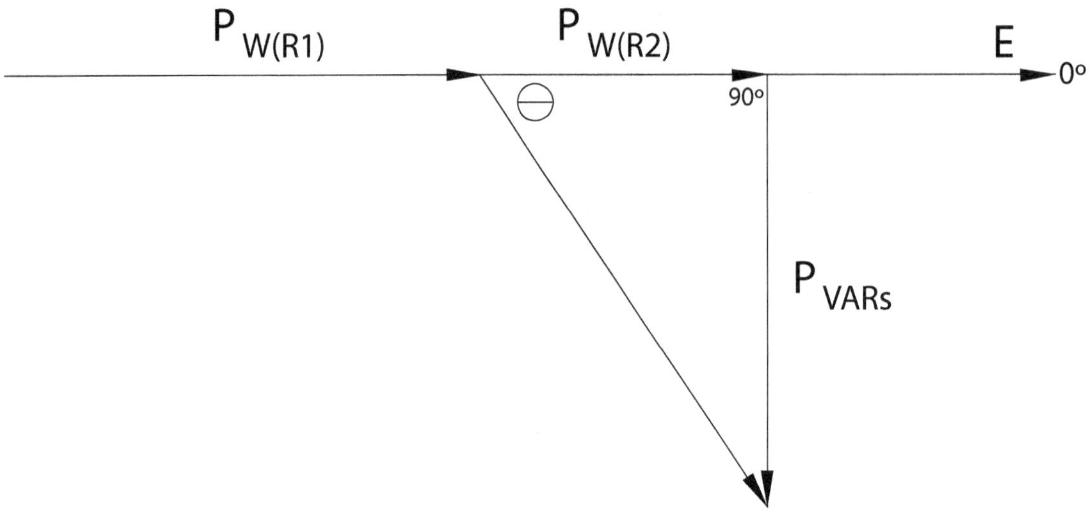

Fig. 74. Resistive/induction loads vector diagram.

Because there is only one voltage source, it is used as the reference (E). Because the resistive load operates at a unity power factor, there will be no reactance power. As for the inductive load, because voltage will lead the current about the load, the reactance power vectors must be drawn below the true power vector at 90° as illustrated. Because the true power consumed by both loads are in phase, they can be totaled to determine the circuit's true power. As a result, the circuit's true power is

$$P_{W(T)} = 1585W + 798.73kW = 2383.73W(2.3837kW)$$

<u>The Circuit's Reactance Power (VARs)</u>
Again, because the resistive load will not contribute any reactance power to the circuit, the circuit's total reactance power will come totally from the inductive load. Because the load's power factor was given, it can be used to determine the load's phase angle. With the phase angle, the resulting reactance power can be determined. Using the formula ($\theta = \cos^{-1}$ pf), the phase angle is

$$\theta = \cos^{-1} .86 = 30.68° \text{ lagging}$$

With the phase angle, the circuit's reactance power amounts to

$$\begin{aligned} P_{VARs} &= 798.73W \times \tan 30.68° \\ &= 798.73W \times -.5933 \\ &= -473.89VARs \end{aligned}$$

Because the reactance power vector was required to be drawn below its true power vector, the position of the reactance power vector is so indicated by the negative sign preceding the power reactance quantity. The negative sign is also an indication of a lagging reactance current.

The Circuit's Apparent Power (VA)

Because the resistive load has a unity power factor, the load's true power and apparent power would be the same: VA_R = 1585VA(1.585kVA). Unlike the resistive load, the inductive load's apparent power can be derived by using the load's power factor. Therefore, the inductive load's apparent power is

$$VA_L = 798.73\text{kW}/.86 = 928.76\text{VA}$$

Because there is no phase angle to consider about the resistive load, the same thing cannot be said for the inductive load. Such phase angle differences will not allow the two apparent power quantities to be simply added together to determine the circuit's apparent power. Therefore, to determine the circuit's apparent power, the circuit's true power and reactance power must be calculated vectorially, which leads to

$$\begin{aligned} VA_T &= \sqrt{(2383.73)^2 + (-473.89)^2} \\ &= \sqrt{(5{,}682{,}168.71) + (\ 224{,}571.73)} \\ &= \sqrt{5{,}906{,}740.44} \\ &= 2430.38\text{VA} \end{aligned}$$

The Circuit's Power Factor

The circuit's power factor can be determined by considering the ratio between the circuit's true power and apparent power with respect to the cosine (cos) function.

Because the reactance current is lagging, the power factor will take on the same identity.

$$PF(\cos\theta) = \frac{\text{true power}}{\text{apparent power}} = \frac{2383.73\text{W}}{2430.38\text{VA}} = .981 \text{ lagging}$$

The Circuit's Operating Current

With the circuit's given voltage and calculated apparent power, the operating current results in

$$I_T = \frac{2430.38\text{VA}}{125\text{V}} = 19.44\text{A}$$

3. Refer to figure 75. Based on the given information, determine the circuit's true power, reactance power, phase angle, power factor, apparent power, and operating current.

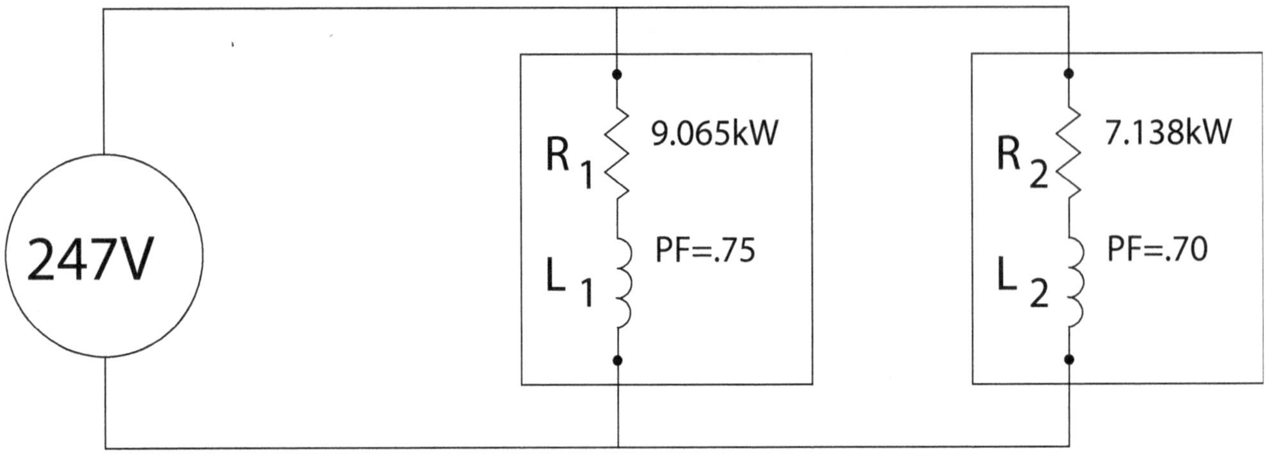

Fig. 75. Induction loads.

The Circuit's True Power (W)

To determine the circuit's true power, the power consumed by both loads is added together as is.

$$P_{(WT)} = 9.065\text{kW} + 7.138\text{kW} = 16.203\text{kW}$$

Just as for the previous circuit, vectors diagrams will be used to illustrate the calculated power quantities. Refer to figure 76.

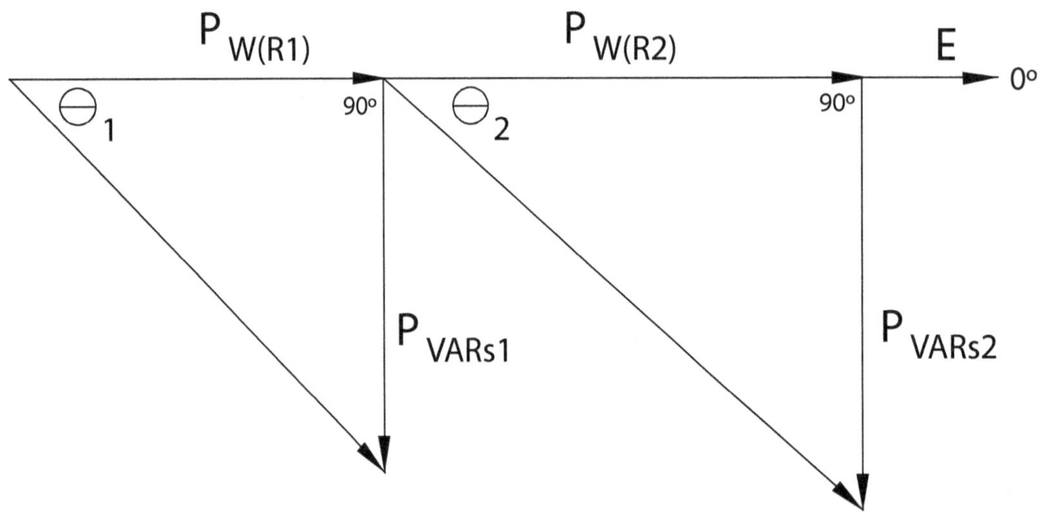

Fig. 76. Induction loads vector diagram.

Again, because there is only one voltage source, it too will be used as the reference (E). Because both loads are inductive, the voltage across the loads will lead the flow of current through the loads. This is why both loads' reactance power vectors are drawn below their respective true power vectors at 90° angles.

The Circuit's Reactance Power (VARs)

Since the loads' power factors were given, they can be used to determine each load's phase angle. With the phase angles, the resulting reactance power quantities can be determined. Observe:

$$\theta_1 = \cos^{-1} .75 = 41.41° \text{ lagging} \qquad \theta_2 = \cos^{-1} .70 = 45.57° \text{ lagging}$$

With the phase angles, the quantity of the each load's reactance power is

$$\begin{aligned} P_{VARs1} &= 9.065\text{kW x tan } 41.41° \\ &= 9.065\text{kW x } -.8819 \\ &= -7.994\text{kVARs} \end{aligned} \qquad \begin{aligned} P_{VARs2} &= 7.138\text{kW x tan } 45.57° \\ &= 7.138\text{kW x } -1.020 \\ &= -7.281\text{kVARs} \end{aligned}$$

Totaling the reactance power quantities of both loads, the circuit's reactance power amounts to

$$P_{VARsT} = -7.994\text{kVARs } (+) -7.281\text{kVARs} = -15.275\text{kVARs}$$

The Circuit's Phase Angle

The circuit's phase angle can be determined by considering the ratio between the circuit's true power and reactance power with respect to the tangent (tan) function.

$$\tan\theta = \frac{\text{reactance power}}{\text{true power}} \quad \text{where} \quad \theta = \tan^{-1}\frac{\text{reactance power}}{\text{true power}}$$

Using the latter formula, the circuit's phase angle is

$$\theta = \tan^{-1}\frac{15.276\text{kVARs}}{16.203\text{kW}} = \tan^{-1}.943 = 43.32° \text{ lagging}$$

The Circuit's Power Factor

Applying the circuit's phase angle, the power factor is

$$PF = \cos 43.32° = .728 \text{ lagging}$$

The Circuit's Apparent Power (VA)

The circuit's apparent power can be determined by dividing the derived power factor into the circuit's true power.

$$P_{VA} = 16.203\text{kW} / .728 = 22.257\text{kVA } (22{,}257\text{VA})$$

The Circuit's Operating Current

With the circuit's given voltage and derived apparent power, the operating current results in

$$I_T = \frac{22.257\text{kVA}}{247\text{V}} = 90.11\text{A}$$

4. A three-phase, 460V synchronous motor operating at 83.7A with an 81.8% leading power factor is connected in parallel with a three-phase, 460V induction motor that pulls 106.4A at a lagging power factor of 92%. Based on the given information, determine the true power, reactance power, phase angle, power factor, apparent power, and operating current produced by both motors when operating at the same time.

The Motor's True Power (kW)
To determine the circuit's true power, the power consumed by both loads is added together as is.

$$kW_{SYN} = \sqrt{3} \text{ x } 460V \text{ x } 83.7A \text{ x } .818 = 54.549kW$$

$$kW_{IND} = \sqrt{3} \text{ x } 460V \text{ x } 106.4A \text{ x } .92 = 77.989kW$$

$$kW_T = 54.549kW + 77.989kW = 132.54kW$$

The Circuit's Reactance Power (VARs)
Because the load's power factors were given, they can be used to determine each load's phase angle. With the phase angles, the resulting reactance power quantities can be determined. Observe:

$$\theta_{SYN} = \cos^{-1} .818 = 35.11° \text{ leading} \qquad \theta_{IND} = \cos^{-1} .92 = 23.07° \text{ lagging}$$

With the phase angles, the quantity of the each load's reactance power is

$$\begin{aligned} P_{VARs\ SYN} &= 54.549kW \text{ x } \tan 35.11° \\ &= 54.549kW \text{ x } .703 \\ &= 38.35kVARs \end{aligned} \qquad \begin{aligned} P_{VARsIND} &= 77.989kW \text{ x } \tan 23.07° \\ &= 77.989kW \text{ x } -.426 \\ &= -33.22kVARs \end{aligned}$$

Totaling the reactance power quantities of both loads, the circuit's reactance power amounts to

$$P_{VARsT} = 38.35kVARs - (-)33.22kVARs = 5.13kVARs \text{ leading}$$

Because the synchronous motor will produce a leading reactance current opposed to a lagging reactance current produced by the induction motor, both loads' reactance power will result to being 180° out of phase and, therefore, the smallest reactance power is subtracted from the larger reactance power. As a result, the total reactance power will have a leading effect. Just remember, the negative sign appearing before the induction motor's reactance power [(–)33.22kVARs] in parentheses does not represent a minus sign. Its only purpose is to indicate a lagging reactance current about the induction motor.

The Circuit's Phase Angle
The circuit's phase angle can be determined by considering the ratio between the circuit's true power and reactance power with respect to the tangent (tan) function.

$$\tan \theta = \frac{\text{reactance power}}{\text{true power}} \qquad \text{where} \qquad \theta = \tan^{-1} \frac{\text{reactance power}}{\text{true power}}$$

Using the latter formula, the circuit's phase angle is

$$\theta = \tan^{-1}\frac{5.13\text{kVARs}}{132.54\text{kW}} = \tan^{-1}.039 = 2.23° \text{ leading}$$

The Circuit's Power Factor

Applying the circuit's phase angle the power factor is

$$PF = \cos 2.23° = .999 \text{ leading}$$

The Circuit's Apparent Power (VA)

The circuit's apparent power can be determined by dividing the derived power factor into the circuit's true power.

$$P_{VA} = 132.54\text{kW} / .999 = 132.67\text{kVA}\ (132{,}670\text{VA})$$

As a reminder, an approximate result of the circuit's apparent power could have been determined earlier by applying the circuit's true power (kW_T) and reactance power (P_{VARsT}) where,

$$\begin{aligned} P_{VA} &= \sqrt{(132.54)^2 + (5.13)^2} \\ &= \sqrt{17{,}593.17} \\ &= 132.64\text{kVA}(132{,}640\text{VA}) \end{aligned}$$

The Circuit's Operating Current

With the circuit's given voltage and derived apparent power, the operating current results in

$$I_T = \frac{132.67\text{kVA}}{460\text{V x } 1.732} = 166.52\text{A}$$

Practical Problems Pertaining to Power Factor (PF) Corrections with Capacitors

5. A three-phase 480V (60Hz), 100HP motor has a 65% power factor. The full load current of the motor is 124 amps. Determine the amount of reactance power needed to improve the motor's power factor to 90%.

   Determine the existing and improved phase angles.

$$PF_{\text{(existing)}} = \cos\theta$$

$$\cos\theta = .65_{\text{(existing)}} \quad \theta = \cos^{-1}.65 \quad \theta = 49.46°$$

$$PF_{\text{(improved)}} = \cos\theta$$

$$\cos\theta = .90_{\text{(improved)}} \quad \theta = \cos^{-1}.90 \quad \theta = 25.84°$$

The existing phase angle of the motor is 49.46°, and once the power factor is improved, the phase angle of the motor will be reduced to 25.84°.

Now, to determine the size of the capacitor needed to improve the power factor, use the following formula:

$$kVAR_{(needed\ capacitor)} = kW\ (\tan\theta_1 - \tan\theta_2)$$

$$\tan\theta_1 = \tan 49.46° \qquad \tan\theta_2 = \tan 25.84°$$

$$= 1.1692 \qquad = .4843$$

Substitute values into the formula:

$$kW_{(motor)} = \frac{E \times I \times PF \times 1.732}{1000}$$

$$kW_{(motor)} = \frac{480V \times 124A \times .65 \times 1.732}{1000} = 67kW$$

$$kVARs_{(needed\ capacitor)} = 67kW\ (1.1692 - .4843)$$

$$= 67kW \times .6849$$

$$= 45.89kVARs$$

A reactance power of 45.89kVARs is required to improve the motor's power factor from 65% to 90%.

6. In question No. 5, what size capacitor in units of micro-farads (μf) is needed to produced 45.89 kVARs?

The first thing to do is to determine the reactance current produced by 45.89 kVARs.

$$I_{kVARs} = \frac{45{,}890\ VARs}{480V \times 1.732} = 55.2A$$

With the reactance current and the operating voltage of the motor (480V at 60Hz), the capacitance reactance (X_C) can be calculated, where

$$X_C = \frac{480V}{55.2A} = 8.7\Omega$$

Using the formula

$$X_C = 1 / 2 \times \pi \times f \times C$$

to derive the formula

$$C = 1 / X_C \times 2 \times \pi \times f$$

the value of the capacitor can be calculated.

$$C = 1 / 8.7\Omega \times 2 \times 3.14 \times 60$$
$$= .00030505F \text{ or } 305\mu f \text{ (rounded off)}$$

A capacitor rated for 305µf is needed to produce 45.89 kVARs.

7. A small manufacturing company has a 600A, 480/277V three-phase service. The service is being fed from a 750kVA transformer. On average, the company consumes 311.54kW per month and has been assessed a power factor penalty by the local utility company for operating with a power factor less than 80%. Determine the actual power factor of the service and the phase angle, the active and reactance (magnetizing) currents, the initial and final apparent power and the reactance current if the power factor is increased to 80%.

   Determine the actual power factor:

$$\text{PF (power factor)} = \frac{\text{kW}}{\text{kVA}} = \frac{311{,}540\text{W}}{480\text{V} \times 600\text{A} \times 1.732} = .625$$

$$\text{PF} = \cos\theta, \quad \theta = \cos^{-1}.625 \quad \theta = 51.32°$$

   The actual power factor is .625, and the phase angle is 51.32°.

   The active current

$$I_{kW} = \frac{311{,}540\text{W}}{480\text{V} \times 1.732} = 374.74\text{A}$$

   The reactance power$_1$

$$\text{kVARs}_1 = 311.54\text{kW} \times \tan 51.32°$$
$$= 389.14\text{kVARs}$$

   The apparent power$_1$

$$\text{kVA}_1 = \frac{389.14\text{kVARs}}{\sin 51.32°} = 498.48\text{kVA (498,480VA)}$$

   The reactance current$_1$

$$I_{kVARs1} = \frac{389{,}140\text{VARs}}{480\text{V} \times 1.732} = 468\text{A}$$

The 468 amps of reactance current is based on the service existing power factor (.625). Notice that the active current is lower than the reactance current, which is a product of the reactance power. Reactance power is considered unproductive or wasted power because it is not converted into useful work, similar to active power. Reactance power only serves to produce the magnetic fields needed to operate induction equipment such as motors, transformers, and inductors. It causes a low power factor and a high apparent power demand. As a result, the utility company assessed a power factor penalty against the company to be reimbursed for providing an apparent power that basically involves less active power than reactance power. Therefore, if the reactance power is decreased, so is the reactance current. If a capacitor is added to the system to improve the power factor, what it actually does is produce a supplemental reactance power, which supplies a large amount of reactance current to the service, which reduces the reactance power demands supplied by the apparent power.

If the power factor is improved to 80% (by using power factor correction capacitor), the improved phase angle is

$$\cos\theta = .80_{\text{(improved)}} \qquad \theta = \cos^{-1}.80 \qquad \theta = 36.87°$$

The reactance power$_2$

$$\text{kVARs}_2 = 311.54\text{kW} \times \tan 36.87°$$

$$= 233.66\text{kVARs}$$

If the power factor is improved to 80%, the reactance power is reduced to 233.66kVARs.

The apparent power$_2$

$$\text{kVA}_2 = \frac{233.66\text{kVARs}}{\sin 36.87°} = 389.43\text{kVA (389,430VA)}$$

If the power factor is improved to 80%, the reactance and apparent powers are both reduced.

The reactance current$_2$

$$I_{\text{kVARs2}} = \frac{233.66\text{kVAR}}{480\text{V} \times \sqrt{3}} = 281\text{A}$$

By improving the power factor to 80%, the reactance current is reduced from 468A to 281A, a difference of 187A. The demand of the transformer is also reduced from 498.48kVA to 389.43kVA, a difference of 109kVA, while the operating current is reduced from 600A ($I_{T1} = \sqrt{[374.74A]^2 + [468A]^2}$) to 468A ($I_{T2} = \sqrt{[374.74A]^2 + [281A]^2}$), a difference of 132A.

ABOUT THE AUTHOR

During the past thirteen years, Alvin Walker, a native of Shreveport, Louisiana, owned and operated a small yet successful electrical contracting business. He now works as an author and instructor specializing in electrical and NEC training, and his services are available throughout the United States. In his over twenty-seven years of experience, he has developed a strong background in electrical maintenance, construction, and design engineering. He has taught Business and Law for Contractors; the National Electrical Code; Electrical Theory; Electrical Service, Motors, and Transformer Load Calculations; and other basic and advanced electrical classes at Bossier Parish Community College, Louisiana State University-Shreveport, Louisiana Technical College (Forcht Wade Correction Center), and Southern University-Shreveport. He also served as the department head of Industrial Electricity at Houston Community College-Stafford. Mr. Walker is best known for his hands-on approach and his ability to simplify and explain the most difficult electrical subject matters. He is a master electrician who holds a Louisiana state license as an electrical contractor. He has a degree in electrical engineering from the University of South Carolina and has worked as an electrical engineer for E.I. DuPont, Westinghouse, and M.W. Kellogg. Currently, he has written two study guides, both based on the two most recent editions of the *National Electrical Code*, as well as an electrical theory book. As an author, his primary objective is to produce electrical material that's thoroughly explained, easily understood, and user friendly. For more about his services and other pending books, visit his Web site at www.alvinwalker.com.

INDEX

R

www.ingramcontent.com/pod-product-compliance
Lightning Source LLC
LaVergne TN
LVHW080033210926
848481LV00011B/3175